Earthquak
Resistant
Structures
Assessmen
Rehabilitat

Earthquake-Resistant Structures - Design, Assessment and Rehabilitation

Editor

Marek Lagunov

scitus
academics

Earthquake-Resistant Structures - Design, Assessment and Rehabilitation
Edited by **Marek Lagunov**

ISBN: 978-1-68117-203-3
Library of Congress Control Number: 2016934750

www.scitusacademics.com
Box No. 4766, 616 Corporate Way,
Suite 2, Valley Cottage,
NY 10989

Printed in United States of America on Acid Free Paper

Preface

Earthquake-resistant structures are structures considered to withstand earthquakes. While no structure can be fully resistant to earthquake damage, the objective of an earthquake-resistant building is to create structures better than their predictable counterparts during seismic activity. Earthquake-resistant structures are designed to withstand the biggest earthquake in their location. This means reducing the loss of life by preventing the collapse of buildings for rare earthquakes, while reducing the loss of functionality for more frequent ones. Buildings, structures and their foundations must be constructed to withstand side loads to prove an earthquake. They must be powerful enough to take the loads. After the worst shock waves, they must not disintegrate and remain in place to retain the strength for the aftershock. There is currently a number of design philosophies in earthquake engineering using experimental results, computer simulations and observations from past earthquakes to provide the performance required at the site of concern for the seismic threat. These ranges from strong and ductile enough to survive the shaking with acceptable damage, to base insulation or structural vibration control technology in order to minimize forces and deformations. In addition to the safety assessment, reinforcement and rehabilitation of existing structures in contrast to earthquake loads, this book highlights the seismic resistance design of the construction of masonry and reinforced concrete structures. This book focuses on earthquakes–resistant buildings, bridges and dams. It covers issues relating to earthquake engineering. The book presents contemporary themes about the recent progress of the earthquake-resistant structures and a helpful tool for graduate students, researchers and practicing structural engineers.,

Table of Contents

CHAPTER 1

A Review of Seismic Isolation for Buildings: Historical Development and Research Needs

Gordon P. Warn [1,*] and Keri L. Ryan [2]

[1] Department of Civil and Environmental Engineering, Pennsylvania State University, 226B Sackett Building, University Park, PA 16802, USA

[2] Department of Civil and Environmental Engineering, University of Nevada, Reno/MS 0258, Reno, NV 89557, USA;

Abstract

Seismic isolation is a technique that has been used around the world to protect building structures, nonstructural components and content from the damaging effects of earthquake ground shaking. This paper summarizes current practices, describes widely used seismic isolation hardware, chronicles the history and development of modern seismic isolation through shake table testing of isolated buildings, and reviews past efforts to achieve three-dimensional seismic isolation. The review of current practices and past research are synthesized with recent developments from full-scale shake table testing to highlight areas where research is needed to achieve full seismic damage protection of buildings. The emphasis of this paper is on the application of passive seismic isolation for buildings primarily as practiced in the United States, though systems used in other countries will be discussed.

Keywords: isolation; elastomeric; sliding; seismic protection

1. Introduction

Seismic isolation is a technique to shift the fundamental natural period of a structure to the long period range, e.g., two to four seconds, by placing horizontally flexible isolation devices at the base of the structure to physically decouple it from the ground. For earthquake excitation this period shift translates into reduced floor acceleration and inter-story drift demands on the superstructure (structure above the isolation system) by comparison to the equivalent non-isolated structure. The reduced demands allow the superstructure to remain elastic, or nearly elastic, following a design level event. Furthermore, the reduced demands minimize the likelihood of damage to displacement sensitive and acceleration sensitive equipment, nonstructural components, and content. The period shift, however, does result in increased displacement demands concentrated at the isolation interface that must be accommodated by the seismic isolation devices. The simultaneous reduction in acceleration and drift demands achieved with seismic isolation makes it one of the most effective strategies to achieve "Operational" or "Fully Operational" performance following a large and infrequent earthquake event.

Though the concept of seismic isolation dates back more than one hundred years, it has only been practiced in the United States for the last three decades. The first record of seismic isolation is an 1870 U.S. Patent filed in San Francisco for a double concave rolling ball bearing, described as an "Earthquake-proof building" [1,2]. The 1870 double concave rolling ball bearing is remarkably similar to modern double concave Friction Pendulum™ bearings [2,3]. In 1985, one hundred and fifteen years later, the Foothill Communities Law and Justice Center in Rancho Cucamonga, California was the first seismically isolated building constructed in the United States [4]. The Law and Justice Center is isolated on 98 high damping natural rubber bearings. A thorough historical perspective and chronology of seismic isolation can be found in Naiem and Kelly [4]. Construction of seismically isolated buildings has increased at an almost exponential rate since the 1980s in Japan and China, while construction in the United States has remained modest [5] irrespective of the demonstrated effectiveness of this technology for protecting structures, nonstructural components, and content from horizontal earthquake ground shaking through more than thirty years of research. In addition to buildings, seismic isolation has been used for the protection of critical, non-building, structures such as bridges, liquefied natural gas (LNG) tanks, and offshore platforms [6]. However these applications are not discussed further as the scope of this paper is on seismic isolation of buildings.

A number of sources, both recent and dated, have provided comprehensive reviews of various aspects of the development, theory, and application of seismic isolation technology. Early reviews were excellent and thorough. For instance, Kelly [7] provided a historical perspective dating back to the rudimentary beginnings of seismic isolation technology, followed by a complete chronology of research and development efforts. Buckle and Mayes [8] also included a thoughtful historical discussion as well as a comprehensive list of the early applications that paved the way for acceptance and wider adoption. Taylor *et al.* [9] presented a review of the use of elastomers in seismic isolation bearings, with emphasis on their long-term behavior. A mid-1990s report provided information on several subtopics including theory, experiments, and application of sliding bearings, hybrid testing, and development and practice in several countries [10].

As the volume of information on seismic isolation has grown exponentially over the past 10–15 years, the attempts at a comprehensive review have diminished. Yet, several focused reviews have emerged. Kunde and Jangid [11] prepared a comprehensive review of research and application of seismic isolation to bridges, including analytical, experimental and parametric studies. Symans *et al.* [12] reviewed the development and application of seismic isolation and damping systems for wood frame structures, which are uniquely challenging to isolate due to the inherent flexibility of the framing system and relatively light mass. A recent primer, developed by the American Society of Civil Engineers (ASCE) [13], discusses the theory, hardware, analysis, design, and testing requirements specific to the United States. Finally, a collaboration of international experts acting as part of the International Council for Research and Innovation in Building and Construction (CIB) Task Group 44, prepared a report comparing devices, design codes, and current state of seismic isolation practice among countries that have been forward in adoption of seismic isolation technology [14].

Complete protection of a building system poses some unique challenges, which include: (1) protection of the nonstructural components and content, which are more sensitive to vertical excitation than the structural system; (2) mitigation of local uplift or tension demands in the isolation system that may be generated by overturning forces in moderate to slender structures; and (3) shifting of displacement demands from the isolation system to the structure in extreme events that can lead to superstructure yielding. While the vast majority of shake table tests have been conducted to demonstrate the effectiveness of seismic isolation to provide a linear elastic response of the structure under horizontal ground motion, only a few experimental studies have focused on the advanced aspects mentioned above.

The current volume of literature precludes a comprehensive review in a single paper. Instead, the aim of this paper is to summarize the current practice, describe widely used seismic isolation hardware, chronicle the history and development of modern seismic isolation through shake table testing of isolated buildings, review past efforts to achieve three-dimensional isolation, and discuss recently identified research needs. This article focuses on the application of passive seismic isolation for buildings primarily as is practiced in the United States, though systems used in other countries will be discussed. The paper is organized into the following major sections: seismic isolation hardware, experimental demonstration, past efforts to achieve 3D seismic isolation, and research needs.

2. Seismic Isolation Hardware

A variety of seismic isolation bearings have been developed and implemented around the world for the seismic protection of structures. In this section, a brief description of the basic construction, mechanical behavior and analytical modeling is presented for each type of bearing. In general, bearings are classified into two categories: (1) elastomeric and (2) sliding.

2.1. Elastomeric Bearings

Elastomeric bearings are composed of alternating layers of natural, or synthetic, rubber bonded to intermediate steel shim plates as shown in Figure 1. The bearings are constructed by placing un-vulcanized rubber sheets and steel shims in a mold, then subjecting the mold to elevated temperature and pressure to simultaneously vulcanize and bond the rubber. A rubber cover is provided to protect the internal rubber layers and steel plates from environmental degradation due to ozone attack [15] and corrosion, respectively. Elastomeric bearings can be categorized as: (1) low-damping natural or synthetic rubber; and (2) high-damping rubber. Low-damping natural rubber material exhibits nearly linear shear stress-strain behavior up to, approximately, 150% shear strain, after which the material stiffens. Natural rubber with type A durometer hardness of 50 is typically used for seismic applications having a shear modulus (G) that ranges from 0.65 MPa to 0.9 MPa. The equivalent damping ratio ξ, for low-damping natural rubber ranges between 2% and 3% at 100% shear strain. To control, or limit, displacements across the isolation interface, external supplemental damping devices such as yielding steel bars, plates, or viscous fluid dampers are typically used in parallel with low-damping natural rubber bearings. A higher level of damping is achieved through the addition of carbon black and other fillers to the raw rubber during the mixing process to produce high-damping rubber bearings [4]. The equivalent damping ratio of high-damping rubber bearings can range from 10% to 20% at 100% shear strain. Though the range of shear modulus for high-damping rubber is similar to that for low-damping rubber, fillers increase the hardness and thus the shear modulus of the rubber so that it can be difficult to achieve low shear modulus and high levels of damping.

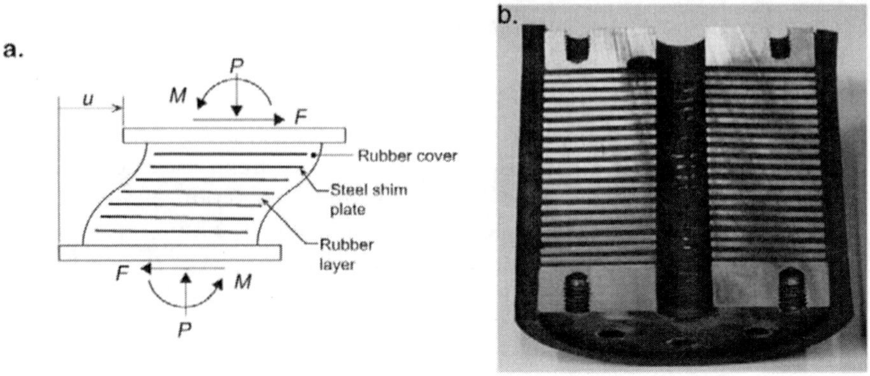

Figure 1. (**a**) Illustration of elastomeric bearing in the horizontally deformed configuration;(**b**) Photograph of elastomeric bearing cross-section.

The total thickness of rubber (T_r) provides the low horizontal stiffness needed to lengthen the fundamental natural period of the system, whereas the close spacing of the intermediate steel shim plates provides a large vertical stiffness and critical load capacity for a given G and bonded rubber area (A_b). However, the steel shim plates have no effect on the horizontal stiffness of the bearing, calculated as:

$$K_h = \frac{GA_b}{T_r} \tag{1}$$

The steel shims restrain the rubber at the bond interface and the spacing of the shims (or individual rubber layer thickness) controls the bulging around the perimeter and thus the compression modulus of the elastomeric layer [16,17]. For example, the compression modulus for an individual, solid, circular rubber layer, assuming the rubber is incompressible, is:

$$E_c = 6GS^2 \tag{2}$$

where S is the shape factor, a dimensionless geometric parameter, defined for a single rubber layer as:

$$S = \frac{\text{Loaded Area}}{\text{Area free to bulge}} \tag{3}$$

Although bulk compressibility was not considered in Equation 2 for simplicity, Chalhoub and Kelly [16] demonstrated that the assumption of incompressibility overestimates the compression stiffness of an elastomeric bearing, where the degree of over-estimation increases with increasing shape factor. Based on their work, Chalhoub and Kelly [16] concluded the effect of bulk compressibility should be considered in the analysis of seismic isolation. The close spacing of steel shim plates, *i.e.*, thin rubber layers, produces a high shape factor that in turn results in a large vertical stiffness:

$$K_v = \frac{E_c A_b}{T_r} = \frac{6GS^2 A_b}{T_r} \tag{4}$$

Inspection of Equations 1,4 reveal the horizontal and vertical stiffness of the elastomeric bearing are related. For example, for circular bearings:

$$\frac{K_v}{K_h} = 6S^2 \tag{5}$$

Traditionally, elastomeric seismic isolation bearings have been designed with shape factors from 15 to 25 [18,19] and as high as 30 [20]. Such high shape factors have been used for the historically perceived benefits of mitigating rocking motion and enhancing stability [21,22], among others. However, according to Equation 5, this range of shape factors results in vertical stiffness that ranges from one to several thousand times greater than the horizontal stiffness. As a result, isolation systems composed of bearings detailed with high shape factors (15–30) typically have vertical isolation periods ranging from 0.03 to 0.15 second. Consequently, the vertical isolation frequencies can align with the dominant frequencies of the vertical spectrum, leading to significant amplification of the vertical acceleration. Therefore, isolation systems composed of elastomeric bearings detailed with high shape factors, e.g., 15–30, provide only horizontal isolation.

Figure 2 presents the shear force-horizontal deformation and vertical force-vertical deformation response of a low-damping natural rubber bearing [23]. While damping in natural rubber bearings is neither hysteretic nor viscous the shear force-horizontal deformation relationship is often modeled using either: (1) a linear viscous representation; or (2) a bilinear hysteretic representation. In Figure 2b, positive vertical force corresponds to tension. The vertical force-deformation behavior is highly nonlinear in tension due to softening or loss of stiffness resulting from multi-chain damage, damage of the micro-structure, and micro-void formation in cross-link polymers subjected to tensile strains [24,25]. Uplift, or tension, in elastomeric bearings is considered undesirable and efforts are made in the design process to avoid uplift on the isolation system. Therefore a simple linear force-deformation relationship is typically used to model the vertical force-deformation behavior where the stiffness is specified as the compression stiffness calculated using Equation 4, or similar, for different bearing plan geometries. If uplift cannot be avoided the vertical force-deformation relationship plotted in Figure 2b can be approximated with user-defined, multi-spring, models or, for example, a self-centering material model available in OpenSees [26], an open source earthquake simulation software.

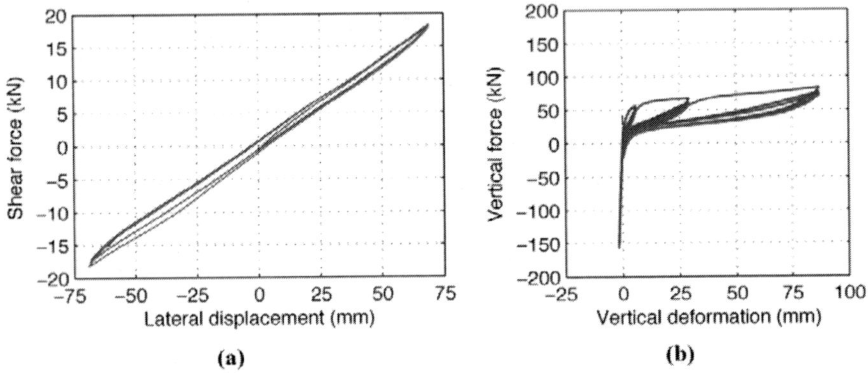

Figure 2. (**a**) Cyclic shear force-horizontal deformation response of low-damping natural rubber bearing; (**b**) Cyclic vertical force-deformation response of low-damping natural rubber bearing.

2.2. Lead-Rubber Bearings

Lead-rubber bearings were first introduced and used in New Zealand in the late 1970s [27,28]. Since then, lead-rubber bearings have been widely used for seismic isolation around the world including the United States and Japan [29]. From a construction perspective, lead-rubber bearings differ from low-damping natural rubber bearings only by the addition of a lead-plug that is press-fit into a central hole in the bearing. The lead-plug deforms plastically under shear deformation, enhancing the energy dissipation capabilities compared to the low-damping natural rubber bearing.

The horizontal force-deformation relationship of a lead-rubber bearing is characterized using bilinear behavior as shown in Figure 3. The zero-displacement force-intercept, Q_d, for a lead-rubber bearing is controlled by the yield strength of the lead in shear, σ_L, and the cross-sectional area of the lead-plug, A_L, or:

$$Q_d = \sigma_L A_L \tag{6}$$

The second-slope stiffness, K_d, is the stiffness of the elastomeric component of the bearing, determined using Equation 1. At a given horizontal displacement, *d*, the effective, or secant stiffness, of the lead-rubber bearing is:

$$K_{eff} = \frac{Q_d}{d} + K_d \tag{7}$$

The vertical stiffness of lead-rubber bearings is calculated from Equation 4 using the effective shear modulus, G_{eff}, calculated from the effective stiffness, K_{eff}, to approximately account for the increase in vertical stiffness from the lead-plug. Similar to elastomeric bearings, the vertical stiffness of a lead-rubber bearing is typically thousands of times larger than the horizontal stiffness so that isolation systems composed of lead-rubber bearings provide isolation only from the horizontal components of ground shaking.

The energy dissipation mechanism is primary hysteretic due to plastic deformation of the lead core. A Bouc-Wen or rate independent plasticity model [30] is typically used for analytically modeling the shear force-horizontal deformation response of lead-rubber bearings. Under bi-directional loading, the bearing model is coupled in the two orthogonal horizontal directions through a circular yield surface. The Bouc-Wen based hysteretic models, however, do not account for the effect of heating in the lead-core with repeated cycling that leads to degradation in the characteristic strength [2,31,32,33]. Theoretical models to account for the effects of heating in lead-rubber bearings have been developed and experimentally verified [31,32]. However, thermo-mechanical models [33] that account for heating and strength degradation in lead-rubber bearings have not been widely implemented for the analysis and design of lead-rubber isolation systems. Kalpakidis and Constantinou [34] developed a theory of scaling based on similitude and provide recommendations for testing reduced scaled bearings to properly account for the effects of heating.

The vertical force-deformation behavior is typically assumed to be linear with stiffness equal to the compressive stiffness of the bearing (Equation 4), though multi-linear models are possible as previously discussed. A model that includes the influence of vertical load on the effective horizontal stiffness and lead-core yield strength was developed by Ryan *et al.* [35] and has been implemented in OpenSees [26]. Models that account for the second-order effects due to vertical load at large horizontal displacement have been developed [36,37,38] and are capable of exhibiting zero or negative tangential horizontal stiffness as has been experimentally demonstrated [39]. However, these models have not been widely adopted due to the extensive experimental data required to calibrate the model parameters.

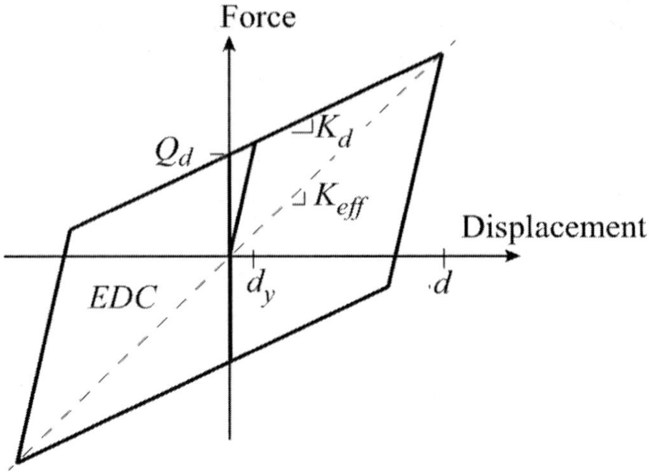

Figure 3. Bilinear horizontal force-displacement characterization of a seismic isolation bearing.

2.3. Sliding Bearings

Sliding bearings support the weight of the structure on a bearing that rests on a sliding interface. The sliding interface is designed with a low coefficient of friction, which limits the resistance to horizontal forces. Most sliding bearings use polytetrafluorethylene (PTFE) type material and stainless steel for the bearing material at the sliding interface. Restoring force is provided either by added springs or through geometry as with the Friction Pendulum™ (FP) bearing [40]. The FP bearing or derivatives such as the multi-spherical Friction Pendulum™ bearings [3,41,42,43] are among the most widely used seismic isolation bearings in the United States. The single FP bearing consists of a base-plate (ductile iron), an articulated slider (ductile iron with bonded PTFE type bearing material) and a spherical concave dish (cast steel with stainless steel overlay) as illustrated in Figure 4. As shown in Figure 4, under horizontal motion the spherical concave dish displaces horizontally relative to the articulated slider and base-plate. Friction between the PTFE type material and stainless steel surface provides frictional resistance and energy dissipation, whereas the radius of curvature of the spherical concave dish provides a restoring force. The FP bearing can be installed upside down from the configuration shown in Figure 4. However the upside down configuration results in the *P-Δ* moment being distributed to the structural element below the FP isolator rather than the element above the FP isolator as would occur with the configuration shown in Figure 4.

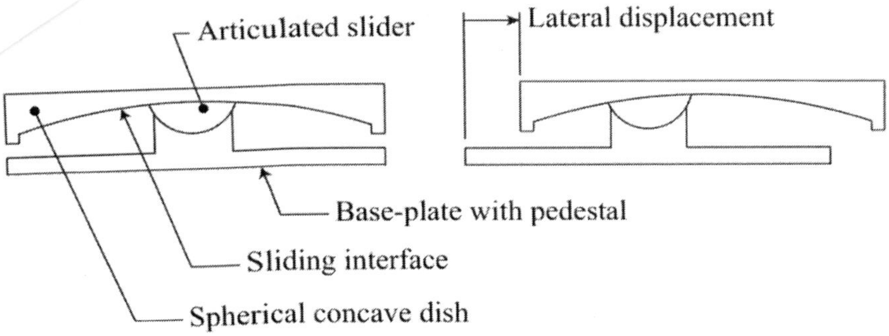

Figure 4. Illustration of Friction Pendulum™ bearing.

The shear force-horizontal deformation behavior of FP bearings is characterized using the bilinear relationship shown inFigure 3. The horizontal strength, or zero-displacement force-intercept, Q_d, is controlled by the bearing material and the weight W carried by the isolators, according to:

$$Q_d = \mu W \tag{8}$$

Where μ is the sliding coefficient of friction of the bearing interface. The sliding coefficient of friction for un-lubricated Teflon type material mated to stainless steel typically ranges between 0.07–0.18 depending on bearing pressure, peak velocity and material [44]. However, a bearing manufacturer reports that sliding coefficients of friction from 0.03 to 0.2 are possible [45]. The second-slope stiffness, K_d, of the FP bearing is controlled by weight acting on the isolator and the radius of curvature, R, of the spherical concave dish according to:.

$$K_d = \frac{W}{R} \tag{9}$$

The effective stiffness of an FP bearing can be determined by substituting Equations 8,9 into Equation 7. The FP bearing is unique in that the period based on the second-slope stiffness, K_d, is controlled only by the radius of the concave dish:

$$T_d = 2\pi\sqrt{\frac{R}{g}} \tag{10}$$

The significance of the weight independent property is that FP bearings can be effective for isolating light-weight structures.

Furthermore, mass irregularities are naturally balanced by corresponding spatial variation in the restoring force, such that torsional response is minimal.

Similar to lead-rubber bearings, a bi-directionally coupled, Bouc-Wen or rate independent plasticity model is commonly used to analytically describe the force-deformation behavior of FP bearing undergoing horizontal planer motion [30]. The influence of bearing pressure and instantaneous velocity on the coefficient of sliding friction are considered using a model proposed by Constantinou *et al.* [46]. Similar to lead-rubber bearings, the characteristic strength of friction pendulum bearings is affected by heating [2]. The coefficient of friction is affected by the ambient, or bulk, temperature of the isolators as well as heating that develops at the isolation interface during high velocity sliding [2]. Heating due to sliding is directly related to the sliding velocity and must be carefully considered when designing reduced scale tests or extrapolating the results of characterization tests performed on reduced scale bearings to those of the prototype [2].

The vertical stiffness of FP bearings is several times larger than high shape factor elastomeric bearings due to the ductile iron, stainless steel and thin layer of PTFE used to construct the bearing. Typically the vertical stiffness of FP bearings is estimated using a simple axial stiffness equation where the modulus is that of ductile cast iron, the area is that of the articulated slider and the length is that of the combined height of the inner pedestal and articulated slider. The resulting vertical isolation period is around 0.03 seconds [45]. Therefore, isolation systems composed of FP bearings with vertical isolation periods around 0.03 seconds provide only horizontal isolation.

More recently, multi-spherical derivatives of the FP bearing have been developed, characterized [3,41,42,43] and implemented, e.g., [47]. The benefit of the multi-spherical sliding bearings over single FP bearings is that they allow an adaptive force-deformation behavior, whereby the stiffness and damping properties of the bearing can change at predetermined displacement amplitudes [41]. Variations of multi-spherical sliding bearings include the Double Pendulum™ (DP) and Triple Pendulum™ (TP), among others. Details of the DP bearing can be found in Fenz and Constantinou [3]. The TP bearing, consisting of four spherical sliding surfaces and three independent pendulum mechanisms, offers the most adaptable behavior of the multi-spherical sliding bearings [41,42,43]. A photograph of a TP bearing used in the NEES TIPS/E-Defense tests [48] is shown in Figure 5a with a cross-section view in Figure 5b. The internal pendulum mechanism with two concave plates and a rigid slider, denoted collectively as the "articulated slider" (Figure 5b), determines the response during low intensity shaking. The outer stainless steel concave surfaces, when designed with different curvatures and friction coefficients, provide two independent pendulum mechanisms that determine the response during medium to high intensity shaking. Details pertaining to the construction and force-deformation behavior of the TP bearing can be found in Fenz and Constantinou [41].

The force-deformation behavior of the TP bearing can be modeled analytically as three single FP elements in series [49]. Each of the individual series elements is modeled using the Bouc-Wen type plasticity model developed for the single FP bearing [30]. The vertical force-deformation behavior of the Triple FP bearings is assumed to be linear [49], and calculated according to the dimensions of the inner-most rigid slider.

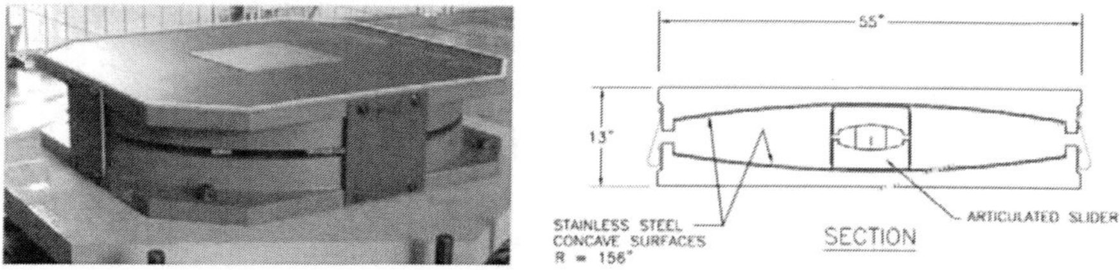

(a) (b)

Figure 5. (**a**) photograph of Triple Pendulum™ bearing; (**b**) view of cross-section.

Both the single FP and the TP bearings provide no resistance to tensile forces and thus are free to uplift. In certain situations uplift in the bearings could occur, e.g., bearings on the perimeter of slender structures or those located under braced frames. For these situations resistance to uplift is considered desirable and a tension-capable sliding bearing, based on the principles of the single FP bearing, was developed [50,51]. The tension-capable device, referred to as the "XY-FP" isolator, consists of two orthogonal concave sliding rails connected through a sliding device that couples the rails in the vertical direction, allowing for the development of tensile forces in the bearing. The XY-FP isolator uniquely provides uplift resistance and an uncoupled bi-directional response [52] due to the separate orthogonal rails.

3. Experimental Demonstration

Extensive shake table testing of base-isolated buildings has been conducted over the past 30 plus years, which has paralleled the development of suitable isolation devices for large scale structures and the evolution of base isolation practice in the United States and other countries. The earliest tests focused on verification of different isolation devices and served mostly as proof-of-concept tests, without rigorous standards for evaluation of the building response. Elastomeric bearings matured more quickly than friction-based sliding systems. Several of the earliest systems were evaluated by shake table testing at the University of California's Earthquake Engineering Research Center (EERC). These systems included lightly damped elastomeric bearings, elastomeric bearings with steel dampers [53], and elastomeric bearings with enhanced damping through a friction fail-safe device [54,55]. The lead-rubber bearing was developed in New Zealand [28], and evaluated also by shake table testing at EERC [56]. All systems were tested on a 5-story frame structure, allowing the development of higher mode response, which enabled the effectiveness of various implementation approaches for seismic isolation to be assessed. These research reports frequently observed that isolation systems with higher levels of damping, especially nonlinear damping, were effective at controlling isolator displacements, but increased floor accelerations and high frequency response.

While the concept of a friction-based system was very simple and attractive, the lack of a suitable restoring force delayed the development of sliding systems. Kelly and Chalhoub [57] tested an isolation system utilizing a combination of elastomeric bearings and flat sliders. The now mature FP system was first tested on a 2-story frame with the isolators located at the top of the first story columns; variations in structural height and aspect ratio were considered [40]. Around this time, several concepts for incorporating restoring force into friction base isolation systems were proposed [58], and an era of shake table testing of friction-based systems was launched at the State University of New York at Buffalo's National Center for Earthquake Engineering Research. A 6-story frame structure was used in evaluation tests of FP isolators [59,60,61] and teflon-disc bearings with helical steel springs [62,63]. Also, a hybrid elastomeric-friction device called resilient friction base-isolator (R-FBI) was tested at EERC [64,65]. All these systems were found to provide effective horizontal isolation.

Alongside the research and development in the United States, a major interest in base isolation developed in Japan, with research carried out mainly in the private sector by the most reputable construction companies [66]. Since Japan is an earthquake-prone region whose construction industry is dominated by a few companies, those companies have and continue to invest in the development of new technologies, including base isolation. Although some private companies and/or government agencies utilized shake table testing (e.g., Kajima, Mitsubishi, Nuclear Power Engineering Test Center), an alternative form of development adopted in Japan was the concept of a demonstration building. Kajima, Oiles, Shimizu, Obayashi, Takenaka, Taisei and Okumura all constructed base-isolation demonstration buildings, many of which were used as laboratory facilities on corporate research campuses. These demonstration buildings were continually monitored by forced vibration, free vibration, and multiple instances of seismic shaking every year, albeit mostly small intensity, as described and referenced in Kelly [66]. Most companies focused on natural rubber bearings with steel dampers, lead-rubber bearings, and eventually the application of rubber compounds with high damping fillers that could provide sufficient damping in a single device to eliminate the need for supplemental damping. Bridgestone emerged as one of the successful developers of the high damping rubber bearing, and also popularized the bolted connection detail, which was shown to improve stability and eventually replaced the doweled connection [20]. Taisei Corporation developed the friction based Taisei Shake Suppression System (TASS) that used TFE elastomeric bearings (bearings with a sliding interface) and Neoprene horizontal springs for a restoring force [67].

As the years have progressed, researchers have continued to invent new concepts for base isolation, many of them taken to the point of successful validation with shake table testing. While an exhaustive treatment of this subject is nearly impossible, a few examples are: (1) suspended pendulum isolator (SPI) [68]; (2) rolling ball isolation [69]; (3) almost lifted structure concept (ALSC) [70]; (4) stable unbonded fiber reinforced elastomeric isolator (SU-FREI) [71]; and (5) core suspended isolation system (CSI) [72]. Few of these systems have advanced to the point of widespread practical implementation. However, the new variations of the original FP isolator described earlier, including the tension capable bearing and the TP bearing, have achieved market acceptance and international implementation. The tension capable bearing was tested and validated by Roussis and Constantinou [50], and two comprehensive programs have evaluated the performance of multi-stage FP devices [73,74]. Some specifics of these test programs are mentioned below.

A few experimental studies focused specifically on evaluation of secondary system response [75,76,77,78]. The studies by Kelly [75] and Juhn *et al.* [77] designed oscillators (vertical cantilevers with mass) mounted on various floors of the frame to simulate the response of light equipment. Kelly [76] found that supplementary damping reduced the benefit of isolation on the secondary system response (floor acceleration and especially floor spectra), while Juhn *et* al. [77] found that the isolation system was very effective in reducing the response of an oscillator tuned to the natural frequency of the fixed-base structure, but could slightly increase the response of the high frequency detuned oscillator. Wolff and Constantinou [78] focused on secondary system response by evaluating floor accelerations and floor spectra for 8 different isolation systems, including combinations of low damping elastomeric bearings, lead-rubber bearings, friction pendulum bearings, and flat sliders with and without viscous dampers. The authors concluded that highly nonlinear systems incorporating friction pendulum bearings or flat sliders were not as detrimental to content response as suggested by earlier research, e.g., [29]. However, meaningful evaluation of the effect of nonlinearity is difficult when the isolation systems considered have significantly different isolation periods, thus providing different levels of force reduction, which was the case in these experiments. While useful, these studies lack a connection to the physical performance of realistic nonstructural components and content that appears in building structures.

Eventually, shake table experiments extended beyond basic validation to examine performance of the overall isolated building under more challenging loading conditions. One consideration was the possibility of uplift in slender structures under large horizontal loading. The effect of column uplift was examined in a 1/5 length scale 7-story R/C isolated structure [79] and a 1/4 length scale 9-story braced-steel frame isolated structure [80]. Different types of elastomeric bearings (natural rubber or lead-rubber) were designed for each test, and doweled connections were used to allow shear transfer while providing no restraint against uplift. The studies found that an uplift occurrence caused a vertical pulse to propagate through the structure, and likewise increased the higher frequency peaks in the horizontal floor spectra. The uplift occurrence sometimes produced a temporary instability in one or more bearing hysteresis loops, but this was a localized effect that was balanced by the other bearings and did not lead to global instability. In one case, the shear key disengaged from the bearing since the uplift of 1.25 cm, which exceeded the height of the shear key, occurred at the same time as the peak bearing displacement [79]. In tests utilizing softer bearings, a combination of moderate uplift and bearing roll out failure was observed, which occurred at a horizontal displacement less than the predicted roll out displacement. The tests suggested that uplift should be anticipated and accounted for in superstructure response and bearing connection details.

To minimize the effects of building overturning, researchers have developed and tested various uplift restraint mechanisms or mechanisms that carry tension. A thorough review of uplift restraint mechanisms studied to date is provided by Roussis [52]. Occasionally, uplift restraint has been provided by a mechanism independent of the isolation device. For example, Sumitomo Construction developed a mechanism—untested but implemented—to effectively tie the building down through interconnected steel arms attached to the structure and the foundation [52]. More commonly, uplift restraint mechanisms have been incorporated into the isolation device. An uplift restraint device embedded in the central hole of an annular elastomeric bearing was proposed and tested [81,82]. The restraint consisted of high strength bolts contained within a sleeve, wherein the bolts were detailed to allow vertical slip. The device could engage in tension and/or during large horizontal displacements to act as a fail-safe displacement restraint. The combined horizontal-vertical restraint provided enhanced functionality to the isolator, but ultimately limited the horizontal displacements that could be accommodated. An uplift restraint for a flat slider or FP isolator was tested by Nagarajaiah *et al.* [83]. The restraint was provided by L-shaped arms that extended from and engaged the underside of the isolator on two sides. The approach was demonstrated for unidirectional excitation, but is difficult to extend to bidirectional excitation and thus has limited application. The use of pre-stressing tendons was evaluated by Kasalanati and Constantinou [84]. Another solution is the tension capable XY-FP bearing mentioned earlier [50,51] that has been implemented in several applications. An alternative practical solution is to pair elastomeric bearings with tension capable cross-linear bearings [85]. The cross-linear bearings, developed in Japan, are essentially low friction sliders that use a double rail mechanism for tensile resistance. The friction coefficient in the cross-linear bearing is minimized by use of linear guides with recirculating ball technology. Shake table verification of this technology was a component of recent tests performed at E-Defense [48], that are described below.

Another project used shake table testing to engage the ultimate capacity of the isolated structure [86]. A three-story R/C building was designed and built to replicate the performance of a base-isolated demonstration building in Sendai, Japan (length scale factor = 0.4). The high damping rubber isolation system stiffened at large displacements, allowing demands to shift to the superstructure to develop the typical ductile degradation modes expected under large earthquakes. The experiment showed that design strategies can be adopted to ensure that the isolation system is not the weak link. In these tests, the isolators withstood significant tension due to the structure overturning.

At least two studies performed shake table tests where the intensity of the excitation was increased until the bearings ruptured [87,88]. Sato *et al.* [87] tested three different types of isolation devices (natural rubber bearings with steel dampers, lead rubber bearings, and high damping rubber bearings) to failure. Several of the tests were stopped only after all 4 bearings had failed; however, in each case the first failure was reported as rupture of the rubber compound under a combination of tension and shear. The failure shear strains in dynamic tests were found to be greater than the failure strains under static, monotonic loading. Takaoka *et al.* [88] performed shake table tests of a slender 1/9 length scale structure with lead rubber bearings, varying the structure slenderness ratio and the bearing aspect (diameter to height) ratio to engage different bearing failure modes. Bearing rupture was reported under extreme cases of buckling, tension, and combinations of tension and shear. Buckling rupture, accompanied by a "sinking" deformation of the bearing, occurred after the horizontal deformation had exceeded the diameter of the bearings. Large tension occurrences were accompanied by a notable increase in horizontal acceleration at the roof level when the uplifted structure landed on the bearings.

Limit state response of single and multi-spherical FP isolators has been observed in several of the previous test programs, e.g., [40,73,74], although engaging the limit states has never been a main focus of these programs. A short duration uplift occurrence of the bearings on one side of the building was observed in Morgan and Mahin [74], during which the slider assembly remained stable and the bearing hysteresis was not affected other than a localized (single bearing) reduction in horizontal shear due to the coupling of shear and supported axial load through friction. In Fenz and Constantinou [73], an accidental aggressive contact of the slider with the displacement restrainer in a DP bearing appeared to cause the building to rock off its supports on one side. The impact increased the base shear of the system, but not to the point of negating the effectiveness of base isolation relative to a non-isolated building; and the authors hypothesized that the model might have collapsed if a displacement restraining ring had not been present. The authors also suggested that an uplift occurrence could be particularly serious if the separation distance exceeds the height of the restrainer ring, causing the top concave plate to rise over the edge of the intermediate slider. However, such behavior has never been observed, and uplift occurences have typically been short duration, limiting the consequences of bearing separation.

In all tests described above, evaluation of the influence of vertical excitation was not a primary focus. However, many of the tests included a vertical component of excitation, and in some cases the test protocol included direct comparisons of shaking with and without vertical excitation. Most test programs have concluded that vertical excitation has no influence or very minor influence on the horizontal response and the effectiveness of base isolation. In tests with elastomeric bearings, Griffith *et al.* [79] noted a very minor (5%–11%) increase in horizontal floor accelerations with the inclusion of vertical excitation, and suggested that a very light coupling may have occurred, but the data was inconclusive. Responses due to horizontal only and horizontal plus vertical excitation were found to be similar in tests of the R-FBI system [65]. For multi-stage FP bearings, the influence of vertical excitation on the floor response spectra appeared to be negligible in Morgan and Mahin [74], and the influence of vertical excitation was also concluded to be "very minor" in Fenz and Constantinou [73]. Some instances of coupling between the system horizontal response and vertical ground excitation have been observed. For instance, in tests of a 3-story building isolated with lead-rubber and natural rubber bearings, Hwang and Hsu [89] found that for a structure with an asymmetric configured isolation system, vertical ground acceleration significantly increased the horizontal floor accelerations in the superstructure. No attempt was made to explain why the floor accelerations were amplified in the asymmetric configuration. Furthermore, data presented in Fenz and Constantinou [73] showed that peak horizontal floor accelerations were amplified by a factor of 2 in some of the tests with the largest peak vertical ground acceleration, which is in contrast to their conclusions.

All of the aforementioned studies involved reduced scale structural models and reduced scale isolation bearings. The multi-story structural models utilized in previous test programs employed length scale factors ranging from about 0.1 to 0.4, with a maximum supported structural weight on the order of 500 kN. Recently, two independent test programs on full scale isolated buildings with weights of 5,000–10,000 kN have been conducted at National Institute for Earth Science and Disaster Prevention (NIED) E-Defense shaking table of Japan. The first of these test programs evaluated the response of a 4-story R/C hospital building isolated with 4 high damping rubber bearings and 4 natural rubber bearings paired with steel dampers [90]. The second of these test programs, a NEES/E-Defense collaboration, evaluated the response of a 5-story steel moment frame building isolated with (1) 9 triple pendulum bearings (TPBs); and (2) 4 lead-rubber bearings (LRBs) paired with 5 cross linear bearings (CLBs). The results of the NEES/E-Defense collaboration have been reported only preliminarily to date [48].

Several features offered by full scale testing have led to significant new observations from the E-Defense tests even considering the wealth of data on reduced-scale systems. First and foremost, each full scale building served as a testbed for evaluation of physical nonstructural components and content. Thus, the effectiveness of seismic isolation to protect a variety of building content was evaluated directly, rather than by speculation of the influence of recorded responses, such as floor accelerations. In fact, the primary objective of the hospital test program [90] was to examine the performance of hospital equipment, and evaluation of the seismic performance of interacting partition walls, suspended ceilings, and sprinkler piping was a key component of the NEES/E-Defense tests [91]. Second, full scale models are required to realistically reproduce the effects of vertical vibration of the floor system and thus assess the influence of vertical excitation on building content. In most prior reduced scale tests, supplementary mass was bolted directly to a bare frame specimen, causing significant distortion of the vertical vibration modes and frequencies.

Based on observations from the hospital test program, functionality of base-isolated hospitals utilizing elastomeric isolation systems can be achieved following a near-fault motion. However, the same cannot be said for a long duration, long period ground motion generated from a subduction earthquake, due to significant motion of furniture items and medical appliances supported by casters. In such motions, velocities measured in the floors of the isolated building were larger than those in the comparable fixed-base building. The tests have generated serious discussion in Japan about how to maintain the function of a base-isolated hospital after a large, long-duration, earthquake. The influence of vertical excitation in this test program has not yet been reported [90].

The significant influence of vertical excitation was a dominant outcome of the NEES/E-Defense test program [48]. Input table excitations with peak vertical accelerations greater than about 0.5g resulted in some damage to suspended ceilings, causing ceiling panels to fall, or in the most extreme cases localized failures of the ceiling grid support system. The performance of the suspended ceilings was similar in all configurations (isolated with TPB, isolated with LRB/CLB, or fixed-base), and the damage seemed to be a direct result of the vertical acceleration recorded in the slabs, which was also similar in all configurations. In the building isolated with TPB, an occurrence of near simultaneous uplift of all 9 TPBs (*i.e.*, bouncing of the entire building on the shake table) induced by a large vertical pulse with PGA = 1.3g. While the input sent a high frequency shock wave up through the columns, it was not transmitted to the slabs, as the slab vibration was dominated by single frequency vibration at the slab natural frequency. This suggests that the vertical motion transmitted to the nonstructural components and content is relatively insensitive to the base conditions (complete fixity, isolators with tensile restraint, or isolators free to uplift). Amplification of horizontal floor accelerations was observed in all configurations when subjected to combined horizontal/vertical table acceleration, although the extent to which this contributed to ceiling damage and content disruption is unclear. The amplification of horizontal acceleration was caused by at least two sources of horizontal-vertical coupling: coupling in the mode shapes due to building irregularities, and coupling in the TPB isolators (horizontal shear is proportional to vertical force). Though investigations are ongoing, the observations overwhelmingly suggest that mitigation of the effect of vertical component of excitation (e.g., 3D isolation, engineered floor systems, improved anchorage of nonstructural components and equipment) may be required to maintain functionality in essential base-isolated buildings.

4. Development and Testing of 3-Dimensional Isolation Systems

Efforts to combine horizontal seismic isolation with vertical vibration isolation date back to the early development of seismic isolation systems, driven by the power and utility industries focused on protection of large equipment. Multiple researchers have independently developed 3-

dimensional isolation systems by modifying the design parameters of laminated rubber bearings. As mentioned previously, the bearing shape factor, controlled by the thickness of the individual rubber layers, determines the vertical stiffness of the system [92]. Hence, a natural approach to provide vertical isolation is to decrease the bearing shape factor (see Equation 4) in order to increase the vertical fundamental natural period. The earliest isolation system developed by Kajima in Japan utilized steel laminated natural rubber bearings that provided a vertical frequency = 5 Hz, which is substantially more vertically flexible than a typical modern design. Oil dampers were included to reduce vertical and rocking motions, and a 2-story R/C acoustic laboratory demonstration building was constructed using this approach [66]. The use of laminated rubber bearings for 3-dimensional isolation was also explored for the United States nuclear industry by Aiken *et al.* [93]. The research progressed to the point of designing a 3D isolation system for a nuclear facility and performing characterization tests of individual 1/4 length scale bearings [94]. The target horizontal and vertical periods of the system were 2 seconds and 0.33 seconds, respectively. The investigation concluded that elastomeric isolators could possibly be designed to effectively provide isolation in the horizontal and vertical directions [93,94]. Application of 3D isolation to elastomeric devices is by nature limiting due to stability issues. As the thickness of the rubber layers increases, the critical load capacity of the bearing P_{cr} decreases proportionally. Furthermore, P_{cr} has been experimentally demonstrated to reduce with increasing horizontal displacement [22,95,96,97]. For a lightweight building, balancing the design objectives for displacement capacity and stability in the deformed configuration can be extremely difficult for elastomeric bearings even without consideration of vertical isolation. Thus, successful implementation of 3D isolation using elastomeric bearings to any building would likely constrain the horizontal isolation period to be somewhat smaller than typically used today.

The GERB system for 3-dimensional earthquake protection of structures was developed by a company that specializes in vibration isolation [98]. The system utilizes helical springs—with similar flexibility in all three directions—and viscous dampers, and was developed for seismic and vibration isolation of diesel and turbo generators. Shake table testing was performed on a 5-story building model in Yugoslavia [98], and the system was also tested by Mitsubishi Industries in Japan [66]. The system can be designed with a vertical fundamental frequency as low as 1.2 Hz, leading to a flexible rocking mode around 0.65 Hz. The authors acknowledged that horizontal floor accelerations increased with the 3D isolation system compared to horizontal only seismic isolation, but argued that the responses were acceptable. A residential building in California isolated with the GERB system was shaken strongly in the

1994 Northridge Earthquake [99]. The maximum recorded floor acceleration was about 0.63g, relative to input PGA < 0.5g; thus the horizontal isolation was less effective than a typical isolation system. Observed vertical ground acceleration was not significant (PGA ≈ 0.1 g).

Prior to the disaster at Fukushima in the 2011 Great East Japan Earthquake and Tsunami, Japan actively pursued 3D seismic isolation approaches for nuclear facilities. Several approaches have been developed that use air springs or pressurized air in the vertical direction [100]. The rolling seal type air spring is a steel/concrete cylinder lowered into an air cavity and attached with a rolling rubber seal, and is configured in series with a laminated rubber bearing for horizontal isolation [101,102]. The cable reinforced air spring consists of an inner cylinder attached to the base and an outer cylinder attached to the structure separated by an air cavity bounded by a flexible rubber sheet [103,104]. The distance between the cylinders allows the device to move both horizontally and vertically and thus accommodate horizontal and vertical motion. Also used in series with laminated rubber bearings, the hydraulic system consists of load carrying hydraulic cylinders filled with nitrogen gas, to which fluctuating pressure can be transmitted by the attached accumulator units [105]. As the flexibility of the isolation system in the vertical direction increases, the structure tends to develop a rocking mode with a substantial modal mass participation factor due to increased coupling in the horizontal and vertical directions. The proposed systems have vertical isolation periods on the order of 1–2 seconds, and generally utilize dampers (oil dampers or viscous wall dampers) and rocking suppression devices to control both vertical and rocking displacements.

A commercial solution for 3D isolation is available through Shimizu Corporation in Japan, and has been implemented in at least one 3-story apartment building [106,107]. Each device consists of a laminated rubber bearing on a steel frame that transmits the loads to three air springs and three shear force transmitting vertical sliders. An oil damper system—two oil dampers connected by cross-coupled pipes—provides both vertical and rocking suppression. The system is quite complicated and costly.

5. Conclusions

This review of shake table testing of base-isolated building has shown that a multitude of devices have been experimentally validated to perform as intended, which is to provide high attenuation of the input excitation to allow the structural system to remain elastic under large horizontal ground shaking. Of the many devices that have been tested, a handful of devices (elastomeric bearings, lead-rubber bearings, Friction Pendulum™, and Triple Pendulum™) claim the majority of the market share in the United States and around the world. Sufficient research has been conducted to suggest that base-isolated buildings can be detailed to survive earthquake events larger than anticipated in design, through activation of one or more ultimate limit state behaviors such as: unrestrained uplift, rupture of elastomeric bearings under tension and shear, buckling of elastomeric bearings, engagement of a displacement restraint or large displacement hardening. Many of these ultimate behaviors of the isolation system will redistribute demands to the superstructure, which should ultimately be detailed for ductile response.

The recent shake table tests of full-scale base isolated buildings at E-Defense have provided insight into the performance of the LRB-CLB and TP isolation systems subjected to tri-directional earthquake ground shaking and have underscored existing knowledge gaps regarding the performance of secondary systems in isolated buildings during representative earthquake ground shaking. Specifically, nonstructural components and content were shown to be vulnerable to damage or disruption under long duration motions generating large floor velocities and motions with large vertical components of excitation. Damage and disruption to these systems can adversely affect the post-event functionality of critical base-isolated facilities, and prevent the performance objectives from being achieved. While the tests at E-Defense provided some data on the demands imposed on these systems, the data is limited due to the large cost associated with performing these tests.

The results from testing at E-Defense have several implications. First, improved understanding of the response of nonstructural components and sensitive equipment is needed for analysis/design of base-isolated buildings. Seismic fragilities for nonstructural systems has generally associated damage with a single demand parameter, such as horizontal floor acceleration. However, the testing suggests that observed damage may be more complex, relating to multiple factors such as peak horizontal acceleration, peak vertical acceleration, and frequency or spectral content of the accelerations. The demands imposed on nonstructural components in base-isolated buildings have different proportions of horizontal to vertical response and different frequency content than conventional buildings, and seismic fragility data of nonstructural systems should be developed in such a way that the results are broadly applicable to all types of buildings.

Second, for buildings that aim to provide post-event functionality following a large earthquake, it is clear that vertical ground acceleration needs to be considered in design. Currently, vertical ground acceleration is rarely incorporated into the response history analysis methods that are often used for the design of critical facilities. To effectively consider vertical excitation, the engineering profession is in need of guidance on: (1) selection and scaling of vertical ground motions that are compatible with the horizontal motions; (2) analytical modeling techniques that can be employed in commercial software to accurately reflect local vertical modes and frequencies related to vibration of the floor system; and (3) accurate analytical models for vertical force-deformation behavior. Floor system modeling techniques should be verified with vibration data of full-scale slabs in realistic structural systems under seismic demands. In particular, for steel framing with concrete floor slabs, the influence of slab to girder connection configuration (fully-composite, partially composite, and non-composite) should be considered. With regard to modeling of the bearings, analytical models for the horizontal force-displacement behavior have outpaced the analytical models for the vertical force-displacement behavior. For example, a simple linear model is typically used to represent the vertical force-displacement behavior of elastomeric bearings while the true behavior is highly nonlinear (see Figure 2b). For FP class of bearings, techniques for evaluating compressive stiffness should be verified and techniques to account for impact upon uplift should be developed and experimentally verified.

Finally, closely associated with the need to evaluate the influence of vertical excitation in the design process is the desire to mitigate the influence of vertical excitation through design. Multiple approaches may be possible. For instance, vertical floor vibrations may be reduced in part through improved slab design. A fundamental understanding of the influence of slab vibration characteristics on the vertical vibration demands and nonstructural component response is needed. Furthermore, limited evidence—described in this review—has suggested that structural configuration irregularities (torsional and vertical) have contributed to horizontal-vertical coupling phenomena observed in some tests.

An alternative is to develop a packaged approach to base isolation that can provide both horizontal and vertical attenuation. To support the revival of the United States nuclear industry, base isolation is being investigated as a viable solution to satisfy Nuclear Regulatory Commission (NRC) guidelines [108] for seismic design based on a 100,000 year return period [109], and a packaged 3-dimensional base-isolation solution is considered highly desirable [109]. With vertical stiffness that is several thousand times the horizontal stiffness, current widely used seismic isolation devices provide isolation only in the horizontal direction, and may actually align the system with dominant frequencies of the vertical ground shaking. Past efforts to achieve 3-dimensional isolation have not produced a viable or cost-effective system. Stability considerations preclude low shape factor bearings from being a viable solution as only modest vertical periods, e.g., 0.33 seconds [94], can be achieved. While the Japanese systems might be effective, they are complex, have not been independently verified and their high cost and proprietary nature reduce the likelihood of adoption outside of Japan.

In summary, more research is needed to: (1) develop data on the 3-dimensional response of buildings isolated on current isolation systems; (2) evaluate existing analytical and numerical models of conventional isolation hardware for predicting demands imposed on the primary structural system as well as secondary and nonstructural systems; and (3) develop improved data on the seismic fragility of secondary and nonstructural systems under a variety of demands. Further research could show that complete protection of secondary and nonstructural systems using conventional isolation approaches might not be possible. Recognizing that the integrity of these systems is paramount to the post-event functionality of critical facilities, alternative or new systems to achieve complete 3-dimensional protection might be necessary.

References

1. Touaillon, J. Improvement in Buildings. U.S. Patent 99,973, 15 February 1870.
2. Constantinou, M.C.; Whittaker, A. *Performacne of Seismic Isolation Hardware under Service and Seismic Loading; Technical Report MCEER-07-0012*; Multidisciplinary Center for Earthquake Engineering Research, State University of New York at Buffalo: Buffalo, NY, USA, 2008.
3. Fenz, D.; Constantinou, M.C. Behaviour of the double concave Friction Pendulum bearing. *Earthq. Eng. Struct. Dyn.***2006**, *35*, 1403–1424.
4. Naeim, F.; Kelly, J.M. *Design of Seismic Isolated Structures: From Theory to Practice*, 1st ed; John Wiley and Sons: Hoboken, NJ, USA, 1999.
5. Taylor, A.; Aiken, I. What's Happened to Seismic Isolation of Buildings in the U.S.? *Structure*, 11 March 2011. Available online: http://www.structuremag.org/article.aspx?articleID=1404 (a ccessed on 30 July 2012).
6. Clarke, C.S.J.; Buchanan, R.; Efthymiou, M. Structural platform solution for seismic arctic environments-Sakhalin II offshore facilities. In Proceedings of the Offshore Technology Conference, Houston, TX, USA, 2-5 May 2005.
7. Kelly, J.M. Aseismic base isolation: Review and bibliography. *Soil Dyn. Earthq. Eng.* **1986**, *5*, 202–216.
8. Buckle, I.B; Mayes, R.M. Seismic isolation: History, application, and performance—A world view. *Earthq. Spectra* **1990**,*6*, 161–201.
9. Taylor, A.W.; Lin, A.N.; Martin, J.W. Performance of elastomers in isolation bearings: A literature review. *Earthq. Spectra* **1992**, *8*, 279–303.

10. Soong, T.T.; Constantinou, M.C. *Passive and Active Structural Vibration Control in Civil Engineering*; Springer-Verlag: New York, NY, USA, 1994.
11. Kunde, M.C.; Jangid, R.S. Seismic behavior of isolated bridges: A-state-of-the-art review. *Electron. J. Struct. Eng.* **2003**,*3*, 140–170.
12. Symans, M.D.; Cofer, W.F.; Fridley, K.J. Base isolation and supplemental damping systems for seismic protection of wood structures: Literature review. *Earthq. Spectra* **2003**, *18*, 549–572.
13. Taylor, A.W.; Igusa, T. *Primer on Seismic Isolation*; ASCE and Task Committee on Seismic Isolation: Reston, VA, USA, 2004.
14. Higashino, M.; Okamoto, S. *Response Control and Seismic Isolation of Buildings*; SPON Press: London, UK, 2006.
15. Lake, G.J.; Lindley, P.B. *Ozone Attack and Fatigue life of Rubber*; Maclaren and Sons LTD: London, UK, 1967; pp. 56–71.
16. Chalhoub, M.S.; Kelly, J.M. Effect of bulk compressibility on the stiffness of cylindrical base isolation bearings. *Int. J. Solids Struct.* **1990**, *26*, 743–760.
17. Constantinou, M.C.; Kartoum, A.; Kelly, J.M. Analysis of compression of hollow circular elastomeric bearings. *Eng. Struct.* **1992**, *14*, 103–111.
18. Highway Innovative Technology Evaluation Center, *Evaluation Findings for Skellerup Base Isolation Elastomeric Bearings*; Civil Engineering Research Foundation: Washington, DC, USA, 1998.
19. Highway Innovative Technology Evaluation Center, *Evaluation Findings for Dynamic Isolation Systems, Inc. Elastomeric Bearings*; Civil Engineering Research Foundation: Washington, DC, USA, 1998.
20. Kelly, J.M. *Dynamic Failure Characteristics of Bridgestone Isolation Bearings; Report No. UCB/EERC–91/04*; University of California: Berkeley, CA, USA, 1991.
21. Kelly, J.M. *Earthquake Resistant Design with Rubber*, 2nd ed; Springer Verlag: New York, NY, USA, 1997.
22. Nagarajaiah, S.; Ferrell, K. Stability of elastomeric isolation bearings. *J. Struct. Eng.* **1999**, *125*, 946–954.
23. Warn, G.P.; Whittaker, A.S. *A Study of the Coupled Horizontal-Vertical Behavior of Elastomeric and Lead-rubber Seismic Isolation Bearings; Technical Report MCEER-06-0011*; Multidisciplinary Center for Earthquake Engineering Research: Buffalo, NY, USA, 2006.
24. Gent, A.N. Fracture of glassy polymers. *ASME Appl. Mech. Div.* **1976**, *19*, 55–68.

25. Kramer, E.J. Microscopic and molecular fundamentals of crazing. *Adv. Polym. Sci.* **1983**, *52-53*, 1–56.
26. *26. Open System for Earthquake Engineering Simulation Framework, version 2.1.0*; University of California: Berkeley: CA, USA, 2009. Available online: http://opensees.berkeley.edu/ (accessed on 20 July 2012).
27. Robinson, W.H. A lead-rubber shear damper. *Bull. N. Z. Natl. Soc. Earthq. Eng.* **1977**, *3*, 93–101.
28. Robinson, W.H. Lead-rubber hysteretic bearings suitable for protecting structures during earthquakes. *Earthq. Eng. Struct. Dyn.* **1982**, *10*, 593–604.
29. Skinner, R.I.; Robinson, W.H.; McVerry, G.H. *An Introduction to Seismic Isolation*; John Wiley and Sons Inc.: New York, NY, USA, 1993.
30. Nagarajaiah, S.; Reinhorn, A.M.; Constantinou, M.C. Nonlinear dynamic analysis of 3-D-base-isolated structures. *J. Struct. Eng.* **1991**, *117*, 2035–2054.
31. Kalpakidis, I.V.; Constantinou, M.C. Effects of heating on the behavior of lead-rubber bearings. I: Theory. *J. Struct. Eng.* **2009**, *135*, 1440–1449.
32. Kalpakidis, I.V.; Constantinou, M.C. Effects of heating on the behavior of lead-rubber bearings. II: Verification of theory. *J. Struct. Eng.* **2009**, *135*, 1450–1461.
33. Kalpakidis, I.V.; Constantinou, M.C.; Whittaker, A.S. Modeling strength degradation in lead-rubber bearings under earthquake shaking. *Earthq. Eng. Struct. Dyn.* **2010**, *39*, 1533–1549.
34. Kalpakidis, I.V.; Constantinou, M.C. Principles of scaling and similarity for testing of lead-rubber bearings. *Earthq. Eng. Struct. Dyn.* **2010**, *39*, 1551–1568.
35. Ryan, K.L.; Kelly, J.M.; Chopra, A.K. Nonlinear model for lead-rubber bearings including axial-load effects. *J. Eng. Mech.* **2005**, *131*, 1270–1278.
36. Iizuka, M. A macroscopic model for predicting large-deformation behaviors of laminated rubber bearings. *Eng. Struct.* **2000**, *22*, 323–334.
37. Yamamoto, S.; Kikuchi, M.; Ueda, M.; Aiken, I.D. A mechanical model for elastomeric seismic isolation bearings include the influence of axial load. *Earthq. Eng. Struct. Dyn.* **2009**, *38*, 157–180.
38. Kikuchi, M.; Nakamura, T.; Aiken, I.D. Three-dimensional analysis for square seismic isolation bearings under large shear deformations and high axial loads. *Earthq. Eng. Struct. Dyn.* **2010**, *39*, 1513–1531.

39. Sanchez, J.; Masroor, A.; Mosqueda, G.; Ryan, K. Static and dynamic stability of elastomeric bearings for seismic protection of structures. *J. Struct. Eng.* **2012**. accepted for publication.
40. Zayas, V.; Low, S.; Mahin, S. *The FPS Earthquake Resisting System; Report No. UCB/EERC-87/01*; University of California: Berkeley, CA, USA, 1987.
41. Fenz, D.; Constantinou, M.C. Spherical sliding isolation bearings with adaptive behavior: Theory. *Earthq. Eng. Struct. Dyn.* **2008**, *37*, 163–183.
42. Fenz, D.; Constantinou, M.C. Spherical sliding isolation bearings with adaptive behavior: Experimental verification.*Earthq. Eng. Struct. Dyn.* **2008**, *37*, 185–205.
43. Becker, T.C.; Mahin, S.A. Experimental and analytical study of the bi-directional behavior of the triple friction pendulum isolator. *Earthq. Eng. Struct. Dyn.* **2012**, *41*, 355–373.
44. Mokha, A.; Constantinou, M.; Reinhorn, A. Teflon bearings in base isolation I: Testing. *J. Struct. Eng.* **1990**, *116*, 438–454.
45. Earthquake Protection Systems. Available online: http://www.earthquakeprotection.com/TechnicalCharacteristicsofFPBearngs.pdf (accessed on 20 April 2012).
46. Constantinou, M.; Mokha, A.; Reinhorn, A. Teflon bearings in base isolation II: Modeling. *J. Struct. Eng.* **1990**, *116*, 455–474.
47. Sarkisian, M.; Lee, P.; Hu, L.; Doo, C.; Zayas, V.; Constantinou, M.; Bachman, R. Property verification of triple pendulumtm seismic isolation bearings. In Proceedings of the 20th Analysis and Computation Specialty Track (2012 ASCE Structures Congress), Chicago, IL, USA, 29-31 March 2012.
48. Ryan, K.L.; Dao, N.D.; Sato, E.; Sasaki, T.; Okazaki, T. Aspects of isolation device behavior observed from full-scale testing of an isolated building at E-defense. In Proceedings of the 20th Analysis and Computation Specialty Track (2012 ASCE Structures Congress), Chicago, IL, USA, 29-31 March 2012.
49. Fenz, D.; Constantinou, M.C. Modeling triple friction pendulum bearings for response-history analysis. *Earthq. Spectra***2008**, *24*, 1011–1028.
50. Rousis, P.C.; Constantinou, M.C. Uplift-restraining friction pendulum seismic isolation system. *Earthq. Eng. Struct. Dyn.* **2006**, *35*, 577–593.
51. Roussis, P.C.; Constantinou, M.C. Experimental and analytical studies of structures seismically isolated with an uplift-restraining friction

pendulum bearing system. *Earthq. Eng. Struct. Dyn.* **2006**, *35*, 595–611.

52. Roussis, P.C. Study on the effect of uplift-restraint on the seismic response of base-isolated structures. *ASCE J. Struct. Eng.* **2009**, *135*, 1462–1471.
53. Kelly, J.M.; Skinner, M.S.; Beucke, K.E. *Experimental Testing of an Energy Absorbing Seismic Isolation System; Report No. UCB/EERC-80/35*; University of California,Berkeley: CA, USA, 1980.
54. Kelly, J.M.; Beucke, K.E.; Skinner, M.S. *Experimental Testing of a Friction Damped Aseismic Base Isolation System with Fail-safe Characteristics; Report No. UCB/EERC-80/18*; University of California: Berkeley, USA, 1980.
55. Kelly, J.M.; Beucke, K.E. A friction damped base isolation system withfail-safe characteristics. *Earthq. Eng. Struct. Dyn.***1983**, *11*, 33–56.
56. Kelly, J.M.; Hodder, S.B. *Experimental Study of Lead and Elastomeric Dampers for Base Isolation Systems; Report No. UCB/EERC-81/16*; University of California: Berkeley, CA, USA, 1981.
57. Kelly, J.M.; Chalhoub, M.S. *Earthquake Simulator Testing of a Combined Sliding Bearing and Rubber Bearing Isolation System; Report No. UCB/EERC-87/04*; University of California: Berkeley, CA, USA, 1990.
58. Mokha, A.; Constantinou, M.C.; Reinhorn, A.M. *Teflon Bearings in Aseismic Base Isolation: Experimental Studies and Analytical Modeling; Technical Report NCEER-88-0038*; State University of New York at Buffalo: Buffalo, NY, USA, 1988.
59. Mokha, A.; Constantinou, M.C.; Reinhorn, A.M. *Experimental Study and Analytical Prediction of Earthquake Response of a Sliding Isolation System with Spherical Surface; Technical Report NCEER-90-0020*; State University of New York at Buffalo: Buffalo, NY, USA, 1990.
60. Mokha, A.; Amin, N.; Constantinou, M.C.; Zayas, V. Experimental study of friction pendulum isolation system. *ASCE J. Struct. Eng.* **1991**, *117*, 1201–1217.
61. Al-Hussaini, T.M.; Constantinou, M.C.; Zayas, V.A. *Seismic Isolation of Multi-Story Frame Structures Using Spherical Sliding Isolation System; Technical Report NCEER-94-0007*; State University of New York at Buffalo: Buffalo, NY, USA, 1994.
62. Constantinou, M.C.; Mokha, A.S.; Reinhorn, A.M. *Experimental and Analytical Study of a Combined Sliding Disc Bearing and Helical Steel*

Spring Isolation System; Technical Report NCEER-90-0019; State University of New York at Buffalo: Buffalo, NY, USA, 1990.

63. Constantinou, M.C.; Mokha, A.S.; Reinhorn, A.M. Study of sliding bearing and helical-steel-spring isolation system.*ASCE J. Struct. Eng.* **1991**, *117*, 1257–1275.
64. Mostaghel, N.; Khodaverdian, M. Dynamics of resilient-friction base isolator (R-FBI). *Earthq. Eng. Struct. Dyn.* **1987**,*15*, 379–390.
65. Clark, P.W.; Kelly, J.M. *Experimental Testing of the Resilient-Friction Base Isolation System; Report No. UCB/EERC-90/10*; University of California: Berkeley, CA, USA, 1990.
66. Kelly, J.M. *Base Isolation in Japan; Report No. UCB/EERC-88/20*; University of California: Berkeley, CA, USA, 1988.
67. 67. Nagashima, I.; Kawamura, S.; Kitazawa, K.; Hisano, M. Study on a base isolation system. In Proceedings of the 3rd Conference on Soil Dynamics and Earthquake Engineering, Princeton University, Princeton, NJ, 22-24 June 1987.
68. Bakhshi, A.; Araki, H.; Shimazu, T. Evaluation of the performance of a suspension isolation system subjected to strong ground motion. *Earthq. Eng. Struct. Dyn.* **1998**, *27*, 29–47.
69. Guerreiro, L.; Azevedo, J.; Muhr, A. Seismic tests and numerical modeling of a rolling-ball isolation system. *J. Earthq. Eng.* **2007**, *11*, 49–66.
70. Tashkov, L.; Manova, K.; Krstevska, L.; Garevski, M. Evaluation of efficiency of ALSC floating-sliding base-isolation system based on shake table test and floor response spectra. *Bull. Earthq. Eng.* **2010**, *8*, 995–1018.
71. Toopchi-Nezhad, H.; Tait, M.J.; Drysdale, R.G. Shake table study on an ordinary low-rise building seismically isolated with SU-FREIs (stable unbounded-fiber reinforced elastomeric isolators). *Earthq. Eng. Struct. Dyn.* **2009**, *38*, 1335–1357.
72. Nakamura, Y.; Saruta, M.; Wada, A.; Takeuchi, T.; Hikone, S.; Takahashi, T. Development of the core-suspended isolation system. *Earthq. Eng. Struct. Dyn.* **2011**, *40*, 429–447.
73. Fenz, D.M.; Constantinou, M.C. *Development, Implementation, and Verification of Dynamic Analysis Models for Multi-spherical Sliding Bearings; Technical Report MCEER-08-0018*; Multidisciplinary Center for Earthquake Engineering Research, State University of New York at Buffalo: Buffalo, NY, USA, 2008.
74. Morgan, T.A.; Mahin, S.A. *The Use of Base Isolation Systems to Achieve Complex Seismic Performance Objectives; PEER Report No. 2011/06*;

Pacific Earthquake Engineering Research Center, University of California: Berkeley, CA, USA, 2011.
75. Kelly, J.M. *The Influence of Base Isolation Systems on the Seismic Response of Light Secondary Equipment; Report No. UCB/EERC-81/17*; Earthquake Engineering Research Center, University of California: Berkeley, CA, USA, 1981.
76. Kelly, J.M.; Tsai, H.C. Seismic response of light internal equipment in base-isolated structures. *Earthq. Eng. Struct. Dyn.* **1985**, *13*, 711–732.
77. Juhn, G.; Manolis, G.; Constantinou, M.C.; Reinhorn, A.M. Experimental investigation of secondary systems in a base-isolated structure. *ASCE J. Struct. Eng.* **1992**, *118*, 2204–2221.
78. Wolff, E.D.; Constantinou, M.C. *Wolff, E.D.; Constantinou, M.C. Experimental Study of Seismic Isolation Systems with Emphasis on Secondary System Response and Verification of Accuracy of Dynamic Response History Analysis Methods; Technical Report MCEER-04-0001*; Multidisciplinary Center for Earthquake Engineering Research, State University of New York at Buffalo: Buffalo, NY, USA, 2004.
79. Griffith, M.C.; Kelly, J.M.; Coveney, V.A.; Koh, C.G. *Experimental Evaluation of Seismic Isolation of Medium-Rise Structures Subjected to Uplift; Report No. UCB-EERC 88-02*; University of California: Berkeley, CA, USA, 1988.
80. Griffith, M.C.; Aiken, I.D.; Kelly, J.M. *Experimental Evaluation of Seismic Isolation of a 9-story Braced Steel Frame Subjected to Uplift; Report No. UCB-EERC 88-05*; University of California: Berkeley, CA, USA, 1988.
81. Kelly, J.M.; Griffith, M.C.; Aiken, I.D. *A Displacement Control and Uplift Restraint Device for Base-Isolated Structures; Report No. UCB-EERC 87-03*; University of California: Berkeley, CA, USA, 1987.
82. Griffith, M.C.; Aiken, I.D.; Kelly, J.M. Displacement control and uplift restraint for base-isolated structures. *ASCE J. Struct. Eng.* **1990**, *116*, 1135–1148.
83. Nagarajaiah, S.; Reinhorn, A.M.; Constantinou, M.C. Experimental study of sliding isolated structures with uplift restraint. *ASCE J. Struct. Eng.* **1992**, *118*, 1666–1682.
84. Kasalanati, A.; Constantinou, M.C. Testing and modeling of prestressed isolators. *ASCE J. Struct. Eng.* **2005**, *131*, 857–866.
85. Toniolo, R. THK:CLB crossed Linear Bearing Seismic Isolators. In Proceedings of the 2008 Seismic Engineering Conference: Commemorating the 1908 Messina and Reggio Calabria Earthquake, Reggio Calabria, Italy, 8-11 July 2008.

86. Clark, P.W.; Aiken, I.D.; Kelly, J.M. *Experimental Studies of the Ultimate Behavior of Seismically Isolated Structures; Report No. UCB/EERC-97/18*; University of California: Berkeley, CA, USA, 1997.
87. Sato, N.; Kato, A.; Fukushima, Y.; Iizuka, M. Shaking table tests on failure characteristics of base isolation system for a DFBR plant. *Nuclear Eng. Des.* **2002**, *212*, 293–305.
88. Takaoka, E.; Takenaka, Y.; Nimura, A. Shaking table tests and analysis method on ultimate behavior of slender base-isolated structure supported by laminated rubber bearings. *Earthq. Eng. Struct. Dyn.* **2011**, *40*, 551–570.
89. Hwang, J.-S.; Hsu, T.-Y. Experimental study of isolated building under triaxial ground excitations. *ASCE J. Struct. Eng.* **2000**, *126*, 879–886.
90. Sato, E.; Furukawa, S.; Kakehi, A.; Nakashima, M. Full-scale shaking table test for examination of safety and functionality of base-isolated medical facilities. *Earthq. Eng. Struct. Dyn.* **2011**, *40*, 1435–1453.
91. Soroushian, S.; Ryan, K.L.; Maragakis, M.; Sato, E.; Sasaki, T.; Okazaki, T.; Tedesco, L.; Zaghi, A.E.; Mosqueda, G.; Alvarez, D. Seismic response of ceiling/sprinkler piping nonstructural systems in NEES TIPS/NEES nonstructural/NIED collaborative tests on a full scale 5-story building. In Proceedings of the 2012 ASCE Structures Congress, Chicago, IL, USA, 29-31 March 2012.
92. Warn, G.P.; Whittaker, A.S.; Constantinou, M.C. Vertical stiffness of elastomeric and lead-rubber seismic isolation bearings. *ASCE J. Struct. Eng.* **2007**, *133*, 1227–1236.
93. Aiken, I.D.; Kelly, J.M.; Tajirian, F.F. *Mechanics of Low Shape Factor Elastomeric Seismic Isolation Bearings; Report No. UCB/EERC-89/13*; University of California: Berkeley, CA, USA, 1989.
94. Tajirian, F.F.; Kelly, J.M.; Aiken, I.D.; Veljovich, W. Elastomeric bearings for three-dimensional seismic isolation. In Proceedings of the 1990 ASME PVP Conference, Nashville, TN, USA, 17-21 June 1990.
95. Buckle, I.G.; Liu, H. Critical loads of elastomeric isolators at high shear strain. In Proceedings of the 3rd U.S.-Japan Workshop on Earthquake Protective Systems for Bridges, Berkeley, CA, USA, 24-25 January 1994.
96. Buckle, I.G.; Liu, H. Experimental determination of critical loads of elastomeric isolators at high shear strain. *NCEER Bull.* **1994**, *8*, 1–5.
97. Buckle, I.G.; Nagarajaiah, S.; Ferrell, K. Stability of elastomeric isolation bearings: Experimental study. *ASCE J. Struct. Eng.* **2002**, *128*, 3–11.

98. Huffman, G.K. Full base isolation for earthquake protection by helical springs and viscous dampers. *Nuclear Eng. Des.* **1985**, *84*, 331–338.
99. Makris, N.; Deoskar, H.S. Prediction of observed response of base-isolated structure. *ASCE J. Struct. Eng.* **1996**, *122*, 485–493.
100. Inoue, K.; Morishita, M.; Fujita, T. Development of three-dimensional seismic isolation technology for next generation nuclear power plant in Japan. In Proceedings of the ASME/JSME 2004 Pressure Vessels and Piping Conference (PVP2004), San Diego, CA, USA, 25-29 July 2004.
101. Suhara, J.; Tamura, T.; Ohta, K.; Okada, Y.; Moro, S. Research on 3-D base isolation system applied to new power reactor 3-d seismic isolation device with rolling seal type air spring: Part 1. In Proceedings of the 17th International Conference on Structural Mechanics in Reactor Technology (SMiRT 17), Prague, Czech Republic, 17-22 August 2003.
102. Suhara, J.; Matsumoto, R.; Oguri, S.; Okada, Y.; Inoue, K.; Takahashi, K. Research on 3-D base isolation system applied to new power reactor, 3-d seismic isolation device with rolling seal type air spring: Part 2. In Proceeding of 18th International Conference on Structural Mechanics in Reactor Technology, Beijing, China, 7-12 August 2005.
103. Kageyama, M.; Iba, T.; Umeki, K.; Somaki, T.; Moro, S. Development of three-dimensional base isolation system with cable reinforcing air spring. In Proceedings of the 17th International Conference on Structural Mechanics in Reactor Technology (SMiRT 17), Prague, Czech Republic, 17-22 August 2003.
104. Kageyama, M.; Iba, T.; Umeki, K.; Somaki, T.; Hino, Y.; Moro, S.; Ikutama, S. Study on three-dimensional seismic isolation system for next generation nuclear power plant: independent cable reinforced rolling-seal air spring. In Proceedings of the 13th World Conference on Earthquake Engineering, Vancouver, BC, Canada, 1-6 August 2004.
105. Kashiwazaki, A.; Shimada, T.; Fujikawa, T.; Moro, S. Study on 3-dimensional base isolation system applying to new type power plant reactor (hydraulic 3-dimensional base isolation system: No.1). In Proceedings of the 17th International Conference on Structural Mechanics in Reactor Technology (SMiRT 17), Prague, Czech Republic, 17-22 August 2003.
106. Takahashi, O.; Aida, H.; Suhara, J.; Matsumoto, R.; Tsuyuki, Y.; Fujita, T. Construction of civil building using three dimensional seismic isolation system: Part 1, design of building using three

dimensional seismic isolation system. In Proceedings of the 14th World Conference on Earthquake Engineering, Beijing, China, 12-17 October 2008.

107. Suhara, J.; Matsumoto, R.; Torita, H.; Tsuyuki, Y.; Kamei, T.; Takahashi, O.; Kunimatsu, Y.; Aida, H.; Fujita, T. Construction of civil building using three dimensional seismic isolation system: Part 2, tests for three dimensional seismic isolation system. In Proceedings of the 14th World Conference on Earthquake Engineering, Beijing, China, 12-17 October 2008.

108. USA Nuclear Regulatory Commission, Identification and characterization of seismic sources and determination of safe shutdown earthquake ground motion. In *USA Nuclear Regulatory Commission Regulatory Guide 1.165*; USA Nuclear Regulatory Commission: Rockville, MD, USA, 1997.

109. Malushte, S.R.; Whittaker, A.S. Survery of past base isolation applications in nuclear power plants and challenges to industry/regulatory acceptance. In Proceedings of the 18th International Conference on Structural Mechanics in Reactor Technology (SMiRT 18), Beijing, China, 7-12 August 2005.

CHAPTER 2

Seismic Analysis of Historic Masonry Buildings: The *Vicarious Palace* in Pescia (Italy)

Michele Betti [1,*] and Luciano Galano [1]

[1] Department of Civil and Environmental Engineering (DICeA), University of Florence, Via di Santa Marta, 3 - I-50139, Florence, Italy;

Abstract

Recent Italian earthquakes have underlined the need for wide monitoring and safety assessment of architectonical heritage. This has emerged also from requirements of the new Italian Technical Recommendations for buildings. Within this subject the paper investigates the seismic vulnerability of a specific monumental masonry building: the Vicarious Palace (*Palazzo del Vicario*) in Pescia, a small town near Florence. The structural behavior of the Palace was investigated using a finite element model in which the non-linearities of the masonry were considered by proper constitutive assumptions. The seismic behavior was evaluated by the pushover method, according to the Italian Technical Recommendations. The results were compared with the ones obtained by a simplified approach based on the kinematic theorem of limit analysis. Comparisons of the expected seismic demand *vs* the seismic capacity of the Palace confirm the weakness of this type of building to suffer extensive damage under earthquakes, as frequently observed in similar construction typologies. Additionally, the comprehension of the structural behavior under seismic loading allows the identification of a proper retrofitting strategy.

Keywords: historic masonry buildings; seismic vulnerability; finite element modeling; nonlinear analysis; pushover analysis

1. Introduction

The Italian earthquakes of Umbria and Marche in 1997 [1] and Abruzzo in 2009 [2], have highlighted the need for an extensive monitoring and safety assessment of historic Italian construction heritage. Following the debate developed in the scientific community, new Italian Technical Recommendations for buildings have been drawn up: first after 2003 [3,4] and later, after 2008 [5,6]. As is known, ancient masonry buildings, although perfectly able to bear vertical loads, usually are not adequate to sustain the horizontal forces produced by earthquakes. So, they are particularly prone to severe damage under seismic loading. In addition, historic built heritage represents an economic concern especially in contexts where tourism has become one of the major sources of wealth. Therefore, preserving historic constructions is a cultural requirement and an economic and developmental demand [7].

The structural analysis of a new masonry building is a relatively simple task, whereas the assessment of the seismic vulnerability of an ancient masonry building is a challenging task, because of several uncertainties affecting the geometrical and the mechanical characteristics of the structural elements [8]. Each masonry building is characterized by its own history and the actual configuration is the result of fusions, additions and replacements of many structural elements. Thus, a correct structural analysis of an historic building requires a deep knowledge of: (a) the building history and its evolution; (b) the geometry; (c) the structural details; (d) the cracking pattern and the damage map; and (e) the masonry construction techniques [9,10].

The characterization of the material properties is required to develop a proper structural analysis. This knowledge can be achieved combining *in-situ*and laboratory experimental tests. Due to the difficulties in obtaining all information to be used for a proper definition of a numerical model, in some cases it is necessary to perform simplified, often iterative, procedures for evaluation of the static and the seismic reliability. Hence, the structural engineers should be able to address the relevant aspects of the problem and provide guidance in analyses and experiments [11].

Maintenance of historic buildings has become a relevant scientific issue that has attracted the interest of researchers all over the world as demonstrated by the growing number of researches presented in recent decades. These studies represent a wide state-of-the-art of the engineering approach for evaluation of safety of historic buildings. Among the various researches in this field there is the work of Lourenço *et al.* [12]. The authors discuss the case of the Monastery of Jerónimos in Lisbon (Portugal) and show that it is possible to understand the behavior and the damage of a complex historic construction providing valuable information in designing *in-situ*tests and monitoring.

The relevance of proper numerical strategies has been pointed out by Cardoso *et al.* [13,14]. These authors discuss the aseismic provisions included in old masonry buildings in downtown Lisbon, after the Lisbon earthquake in 1755. They propose an iterative method for the seismic assessment in which the damage to the structural elements or connections is identified and the stiffness of the structural model is changed accordingly. Each iteration comprises a linear elastic analysis, so the procedure is useful for current design practice. Betti and Vignoli [15] discussed the seismic vulnerability of the Basilica of Santa Maria of Impruneta (Italy), and proposed the combination of a finite element model of the building with a simplified approach based on the kinematic theorem of limit analysis. The results suggest that the comparison between different techniques of analysis is mandatory to cover the unknowns affecting the mechanic properties of the materials. Relevance of a wide spread analysis approach has been pointed out also by Mallardo *et al.* [16]. The authors discuss the seismic behavior of a Renaissance Palace in Ferrara (Italy). Firstly, a 3D nonlinear model of the Palace is considered in order to understand the large-scale structural performance. Next, a detailed study of the main façade of the Palace is presented, by means of three reduced 2D nonlinear models. The authors show that the use of different modeling strategies allows a critical evaluation of the seismic vulnerability of the main elements. An integrated multilevel approach that combines laboratory and non-destructive testing methods with monitoring systems has been recently presented by Anzani *et al.* [17] to evaluate the state of conservation of historic structures. They discussed the cases of some damaged towers in Italy and suggested an investigation procedure to be adopted for safety assessment of historic towers.

This paper presents an evaluation of the seismic vulnerability of the Vicarious Palace (*Palazzo del Vicario*) in Pescia, a small town near Florence. In the first part, the geometrical and mechanical descriptions of the building are presented, together with a brief report of its history. Then, the seismic assessment approach is illustrated.

Firstly, a structural analysis of the Palace was made using a finite element model. Specific assumptions of the material properties and the nonlinear behavior of the masonry were considered. Static analysis under vertical loads (dead and live loads), and pushover analyses were performed, according to the Italian Technical Recommendations [3,4]. Next, a simplified approach based on the kinematic theorem of limit analysis was applied. With reference to the failure mechanisms activated in similar buildings during past earthquakes [18,19], some macro-elements were individuated and analyzed.

The results provide a description of the building response under seismic loading and offer a representative case which is fruitful in extending the comprehension of the structural behavior of this building typology.

2. The *Vicarious Palace* in Pescia

2.1. Historic Notes

The Vicarious Palace is located in the main square (*Piazza Mazzini*) of the city of Pescia. Figure 1a shows the front view of the building. The research started with the analysis of historic data. The material was found in the archive of the Ministry Institution for Environmental and Architectural Heritage of Florence, Prato and Pistoia (the institution employed to preserve the cultural heritage in the Florence metropolitan area).

The original structure of the Palace dates back to the twelfth or thirteenth century, when the Holy Roman Emperor Frederick II of Swabia (Jesi, Italy, 1194–Fiorentino di Puglia, Italy, 1250) created the institution of the Imperial Vicar, a prince charged with administering part of the Holy Roman Empire on behalf of the Emperor. Soon after its construction the Palace was used for the administrative control of territory and for residence of the local governor. Thus, the history of the Palace intersects with the local history of the city.

Along the centuries the Palace was subjected to several modifications. Some of them were aimed at renewing the building according to the necessities of the time, such as changes in shape, in number of levels, etc., other interventions were aimed at retrofitting the building. In the middle of the fourteenth century a new level was added, so the Palace became a three storey building [20,21].

A further level was added during retrofitting started in 1617, so the Palace then became a four storey building [22]; at the same time a new portico was added. This configuration of the Palace remained almost unchanged until the early years of the twentieth century (Figure 1b) when the building was subject to the final modifications. In the thirties of the twentieth century the original medieval configuration of the Palace was reconstructed, according to the propagandistic styles of Fascism to celebrate the history of Italy. So, the fourth floor was eliminated and the portico was demolished [21]. Interested readers can refer to [20,21,22] for more details.

Figure 1. (a) The Vicarious Palace as is today; (b) the Vicarious Palace at the beginning of XX century.

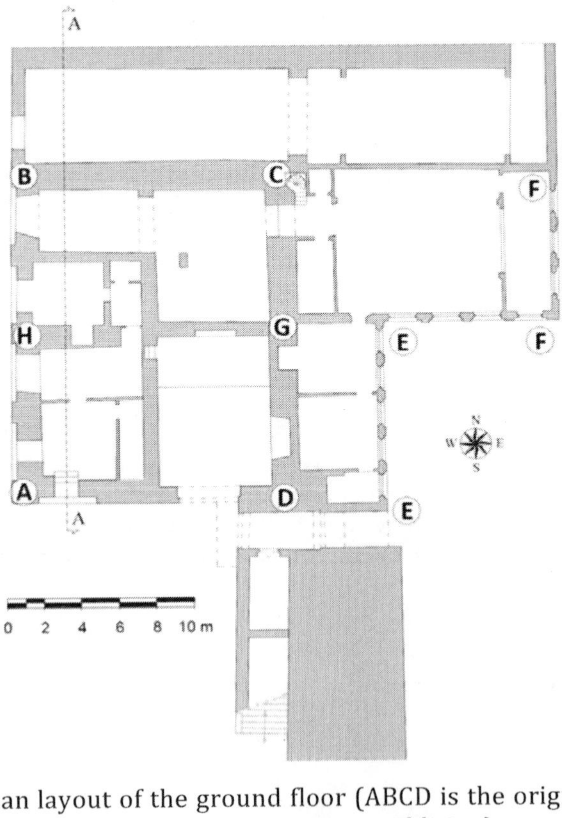

Figure 2. Plan layout of the ground floor (ABCD is the original medieval square nucleus; EE-FF is a twentieth century addition).

2.2. Geometrical and Material Data

The plan layout of the ground floor of the Palace is reported in Figure 2. It is possible to recognize the original medieval square nucleus, delimited by the walls indicated with AB, BC, CD and DA, whose dimensions are about 18 m × 15 m. In particular, the wall CD probably corresponded to the main city walls [20]. This square ground floor was an open space (as a loggia) used for public administrative activities. The adjoining walls EE, EF and FF were added during the last intervention in the thirties of the last century. The actual height of the Palace, after the demolition of the fourth storey, is about 17 m as reported in the section represented in Figure 3. The horizontal structures of the first floor are made of masonry vaults, while the upper floors and the roof are timber structures. Views of the main façades of the building are reported in Figures 4, 5 and 6.

The walls of the oldest nucleus at the first level are stone masonry walls (local sandstone) and have a thickness of about 1 m. The walls of the upper levels are built with bricks and their thickness decreases as the level increases. Masonry was also used for the construction of the walls of the twentieth century addition. A specific plan of *in-situ*tests to assess the mechanical properties of the masonries was not possible in this case. Hence, the strength and deformability of the walls were evaluated considering the Technical Recommendations [3,6] and the results of *in-situ* and laboratory tests on similar masonry materials [23,24], see Table 1. These parameters are conservative estimations of the average values for existing buildings.

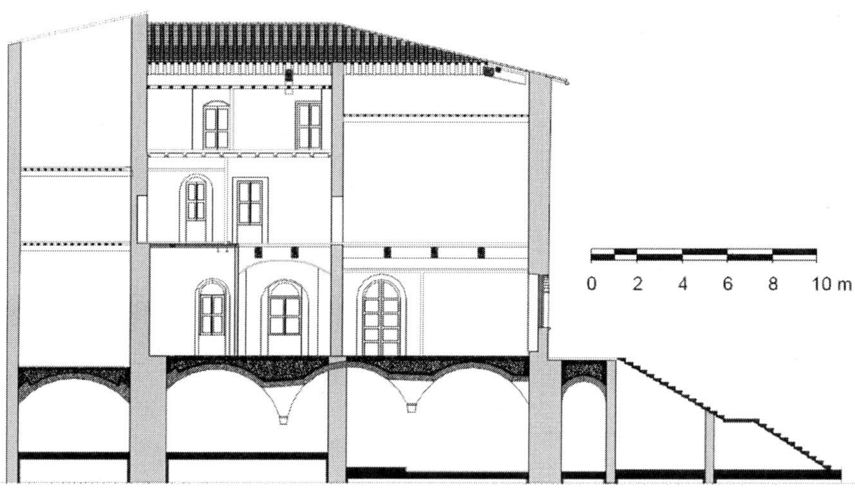

Figure 3. Section A-A of the Palace.

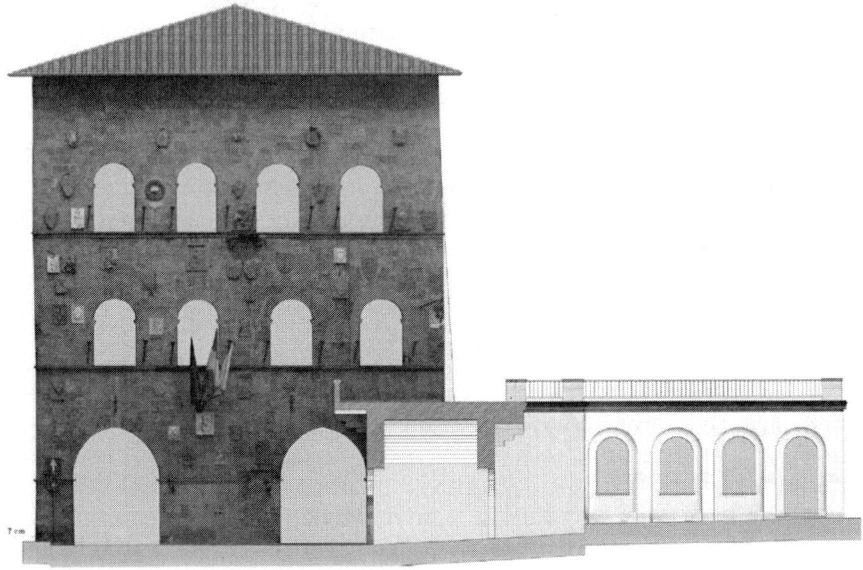

Figure 4. View of the Southern façade.

Figure 5. View of the Western façade.

Figure 6. View of the Eastern façade.

Table 1. Mechanical parameters of the masonries.

		Stone walls	Brick walls
γ (kN/m^3)	Specific weight	21	18
f_m (N/mm^2)	Compressive strength	2.5	3.0
f_t (N/mm^2)	Tensile strength	0.12	0.12
τ_0 (N/mm^2)	Shear strength	0.093	0.100
E (N/mm^2)	Longitudinal elastic modulus	1740	2100
G (N/mm^2)	Tangential elastic modulus	290	350
υ (-)	Poisson coefficient	0.2	0.2

3. 3. Finite Element Model

The structural analysis of the Palace was approached by a 3D finite element model based on the concepts of homogenized material and smeared crack modeling using the ANSYS code v.11.0 [25]. The masonry walls were discretized with *Solid65*elements (isoparametric three-dimensional eight node elements), whereas the ground floor vaults were discretized with*Shell63* elements (isoparametric two-dimensional four node elements). The wooden floors and timber roof were not considered in the model and their own weights were applied on the bearing walls by means of vertical concentrated loads.

An *in-situ* survey of the building was made to accurately reproduce the geometry, the structural details and the irregularities of the Palace. This investigation consisted of a geometrical relief aimed at characterizing the masonry texture and to check up the quality of wall-to-wall, wall-to-floors and wall-to-roof connections. The 3D model is represented in Figure 7. It consists of 37,402 nodes, 36,557 *Solid65* elements and 813 *Shell63* elements, corresponding to 108,399 DOFs.

The nonlinear behavior of the masonry was represented by combining plasticity with a smeared crack approach. The perfectly-elastic plastic behavior with the Drucker-Prager (DP) yield surface was used to reproduce the plasticity properties, whereas the Willam-Warnke (WW) failure surface was considered for model crush and cracking [26].

Both DP and WW models were been extensively used to simulate the inelastic behavior of masonry. Discussing the homogenization approach for masonry, Zucchini and Lourenco [27] adopted the DP model to simulate the plastic deformation in masonry cells. They showed that it is possible to account for degradation of masonry mechanical properties in compression. Cerioni *et al.* [28] adopted the DP criterion to investigate the seismic behavior of the Parma Cathedral Bell-Tower. The WW criterion was used by Adam *et al.* [29] for model cracking and crushing capabilities of masonry materials; comparison between numerical and experimental results shows good agreement. Chiostrini *et al.* [23] combined the DP and the WW criteria to reproduce the results of several diagonal tests on masonry specimens, obtaining good agreement with experimental values. Betti and Vignoli combine the DP and the WW criteria to discuss the seismic vulnerability of a masonry church [15,30].

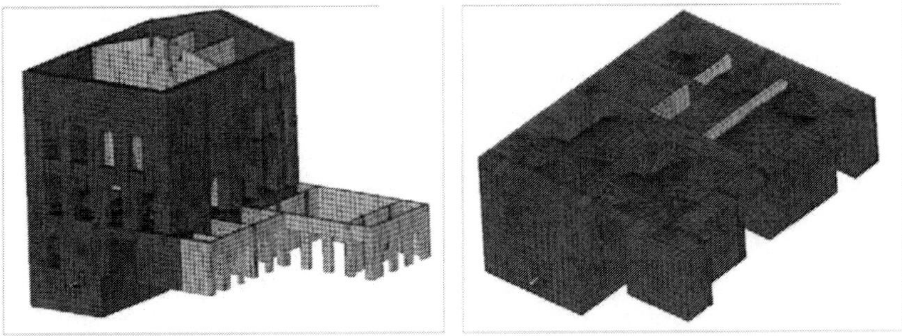

Figure 7. Finite element model of the Vicarious Palace.

The cohesion c and the internal friction angle φ are the two material parameters required to define the DP yield surface. The Willam-Warnke (WW) failure surface is defined by two material constants, *i.e.* the uniaxial compressive strength f_{cWW} and the uniaxial tensile strength f_{tWW} of the masonry. A proper selection of these values allows the introduction of a cut-off to the tensile strength and an upper limit to the biaxial compressive strength. So, the model reproduces effectively the small tensile strength of the masonry, the plasticity behavior in regime of average compression and the crushing phenomenon for high compressive stresses. Table 2 collates the values of the nonlinear material parameters.

Table 2. Constitutive parameters of the masonries adopted for nonlinear FEM analyses (DP = plasticity criterion and WW = failure surface).

	DP yield surface	
c (N/mm^2)	Cohesion	0.09
φ (°)	Friction angle	38°
δ (°)	Dilatancy angle	15°
	WW failure surface	
f_{cWW} (N/mm^2)	Uniaxial compressive strength	4.00
f_{tWW} (N/mm^2)	Uniaxial tensile strength	0.12
β_c (-)	Shear transfer coefficient for closed cracks	0.75
β_t (-)	Shear transfer coefficient for open cracks	0.25

Firstly, the 3D model was used to evaluate the stresses in the masonry walls produced by the vertical loads *i.e.*,: the dead weights, the live floor loads (2.0 kN/m^2) and the snow load on the roof (1.1 kN/m^2). The whole structure was analyzed in the nonlinear range to identify the weak points of possible failure, assuming fixed base restraint conditions.

Results in terms of vertical stresses are reported in Figures 8 and 9. The average compressive stress at the base of the walls is about 0.5 N/mm² with small tensile stresses located on the top surface of the walls. These tensile stresses are produced by the local numerical effect depending on the concentrated loads of the roof. The maximum compressive stress (about 1.7 N/mm²) is reached in a column close to an arch opening in the internal wall (Figure 9). Even if this value is high it is much less than the crushing limit of the masonry (local sandstone). Results of the analysis illustrate that the Palace is adequate to withstand the vertical loads under exercise conditions. This is a common result for this typology of buildings designed by skilled manufacturers.

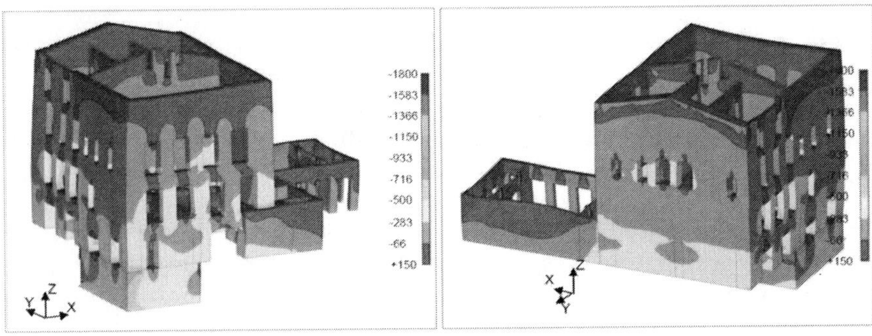

Figure 8. Static analysis: vertical compressive stresses (units are kN/m²).

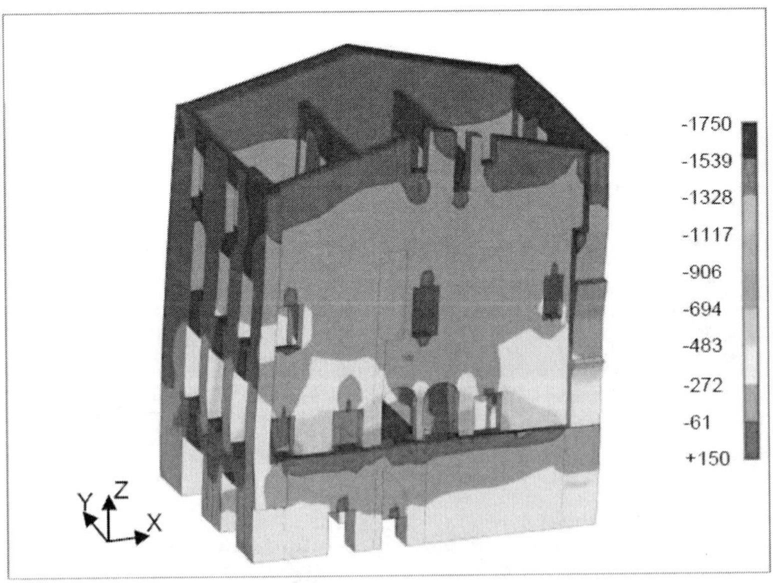

Figure 9. Static analysis: vertical compressive stresses, detail of the internal wall (units are kN/m²).

4. Seismic Analysis

4.1. Kinematic Analysis

The main assumptions of the kinematic method are: (a) masonry is a no-tension material (although this hypothesis may be considered conservative, only very small tension forces are transferred across the mortar joints); (b) masonry has infinite compressive strength (this hypothesis is not conservative, but in most cases collapse of masonry is due to cracks opening rather than to compression failure); and (c) structures are considered as assemblages of macro-elements (rigid bodies). The assessment of these macro-elements is based on structural geometry, shape (*i.e.*, whole façades), details (*i.e.*, the quality of existing connections between walls), and by examining collapses and damages consequent to past earthquakes in similar buildings [11,31].

Each allowable mechanism has a single degree of freedom and it is characterized by the formation of linear hinges among the single macro-elements. The seismic loads are assumed to be static horizontal forces amplified by a kinematic multiplier α; the gravity loads are stabilization forces. Given an allowable mechanism, the corresponding kinematic multiplier at collapse α_0 is evaluated by the Virtual Work Theorem (V.W.T.).

The mechanisms considered in this study concern partial and global overturns of the main façades, and the collapse multipliers α_0 are evaluated according to Eq. (1) (see the general scheme reported in Figure 10; A, B, C and D identify different mechanisms):

$$\alpha_0\left(\sum_{i=1}^{n} W_i\delta_{xi} + \sum_{i=1}^{n} P_i\delta_{xi}\right) + Q\delta_{xq} - \sum_{i=1}^{n} W_i\delta_{yi} - \sum_{i=1}^{n} P_i\delta_{yi} = L_{fi} \qquad (1)$$

In Equation (1) W_i are the dead weights of the elements involved in the mechanism; P_i are the weights transmitted at the selected macro-element by vault, floors or roof; δ_{xi} are the horizontal virtual displacements of the points of application of W_i and P_i; δ_{yi} are the vertical virtual displacements of the points of application of W_i and P_i; Q is the static drift of the vault; δ_{xq} is the corresponding horizontal virtual displacement and L_{fi} is the virtual work of the internal forces, here assumed equal to zero. After the evaluation of the collapse multiplier α_0, the corresponding seismic spectral acceleration a_0^* is given by:

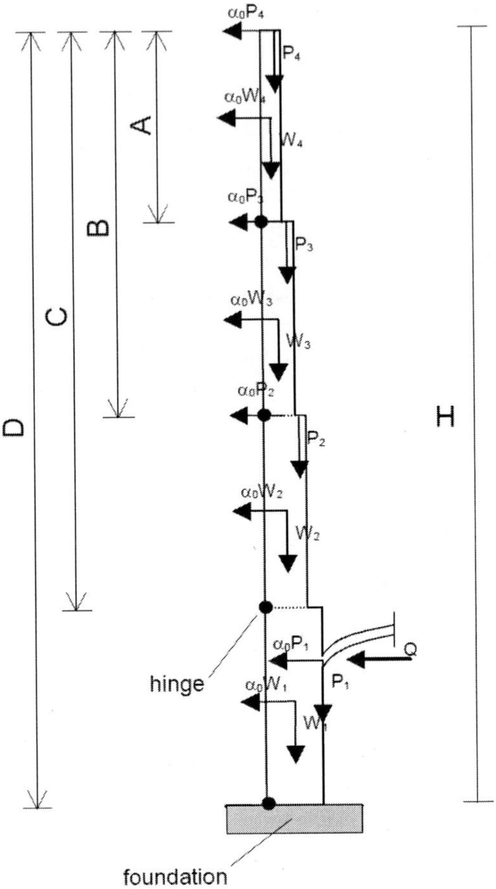

Figure 10. Out-of-plane mechanisms; P_i are the vault or the floor vertical loads; W_i are the own weights of the walls; $\alpha_0 P_i$ and $\alpha_0 W_i$ are the seismic loads.

$$a_0^* = \frac{\alpha_0 \sum_{i=1}^{n}(W_i + P_i)}{M^*}; \qquad M^* = \frac{\left[\sum_{i=1}^{n}\left(W_i\delta_{xi} + P_i\delta_{xi}\right)\right]^2}{g\sum_{i=1}^{n}\left(W_i\delta_{xi}^2 + P_i\delta_{xi}^2\right)} \tag{2}$$

where M^* is the effective participating mass and g is the gravity acceleration. According to [3] the analyzed mechanism will not be possible under design earthquake if:

$$a_0^* \geq a_1^* = \frac{a_g \cdot S}{q} \cdot \left(1 + 1.5 \cdot \frac{Z}{H}\right) \tag{3}$$

where a_gS is the design elastic spectral acceleration at $T = 0$, q is the behavior factor (assumed equal to 2.0), Z is the height to the ground of the linear hinge of the mechanism and H is the total height of the façade. Each macro-element, *i.e.*, each masonry façade, was analyzed considering first the overturning of the last floor only (the fourth floor, case A), then the overturning of the fourth and the third floor (case B), until in the last case where the overturning of the whole façade was analyzed (case D). The results are summarized in Table 3. The more vulnerable walls are the DA, the AH and the HB in which the vulnerability indexes a_1^*/a_0^* are higher. In all cases the overturning minimum collapse multipliers are obtained when the whole façade mechanism is considered; this suggests proper strengthening techniques.

Table 3. Vulnerability indexes obtained with kinematic analysis (α_0 = collapse multipliers; a_0^*= seismic spectral accelerations corresponding to α_0; a_1^*= seismic spectral accelerations demanded for ultimate limit state).

WALL	Mechanism	α_0	a_0^* (m/s²)	a_1^* (m/s²)	a_1^*/a_0^*	Check ($a_0^* \geq a_1^*$)
1-DA	A	0.48	4.75	2.62	0.55	YES
	B	0.08	0.78	1.99	2.55	NO
	C	0.06	0.60	1.62	2.70	NO
	D	0.03	0.33	1.10	3.33	NO
2a-AH	A	0.36	3.52	2.60	0.74	YES
	B	0.09	0.84	2.01	2.39	NO
	C	0.06	0.56	1.58	2.82	NO
	D	0.03	0.27	1.10	4.07	NO

4.2. Pushover Analysis

The seismic behavior of the Palace was also investigated by a non linear static analysis. The FE model was subject to constant gravity loads and, subsequently, to monotonically increasing horizontal forces (pushover analysis [32,33]). Specifically, the effects of the seismic action were evaluated by applying two systems of horizontal forces, not acting simultaneously, perpendicular to one another. The distribution of these forces is directly proportional to the masses by the displacements of the first modal shape in the pertinent direction (modal load).

Soon after its introduction in the late seventies, the pushover analysis became very attractive in earthquake engineering due to its relative simplicity (e.g., it requires quite simple material models, which do not account for progressive damage accumulation) and its reduced computational effort, compared with nonlinear dynamic analysis. Nevertheless, the method is not devoid of disadvantages and critical points, as highlighted by the scientific community. A comprehensive review of criticisms, advantages and disadvantages, together with the theoretical basis of the method, is reported in [34,35]. In this study a conventional pushover is performed, *i.e.*, distribution of the horizontal forces does not change with progressive structural degradation occurring during an earthquake. This means that the analysis does not account for the progressive changes in modal frequencies due to crushing and cracking phenomena. This is a critical point for the application of conventional pushover to historic masonry buildings, because it is predictable that damaging of the building leads to a period elongation. Therefore, different spectral amplifications and modal load distributions should be considered [36,37]. The hypothesis of invariance of the distribution of the horizontal loads could cause an overestimation of the building seismic capacity, especially when non uniform damage or high level of cracking are expected. Yet, also in its conventional form pushover provides an efficient alternative method to expensive computational inelastic dynamic analyses and offers effective information on the damage state.

Results of pushover analysis in terms of displacements and crack and crushing patterns are reported in Figures 11 to 14. Figures 11 and 12 report the final configuration of the Palace for seismic loading acting in the transversal direction (+X and −X). In both cases there is a remarkable out-of-plane deformation of the wall AH, that confirms the results obtained with the kinematic analysis. Figures 13 and 14 show the Palace configuration for seismic loading acting in the longitudinal direction (+Y and −Y). Again it is possible to find a good agreement in terms of failure process between pushover and kinematic analysis, both confirming the overturning of the wall DA.

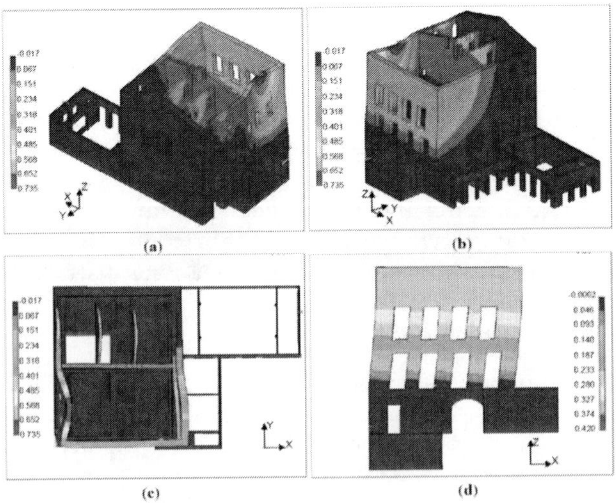

Figure 11. Pushover analysis results (+X direction): (a), (b), (c) and (d) horizontal displacement in X direction (units are m).

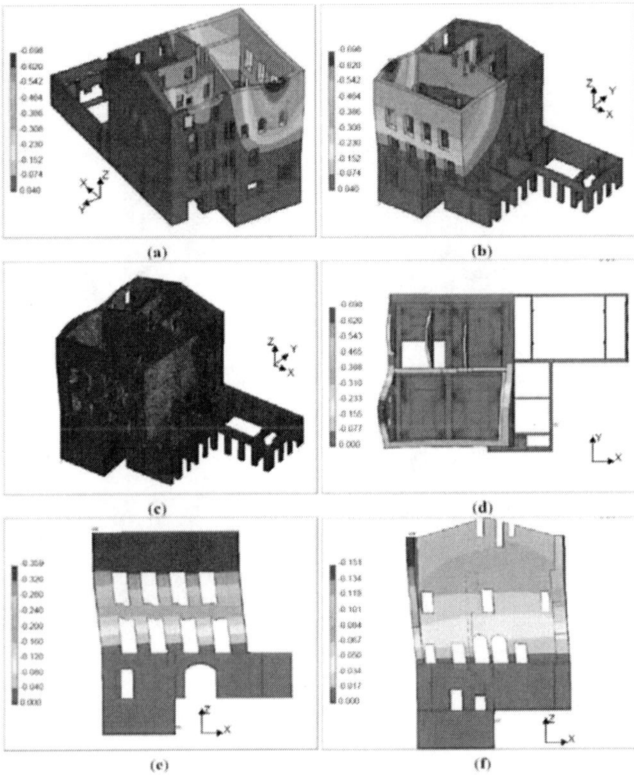

Figure 12. Pushover analysis results (–X direction): (a), (b), (d), (e) and (f) horizontal displacement in X direction (units are m), (c) final cracking and crushing patterns.

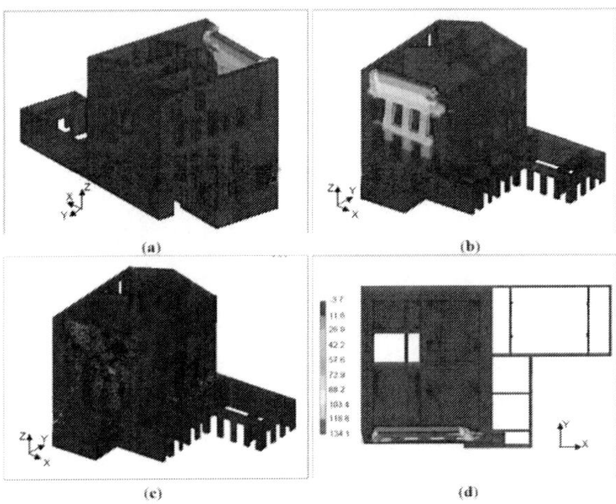

Figure 13. Pushover analysis results (+Y direction): (a), (b) and (d) horizontal displacement in Y direction (units are m), (c) final cracking and crushing patterns.

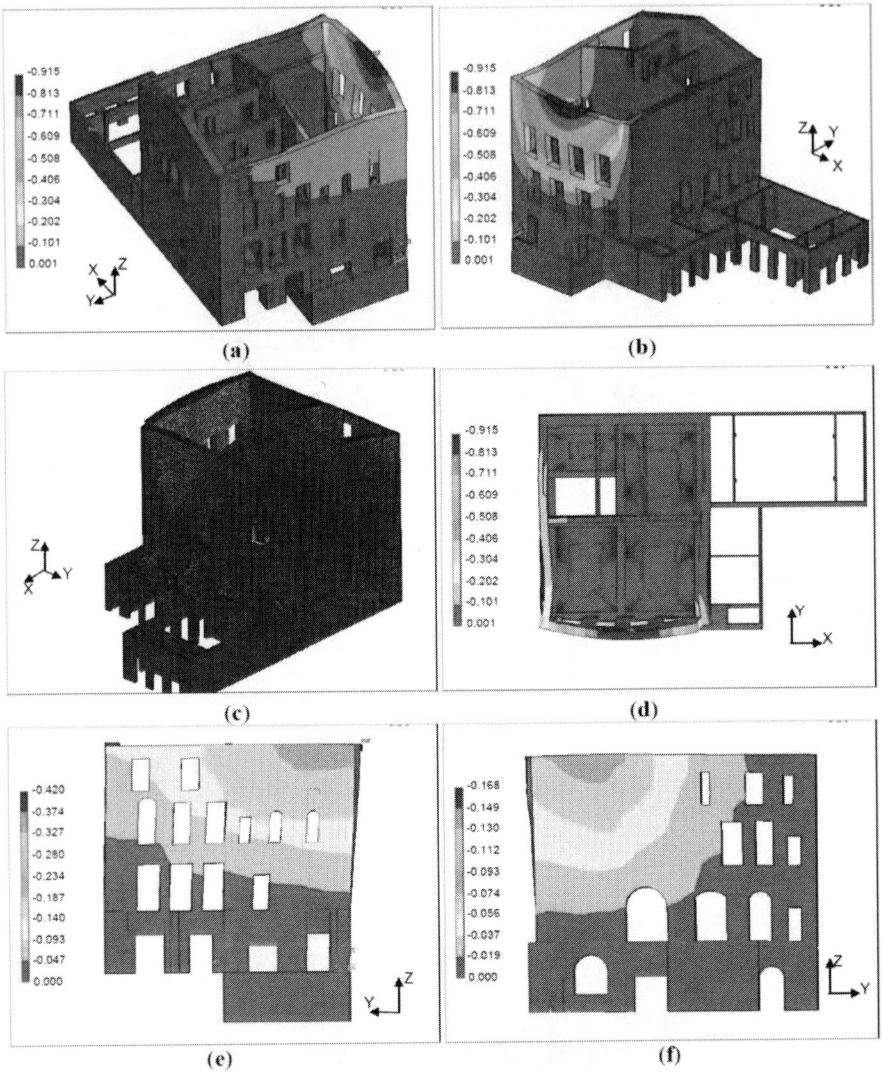

Figure 14. Pushover analysis results (−Y direction): (a), (b), (d), (e) and (f) horizontal displacement in Y direction (units are m), (c) final cracking and crushing patterns.

In case of seismic loading acting in the transversal direction (+X and −X) the analyses stop at a horizontal load corresponding to about 28–30% of the overall weight of the Palace; in case of seismic loading acting in the longitudinal direction (+Y and −Y) the analyses stop at a horizontal load corresponding to about 35–37% of the overall weight of the Palace.

Results obtained by analyses agree well with those obtained by the macro-element approach. The vulnerability of the Palace is mostly due to the absence of adequate connections between the opposite walls of the building due to the typology of the floors and the roof made of wooden beams. This has been already highlighted by other authors [38]. Absence of effective connections between structural components leads to overturning collapses of the perimeter walls under seismic loading due to the lack of box-behavior. Results suggest a proper strategy to improve the strength of the Palace against horizontal loads. A traditional retrofitting by means of steel chains and the creation of effective connections between wooden floors and masonry walls will prevent overturning of the façades [39].

5. Concluding Remarks

The paper has presented a multidisciplinary approach for structural analysis of historic buildings. Engineers involved in the study of monumental heritage are frequently asked to cover a great number of unknowns (materials properties, structural geometry, connections between structural and non-structural elements, stiffness of the horizontal diaphragms, existing damages, *etc.*). The combination of simple but not computationally expensive instruments (e.g., kinematic method) with more sophisticate ones (e.g., finite element approach) can offer proper indications for a subsequent *in-situ* investigation and for preliminary retrofitting solutions. The study on the Vicarious Palace in Pescia leads to the following conclusions. The static analysis confirms that the structural configuration of the Palace is adequate to withstand the vertical loads, as is usual for historic masonry buildings. The seismic analysis was approached by combining a global pushover analysis with a simplified approach based on the kinematic method. Results of the two types of analysis suggest that the more vulnerable parts of the Palace are the walls corresponding to the Southern and the Western façades. Overall, the study offers a screening of the most vulnerable elements of the Palace, suggesting that retrofitting should be planned to improve the connection between macro-elements by means of traditional techniques such as steel chains. Furthermore, connections between wooden floors and masonry walls should be also provided.

Acknowledgements

The authors would like to thank Eng. Paolo F. Bertini and Eng. S. Fiorindi for their contribution to numerical analyses.

References

1. Valluzzi, M.R.; Saisi, A.; Cardani, G.; Binda, L. Vulnerability analysis of the historical buildings in seismic area by a multilevel approach. *Asian J. Civil Eng. (Build. Hous.)* **2006**, *7*, 343–357.

2. Paganoni, S.; D'Ayala, D. Assessment and analysis of damage in L'Aquila historic city centre after 6th April 2009. *Bull. Earthquake Eng.* **2011**, *9*, 81–104.
3. O.P.C.M. 3274 del 20 Marzo 2003. Primi elementi in materia di criteri generali per la classificazione sismica del territorio nazionale e di normative tecniche per le costruzioni in zona sismica, G.U. n. 105, 8 Maggio 2003, S.O. n.72 e O.P.C.M. 3431 del 3 Maggio 2005, Ulteriori modifiche ed integrazioni all'O.P.C.M. 3274.
4. D.M. del Ministero delle Infrastrutture e dei Trasporti 14 Settembre 2005. "Norme tecniche per le costruzioni", G.U. n. 222.
5. D.M. del Ministero delle Infrastrutture e dei Trasporti del 14 Gennaio 2008. Nuove Norme Tecniche per le Costruzioni, G.U. n. 29.
6. Circolare del Ministero delle Infrastrutture del 2 Febbraio 2009. Istruzioni per l'applicazione delle "Nuove norme tecniche per le costruzioni" di cui al D.M. del 14/01/2008, G.U. n. 47.
7. Ibenholt, K.; Bowitz, E. Economic impacts of cultural heritage—Research and perspectives. *J. Cult. Herit.* **2009**, *10*, 1–8.
8. Modena, M.; Casarin, F. Structural assessment and seismic vulnerability analysis of the Reggio Emilia Cathedral, Italy. In Proceeding of Structural Analysis of Historical Constructions, New Delhi, India, 6–8 November 2006.
9. Foraboschi, P.; Barbieri, A.; Siviero, E. *Lettura Strutturale Delle Costruzioni (in Italian)*; 1997; CittàStudiEdizioni: Milano, Italy.
10. Lourenço, P.B. Computations on historic masonry structures. *Prog. Struct. Eng. Mater.* **2002**, *4*, 301–319.
11. Orlando, M.; Bartoli, G.; Betti, M. Evaluation study on structural fault of a Renaissance Italian Palace. *Eng. Struct.***2010**, *32*, 1801–1813.
12. Ramos, L.F.; Fernandes, F.M.; Krakowiak, K.J.; Lourenço, P.B. Failure analysis of Monastery of Jerónimos, Lisbon: How to learn from sophisticated numerical models. *Eng. Fail. Anal.* **2007**, *14*, 280–300.
13. Bento, R.; Lopes, M.; Cardoso, R. Seismic evaluation of old masonry buildings. Part I: Method description and application to a case-study. *Eng. Struct.* **2005**, *27*, 2024–2035.
14. Bento, R.; Lopes, M.; Cardoso, R. Seismic evaluation of old masonry buildings. Part II: Analysis of strengthening solutions for a case study. *Eng. Struct.* **2005**, *27*, 2014–2023.
15. Vignoli, A.; Betti, M. Numerical assessment of the static and seismic behaviour of the Basilica of Santa Maria all'Impruneta (Italy). *Construct. Build. Mater.* **2011**, *25*, 4308–4324.

16. Milani, G.; Milani, E.; Malvezzi, R.; Mallardo, V. Seismic vulnerability of historical masonry buildings: A case study in Ferrara. *Eng. Struct.* **2008**, *30*, 2223–2241.
17. Carpinteri, A.; Binda, L.; Anzani, A.; Invernizzi, S.; Lacidogna, G. A multilevel approach for the damage assessment of historic masonry towers. *J. Cult. Herit.* **2010**, *11*, 459–470.
18. Saisi, A.; Binda, L. Research on historic structures in seismic areas in Italy. *Prog. Struct. Eng. Mater.* **2005**, *7*, 71–85.
19. Giuffrè, A. *Sicurezza e conservazione dei centri storici. Il caso Ortigia*; 1993; Laterza: Bari, Italy.
20. Salvagnini, G. *Pescia, una città. Proposta metodologica per la lettura di un centro antico*; 1975; La Valdera: Firenze, Italy.
21. Merlo, A. La loggia nella città medioevale. Genesi, rilievo e ricostruzione dei processi di trasformazione: l'esempio di Pescia. In *Tesi di dottorato, Sezione Architettura e Disegno*; 2001; Firenze, Italy.
22. Salvagnini, G. Pescia, una comunità nel Seicento (1563–1738). 1989; Granducato Edizioni: Firenze, Italy.
23. Vignoli, A.; Galano, L.; Chiostrini, S. A. In-situ tests and numerical simulations on structural behaviour of ancient masonry. In Proceedings of Monument-98, Workshop on Seismic Performance of Monuments, Lisbon, Portugal, 12–14 November 1998.
24. Vignoli, A.; Galano, L.; Chiostrini, S. Mechanical characterization of stone masonry panels and effectiveness of strengthening techniques. In Proceeding of the 4th International Symposium on Computer Methods in Structural Masonry, London, UK, 3–5 September 1997.
25. *ANSYS v.11.0. ANSYS Manual Set*; 1998; ANSYS Inc: Canonsburg, PA, USA.
26. Willam, K.J.; Warnke, E.P. Constitutive model for the triaxial behaviour of concrete. In. In Proceedings of the IASBE Seminar on Concrete Structures Subjected to Triaxial Stresses, Bergamo, Italy, 17–19 May 1974; International Association for Bridge and Structural Engineering; Zurich, Switzerland; 19, pp. 1–30.
27. Lourenço, P.B.; Zucchini, A. Mechanics of masonry in compression: Results from a homogenisation approach. *Comput. Struct.* **2007**, *85*, 193–204.
28. Donida, G.; Brighenti, R.; Cerioni, R. Use of incompatible displacement modes in a finite element model to analyze the dynamic behavior of unreinforced masonry panels. *Comput. Struct.* **1995**, *57*, 47–57.

29. Calderón, P.A.; Giménez, E.; Pallarés, F.J.; Ivorra, S.; Adam, J.M. Axially loaded RC columns strengthened by steel caging. Finite element modelling. *Construct. Build. Mater.* **2009**, *23*, 2265–2276.
30. Vignoli, A.; Betti, M. Modelling and analysis of a Romanesque church under earthquake loading: Assessment of seismic resistance. *Eng. Struct.* **2008**, *30*, 352–367.
31. Speranza, E.; D'Ayala, D. Definition of collapse mechanisms and seismic vulnerability of historic masonry buildings.*Earthquake Spectra* **2003**, *19*, 479–509.
32. Penna, A.; Lagomarsino, S.; Falasco, A. On the use of pushover analysis for existing masonry buildings. In Proceeding of the 1st European Conference on Earthquake Engineering and Seismology, Geneva, Switzerland, 3–8 September 2006.
33. D'Amore, E.; Kim, S. Push-over analysis procedures in earthquake engineering. *Earthquake Spectra* **1999**, *15*, 417–434.
34. Amr, S.E.; Vassilis, K.P. Evaluation of conventional and adaptive pushover analysis I: Methodology. *J. Earthquake Eng.***2005**, *9*, 923–941.
35. Vassilis, K.P.; Amr, S.E.; Juan, F.P. Evaluation of conventional and adaptive pushover analysis II: Comparative results. *J. Earthquake Eng.* **2006**, *10*, 127–151.
36. Antoniou, S.; Pinho R., *REPLACE*. Advantages and limitations of adaptive and non adaptive force-based pushover procedures. *J. Earthquake Eng.* **2004**, *8*, 497–522.
37. Goel, R.K.; Chopra, A.K. A modal pushover analysis procedure to estimate seismic demands for unsymmetric-plan buildings. *Earthquake Eng. Struct. Dyn.* **2004**, *33*, 903–927.
38. Ramos, L.; Lourenço, P.B. Modeling and vulnerability of historical city centers in seismic areas: A case study in Lisbon. *Eng. Struct.* **2004**, *26*, 1295–1310.
39. ICOMOS. Recommendations for the analysis, conservation and structural restoration of architectural heritage, International Scientific Committee for Analysis and Restoration of Structures of Architectural Heritage, 2001.

CHAPTER 3

Vulnerability of Reinforced Concrete Structures Subjected to Flood

Natarajan Chidambarathanu[1] and Remya Retnan[1]

[1] Department of Civil Engineering, National Institute of Technology Tiruchirappalli, India

1. Introduction

Floods are one of the most widespread and destructive natural disasters occurring in the world (Singh and Sharma 2009), and with the increase in constructions along river courses and concentration of population around floodplain areas, flood-induced damages have been continuously increasing. The annual disaster record reveals that flood occurrence increased about ten folds over the past five decades (Scheuren et al. 2007). Thus, floods are posing a great threat and challenge to planers, design engineers, insurance industries, policy makers, and to the governments.

Structural and non-structural measures can be used to deal with floods (Sagala 2006). Structural measures include a set of works aiming to reduce one or more hydraulic parameters like runoff volume, peak discharge, rise in water level, duration of flood, flow velocity, etc. Non-structural measures involve a wide range of measures to reduce flood risk through flood forecasting and early warning systems, emergency plans, and posing land use regulations and policies. The futuristic reinforced concrete buildings can be considered as a symbol of modern civilization. These buildings are usually constructed based on the guide lines given by the standard code books (like IS:456:2000, for India). Unfortunately, the code provisions consider the seismic loads and wind effects alone, while accounting the dead and live design loads, and exclude the flood loads. This implies the necessity to bring out corrective measures that can be adopted to reduce vulnerability before harm occurrences.

This chapter focuses on both the incorporation of flood loads during the design stage and the assessment of flood vulnerability of reinforced concrete buildings. Vulnerability is expressed as a fraction of ground floor height and assumes that flood water at most immerse the building up to ground floor level. The importance of the outcome arises from the need of a strengthening solution to avoid failure of new or existing structures during floods.

1.1. Forces Due To Flood

The physical forces which act on the buildings include hydrostatic loads (Fig.1.), hydrodynamic loads (Fig.2.), and impact loads, and these loads can be exacerbated by the effects of water scouring soil from around and below the foundation (FEMA, 2001). The *hydrostatic loads* are both lateral (pressures) and vertical (buoyant) in nature. The lateral forces result from differences in interior and exterior water surface elevations. As the floodwaters rise, the higher water on the exterior of the building acts inward against the walls of the building. Sufficient lateral pressures may cause permanent deflections and damage to structural elements within the building. The buoyant forces are the vertical uplift of the structure due to the displacement of water, just as a boat displaces water causing it to float. These uplift forces may be the result of the actual building materials (the floating nature of wood products), or due to air on the interior of a tightly built structure. When the buoyant forces associated with the flood exceed the weight of the building components and the connections to the foundation system, the structure may float from its foundation.

The water flowing around the building during a flood creates *hydrodynamic loads* on the structure. These loads are the frontal impact loads from the upstream flow, the drag on the sides of the building, and the suction on the rear face of the building as the floodwaters flow around the structure. The magnitude of the hydrodynamic loads depends on both the velocity of water and the shape of the structure. Like the hydrostatic pressures, these lateral pressures may cause the collapsing of either structural walls or floors.

Impact loads during floods may be the direct forces associated with waves, as typically encountered during coastal flooding, or the impact of debris floating in the waters, including logs, building components, and even vehicles. Impact loads can be destructive because the forces associated with them may be an order of magnitude higher than the hydrostatic and hydrodynamic. Floating debris can have devastating effects, as they apply large and/or concentrated loads to the structural elements of the building.

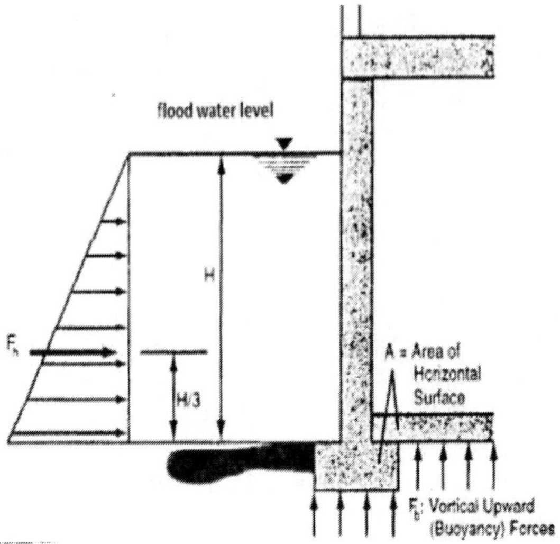

Figure 1. Schematic sketch of hydrostatic force (FEMA, 2001)

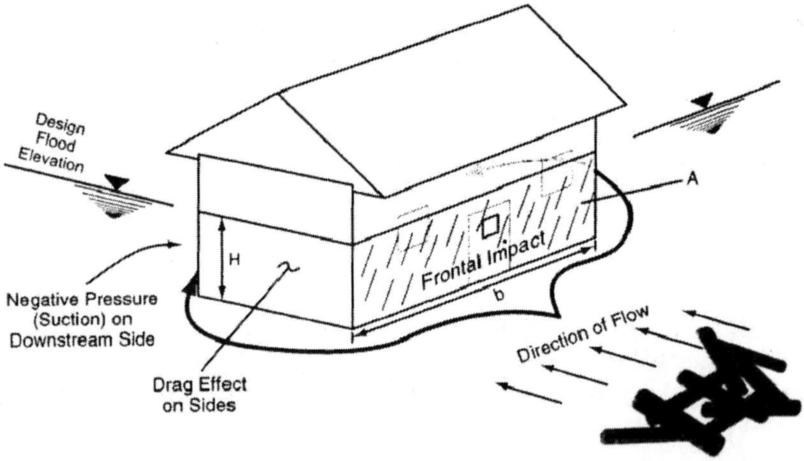

Figure 2. Schematic sketch of hydrodynamic force (FEMA, 2001)

2. Literature Review

FEMA (2001) published a manual focusing on the retrofitting of family residences subject to flooding without wave action. The measures include elevation of the structure in place, relocation of the structure, construction of barriers, dry flood proofing and wet flood proofing. The analyses necessary to determine flood-related hazard factors are also presented.

Kelman (2002), in a dissertation on Physical Flood Vulnerability of Residential Properties in Coastal Eastern England, examined the lateral pressure from flood differential depth between inside and outside a residence.

Kelman and Spenc (2004) categorised flood actions on buildings as energy transfers, forces, pressures, or the consequences of water or contaminant contact.

Messener and Meyer (2005) argued that the challenge consists in understanding the interrelations and social dynamics of flood risk perception, preparedness, vulnerability, flood damage and flood management, and to take this into account in a modern design of damage analysis and risk management.

Sagala (2006) examines the physical vulnerability to flood and people's coping mechanisms in flood prone residential areas in Naga city of Philippines. Six structural types of buildings were chosen and for each type of vulnerability curves (flood depth/damage) were plotted. Results indicate that buildings with plywood walls and wooden floors are the most vulnerable while the type with hollow block walls and concrete floors is the least vulnerable.

Arulselvan et al. (2007) conducted an experimental investigation on the influence of brick masonry infill in a reinforced cement concrete frame and validated outcomes by comparing them with theoretical results obtained by finite element analysis. Until the cracks developed in the infill, the contribution of infill to both stiffness and lateral stiffness was found to be very significant. The strains measured in infilled beams and columns were 20% less than bare frame beams up to failure of brick walls.

Haugen and Kaynia (2008) presented a method for prediction of damage in a structure impacted by a debris flow of known magnitude. The method uses the principles of dynamic response of structures to earthquake excitation, and fragility curves proposed in HAZUS for estimation of the structural vulnerability, by the damage state probability. The model was tested on a debris flow in Italy and it gave probabilities between 34% and 66% for reaching the damage levels which actually occurred for five out of six structures.

Kreibich et al. (2009) investigated the importance of flow velocity, water depth and combinations of these two parameters on various types of damages to buildings and roads. A significant influence of flow velocity on damage to roads was found, in contrast to a minor influence on monetary losses and business interruption. The energy head is suggested as a suitable flood impact parameter for reliable forecasting of structural damage to residential buildings.

Lopez et al. (2010) developed a methodology to estimate flood vulnerability to buildings, in either riverine or coastal settings, based on the aggregated damage to individual building components. Building vulnerability is modelled based on analytical representations of the failure mechanisms of individual building components.

3. Methodology

The present work focuses on the assessment of flood physical vulnerability of building expressed as a factor of ground floor height. The influence of design variation zones or boundary conditions has been also investigated. The methodology is schematized in Fig.3.

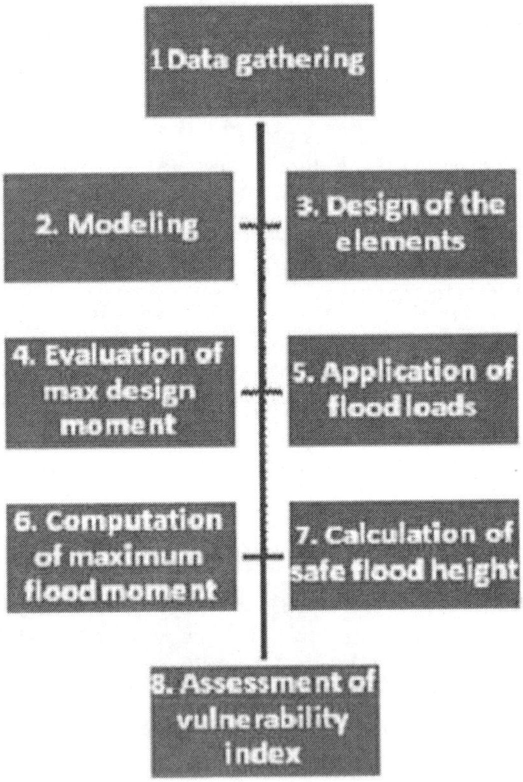

Figure 3. The steps of the methodology

3.1. Building Details

The building configuration used for the study is regular, with plan dimensions 9m×18m. Table 1 lists the data associated with a four storey reinforced concrete building considered for the analysis, while the plan and elevation of the building are shown in Fig.4. and Fig.5., respectively. In Fig.4., the direction of interest refers the perpendicular direction of flood.

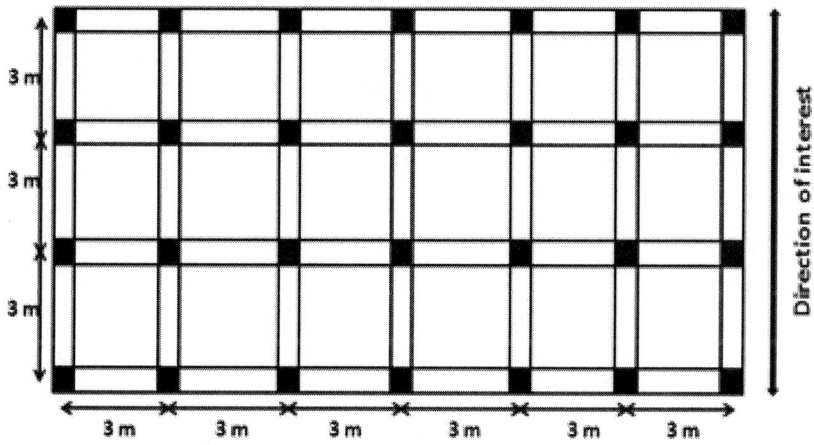

Figure 4. Plan of considered building

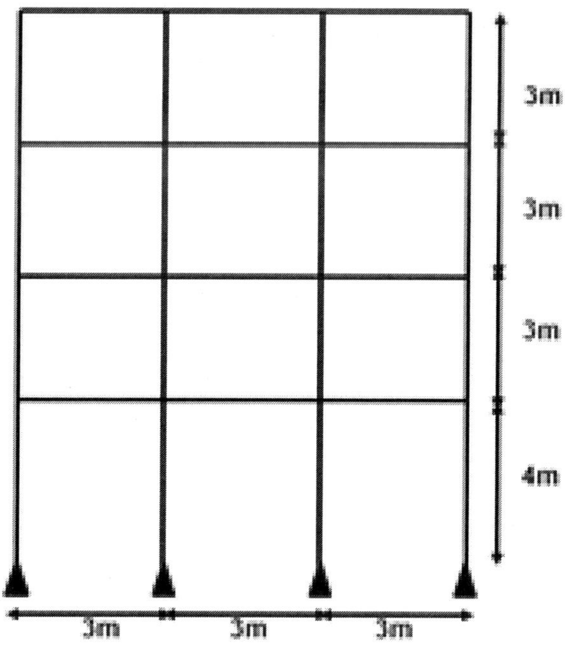

Figure 5. Elevation of frame

Table 1. Reinforced concrete building details

Ground floor height	4m
Remaining floors height	3m
No. of bays in X direction	6
No. of bays in Y direction	3
Bay width	3m in both X and Y directions
Column size	300mmx300mm
Beam size	250mmx300mm
Masonry wall thickness	230mm
Slab thickness	120mm
Unit weight of the concrete	25 kN/m^3
Unit weight of masonry	20 kN/m^3
Elastic modulus of steel	$2\times10^8 kN/m^2$
Yield strength of steel	415 N/mm^2
Young's modulus of concrete	$25\times10^6 kN/m^2$
Poisson ratio of concrete	0.2
Compressive strength of concrete	20 N/mm^2
Young's modulus of masonry	$13.8\times10^6 kN/m^2$
Poisson ratio of masonry	0.25
Floor finish load	$0.5kN/m^2$
Terrace water proofing (TWF) load	$1.5kN/m^2$
Live load on roof	$1.5kN/m^2$
Live load on floor	$3kN/m^2$

3.2. Modelling

To compute the critical effect, the flood was assumed to act along the 18m side and an intermediate 2D frame along 9m side was considered for the study. Three frame models were used, a) bare frame model, without any partition walls (Fig. 6.); b) frame with light weight partition wall; c) frame with structural infill wall (Fig. 7.). The infill walls were modelled as a diagonal strut having width 230mm, very low moment of inertia, modulus of elasticity 13800 N/mm^2 and Poisson ratio 0.25. The weight of light weight partition walls were considered negligible. Hence, frame models for both bare frame and frame with light weight partition walls were similar but the difference will come in to the picture while applying flood load.

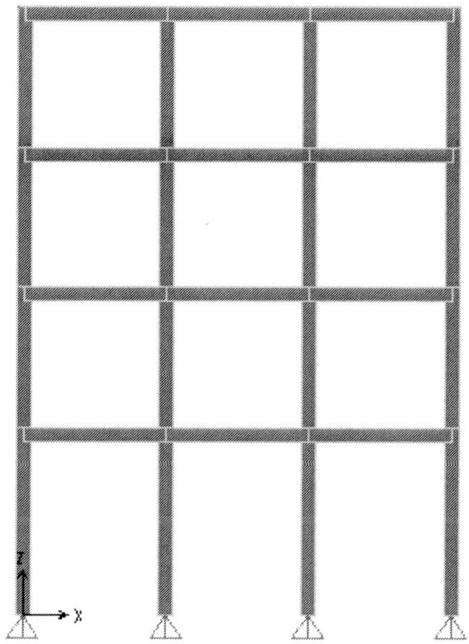

Figure 6. Bare frame SAP model

Figure 7. Frame with structural infill walls

3.3. Analysis

The procedure consists of linear static and linear dynamic analysis. When the linear static or dynamic procedures are used for seismic evaluation, the design seismic forces, the distribution of applied loads over the height of the buildings, and the corresponding displacements are determined using a linearly elastic analysis. The various steps involved in SAP model analysis are the following:

- Modelling of frame sections.
- Defining and assigning material properties and section properties.
- Assigning support conditions.
- Defining and assigning load patterns and load cases.
- Assigning load combinations.
- Setting up of analysis option.
- Running analysis.
- Inferring the results.

The load combinations considered for the study are:
a) 1.5 (DL + IL) b) 1.2 (DL + IL ± EL)

c) 1.5 (DL ± EL) d) 0.9 DL ± 1.5 EL

Analyses were carried out for six different conditions of seismic zones, flood duration, flood water height, flood forces, frame models, and support conditions, to obtain the maximum design moment, flood moment and lateral displacements.

3.4. Calculation of Design Moment

The earthquake load calculations were made for all the zones and all the models analysed, and designed for IS 456:2000. Here, the earthquake zones are considered to demonstrate the different structural variations but not the multi-hazard conditions (Table 2). The design moment is lower for fixed support condition than hinged condition.

Table 2. Zone factor (Ref. IS 1893-2002)

Seismic zone	II	III	IV	V
Seismic intensity	Low	Moderate	Severe	Very severe
Z	0.1	0.16	0.24	0.36

3.5. Calculation of Flood Loads

Flood loads are assumed to act as: a) hydrostatic loads; b) impact loads as equivalent static loads; c) impact loads as dynamic loads, considering the duration of flood. The hydrostatic loads consist of both lateral pressures and buoyancy forces. Lateral pressure is calculated using the formula $P_s = \gamma h_f$ (in kN/m^2), where $\gamma = 9.81 \ kN/m^3$ for water, and h_f is the water depth in meters. Since lateral hydrostatic loads are acting as triangular loads, the resultant hydrostatic load (F_f) acts at $h_f/3$ distance from ground level. Buoyancy force has a significant effect either if the building is surrounded by water or in submerged condition. Here, the flood is considered as slow moving; hence the effect of buoyancy is neglected. Impact loads are velocity dependent loads. As no codes or design books are available for incorporating the impact effects, the magnitude of these loads is arbitrarily considered as a factor of hydrostatic force acting laterally as UDL over the surface. Table 3 shows the magnitude of flood loads acting on the column for the frame models.

Table 3. Flood loads on frame models

h_f (m)	F_f (kN)	Impact UDL (kN/m)	
		$0.1\gamma h_f$	$0.2\gamma h_f$
2	5.89	0.59	1.18
3	13.24	0.88	1.77
4	23.54	1.18	2.35

The flood loads are assumed as dynamic loads by considering the duration of flood t_d. The dynamic displacement and dynamic flood moment are found using a deformation response factor (R). R is the ratio dynamic to static displacement caused by the flood force. The dynamic flood load is assumed as a rectangular pulse (Fig.8.).

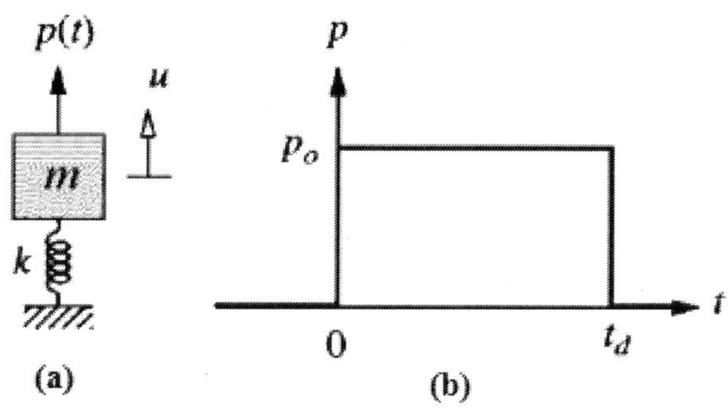

Figure 8. a) SDF system (b) Rectangular pulse load (Chopra, 2009)

The governing equation is:

$$m\ddot{u} + ku = p(t) = \begin{cases} p_0 & t \le t_d \\ 0 & t \ge t_d \end{cases}$$

The R value obtained after solving the equation 2 is (Chopra 2009):

$$R = \frac{u}{u_{st}} = \begin{cases} 2 \sin \pi \frac{t_d}{T_n} & \frac{t_d}{T_n} \le \frac{1}{2} \\ 2 & \frac{t_d}{T_n} \ge \frac{1}{2} \end{cases}$$

Where, u is the dynamic displacement, u_{st} is the static displacement, t_d is the flood duration and T_n is the fundamental natural time period of the structure. The t_d/T_n ratios and corresponding R values used are shown in Table 4. R = 1 indicates the flood as static while R = 2 indicates suddenly applied flood load. Since the flood assumed for the study is slow moving, R will always lies in between 1 and 2.

Table 4. Deformation response factor

t_d/T_n	1/6	1/4	1/3	1/2
R	1.000	1.4142	1.7321	2.000

3.6. Calculation of Flood Moment and Height

Afterwards, analyses have to be carried out for different frame models in each zone with different boundary conditions, and the maximum flood moment in each case must be evaluated. The safe flood height is the height of flood up to which the structure is safe. It is obtained by plotting the moment due to hydrostatic force versus flood height: height corresponding to the design moment gives the safe flood height (h_f, safe).

The vulnerability index is assessed as a factor of ground floor height. It indicates the extent of damage that a flood can cause if the water reaches up to ground floor height. It is calculated using the equation (3).

$$Vulnerability\ index = \frac{ground\ floor\ height\ -\ safe\ flood\ height}{ground\ floor\ height}$$

4. Experimental Results

The analysis was carried out for three frame models under different conditions of:

- Flood loadings: static, equivalent static, and dynamic loads;
- Support conditions: hinged and fixed;
- Seismic zones;
- Flood water height: 2m, 3m and 4m;
- Flood duration: $T_n/6$, $T_n/4$, $T_n/3$ and $T_n/2$.

For each zone, earthquake loads were assessed for all the zones and all the models designed for IS 456:2000. The earthquake zones (Table 5) are considered to demonstrate the different structural variations but not the multi-hazard conditions. The maxima design moments for both the bare frame and the frame with light weight partition walls are similar, since the weight of partition wall is considered as negligible. The sizes of frame sections, selected according to these moments, are given in Table 6. For the frame with structural infill, the infill walls were modelled as diagonal structures. After applying flood loads, for different frame models and in each zone, for hinged support condition, the maximum flood moment in each case was evaluated. Assuming flood heights of 2m, 3m and 4m from ground level, maxima moments were also obtained (Table 7). Because of the free movement of water in between the columns of the bare frame, the flood moment for bare frame model is very low if compared to the other models.

Table 5. Maximum design moments in kN-m

Zone	Bare Frame	Light weight infill	Structural infill
II	33.56	33.56	64.90
III	45.00	45.00	92.66
IV	62.26	62.26	128.58
V	86.40	86.40	184.09

Table 6. Frame cross-sections in mm

Frame model	Column size	Beam size
Bare Frame	300 x 300	250 x 300
Light weight infill	300 x 300	250 x 300
Structural infill	350 x 350	300 x 350

Table 7. Flood moment due to hydrostatic force (without impact factor) in kN-m

h_f (m)	Bare Frame	Light weight infill	Structural infill
2	5.74	32.14	30.09
3	9.45	97.58	83.94
4	20.18	205.84	166.33

Impact force is assumed to act as UDL, and its value is arbitrarily taken as a factor of hydrostatic force. The impact factors considered are 0.1 and 0.2. For all the models, the moments are linearly increasing as impact load increases, because impact force is considered as a factor of hydrostatic load (Table 8). Non-linearity will come only while considering flood duration. Flood is assumed to act as dynamic rectangular load with flood duration t_d and the maximum flood moment obtained in each case is shown in Table 9.

Table 8. Moment due to hydrostatic and equivalent static impact forces in kN-m

h_f (m)	Bare Frame		Light weight infill		Structural infill	
	$0.1\gamma h_f$	$0.2\gamma h_f$	$0.1\gamma h_f$	$0.2\gamma h_f$	$0.1\gamma h_f$	$0.2\gamma h_f$
2	5.74	5.74	36.67	41.20	33.62	37.16
3	10.69	11.92	109.92	122.27	92.5228	101.10
4	22.53	24.89	229.40	252.96	180.05	193.77

Table 9. Flood moment due to dynamic flood forces in kN-m

Zone	Bare frame				Frame with partitions			
	R=1	R=1.4142	R=1.7321	R=2	R=1	R=1.4142	R=1.7321	R=2
II	33.56	47.46	58.13	67.12	64.90	91.78	112.41	129.80
III	45.00	63.64	77.95	90.00	92.66	131.04	160.49	185.32
IV	62.26	88.05	107.84	124.52	128.58	181.84	222.71	257.16
V	86.40	122.18	149.65	172.80	184.09	260.35	318.87	368.19

Table 10. Duration of flood (td) in sec

Frame type	R=1	R=1.414	R=1.732	R=2
Bare frame and Frame with light weight infill	0.0448	0.0673	0.0897	0.1345
Frame with masonry infill	0.0092	0.0139	0.0185	0.0277

The fundamental frequency and duration of flood will be the same for both the frames. Also, the flood moment obtained is the same for frame with structural and non-structural partitions, because the contact area of flood water is the same for both frames. The safe flood height is obtained by plotting the moment due to hydrostatic force versus flood height. For example, for a frame with light weight partition wall in Zone II, design moment is 33.56 kN-m (Table 5) and its maximum moment due to hydrostatic loading is shown in Table 11. From the graph, the safe flood height corresponding to design moment 33.5596 is 2.0276 m.

Table 11. Maximum flood moment for the frame with light weight partition wall in Zone II.

h_f (m)	Max flood moment (kN-m)
2	32.14
3	97.58
4	205.84

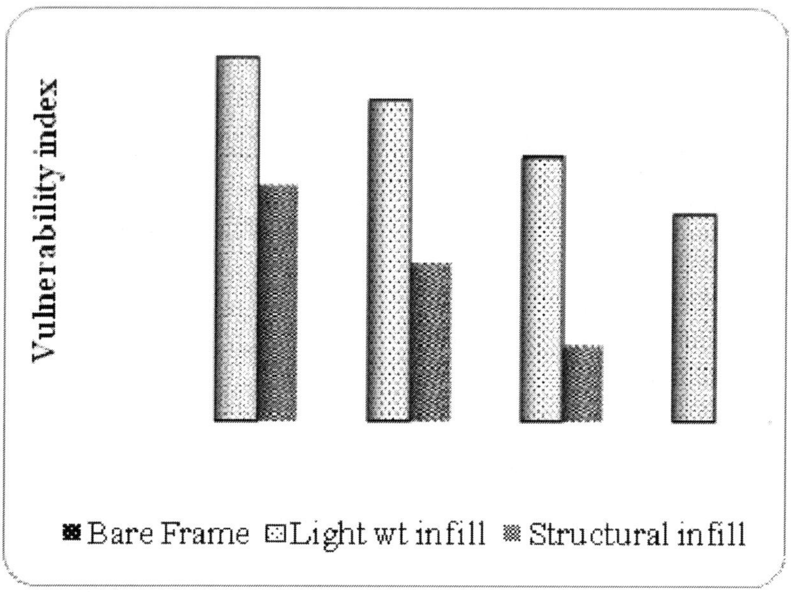

Figure 9. Variation of vulnerability in various zones

The vulnerability index of frame with light weight partition wall is high (49.3%) if compared to the other frames (Fig.8.). For frame with structural infill it only reaches a maximum of 32%, while it is zero for bare frame model. Vulnerability indexes obtained due to hydrostatic and equivalent static impact forces show that the highest values pertain to frame with light weight partition wall (Table 12). Vulnerability indexes obtained due to dynamic flood forces in various zones for different flood duration are shown in Table 13.

Table 12. Vulnerability due to hydrostatic and equivalent static impact forces

Zone	Bare Frame		Light weight infill		Structural infill	
	$0.1\gamma h_f$	$0.2\gamma h_f$	$0.1\gamma h_f$	$0.2\gamma h_f$	$0.1\gamma h_f$	$0.2\gamma h_f$
II	0	0	0.518	0.539	0.347	0.371
III	0	0	0.461	0.484	0.244	0.27
IV	0	0	0.391	0.416	0.134	0.163
V	0	0	0.311	0.339	0	0.027

Table 13. Vulnerability index due to dynamic flood forces

		Bare frame		
R	1	1.414	1.732	2
Zone II	0	0.142	0.300	0.393
Zone III	0	0.061	0.233	0.336
Zone IV	0	0.000	0.160	0.272
Zone V	0	0.000	0.085	0.208
		Frame with light weight infill		
Zone II	0.493	0.642	0.707	0.747
Zone III	0.435	0.600	0.674	0.717
Zone IV	0.362	0.549	0.632	0.681
Zone V	0.279	0.490	0.584	0.639
		Frame with structural infill		
Zone II	0.320	0.519	0.607	0.660
Zone III	0.214	0.444	0.546	0.607
Zone IV	0.103	0.365	0.482	0.551
Zone V	0.000	0.266	0.401	0.481

The storey drifts are evaluated from the lateral joint displacements. According to IS 1893-2002 Cl.7.11.1, the maximum storey drift is 0.004 H, where H is the height of the building. In this study, H = 13 m and hence the maximum allowable storey drift is 52 mm. The frame with structural infill wall has low storey drift if compared to bare frame, because infill walls have significant effect in resisting lateral storey drift (Table 14). For the frame with light weight partition wall, storey drift reaches 71.32mm, which is more than that specified for seismic resistant building (Table 15). Hence a frame with non-structural partitions with hinged support is not preferred in flood prone areas.

Table 14. Storey drifts due to hydrostatic forces

h_f (m)	Bare Frame	Light weight infill	Structural infill
2	0.634	5.939	0.084
3	2.017	19.696	0.205
4	4.665	46.242	0.466

Table 15. Storey drifts due to hydrostatic and equivalent static forces

h_f (m)	Bare Frame		Light weight infill		Structural infill	
	0.1yh_f	0.2yh_f	0.1yh_f	0.2yh_f	0.1yh_f	0.2yh_f
2	0.807	0.98	7.666	9.393	0.1	0.116
3	2.578	3.138	25.372	30.975	0.264	0.324
4	5.918	7.172	58.78	71.317	0.616	0.766

The relative cost for any frame model is calculated with respect to the design moment of bare frame model in zone II (Eq. 4):

$$Cost\ relative_{zone\ III,IV,V} = \frac{DM_{zone\ III,IV,V} - DM_{zone\ II\ (bare)}}{DM_{zone\ II\ (bare)}}$$

Where $DM_{zoneIII,IV,V}$ are the design moments in zones III, IV and V for frame with partitions, and $DM_{zoneII(bare)}$ is the design bending moment of bare frame in zone II.

The relative costs for the three frame models are shown in Table 16. The graph of relative cost versus vulnerability index shows that for the frame with light weight partition wall the cost is increasing but the vulnerability is not reducing that much. Moreover, even though the initial cost is higher for frame with structural partitions, its vulnerability is lower if compared to frame with non-structural partitions (Fig.10.).

Table 16. Relative cost as a factor of design moment for three frame models

Zone	Bare Frame		Light weight infill		Structural infill	
	DM	cost relative	DM	cost relative	DM	cost relative
II	33.560	0	33.560	0	64.898	1
III	45.001	0.341	45.001	0.341	92.658	1.761
IV	62.259	0.855	62.259	0.855	128.579	2.831
V	86.398	1.574	86.398	1.574	184.094	4.486

The vulnerability obtained for different flood loadings is compared with partitions, zones and flood duration (Fig.11. and 12). Dynamic load with R = 1.4142 is used for comparing the results with static results. Frame with light weight infill wall is more vulnerable (64.2%) and bare frame is less vulnerable (14.2%). This is due to the free movement of water in between the columns of the bare frame, so that the contact area of flood water is very low if compared to the other frames. For the frame with masonry infill, vulnerability is less compared to light weight partition, even though the flood moment is the same for both the cases. It is due to the structural action of masonry infill against the lateral flood load.

Comparing vulnerability for different flood loadings to seismic zones (Fig.13.), for the frame with light weight infill, vulnerability is higher in Zone II (64.2%) and it reduces as zone increases (zone V: 49%). For frame with masonry infill, vulnerability is reaching zero as zone varies from II (51.9%) to V (Fig.14.). This is because the design moment of building in zone V is higher if compared to zone II and hence the building in zone V will be more resistive to flood.

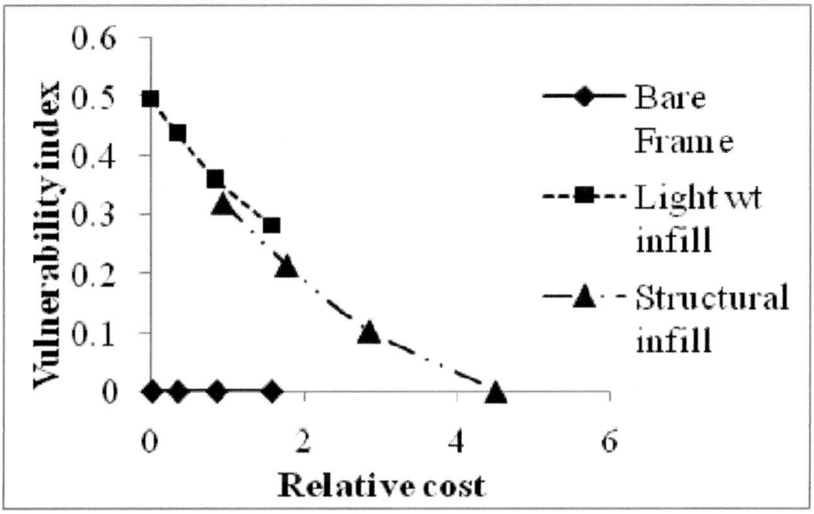

Figure 10. Variation of vulnerability against cost

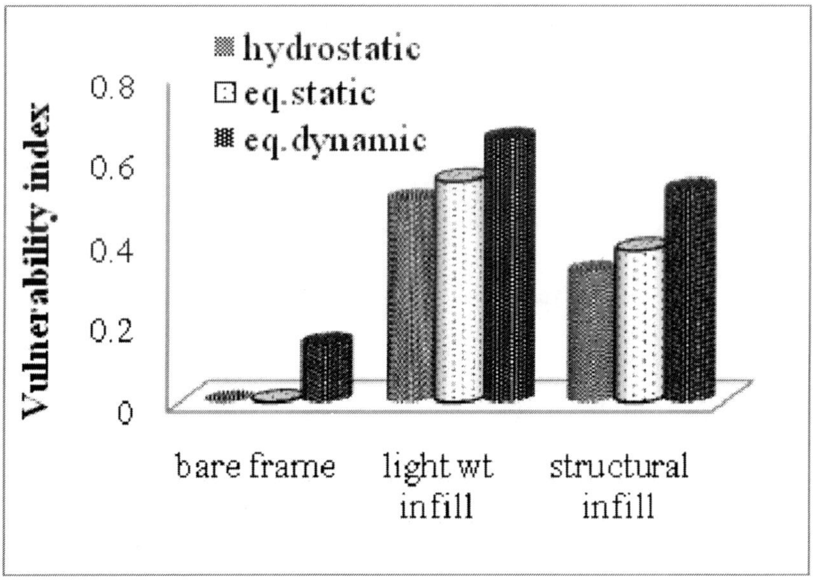

Figure 11. Vulnerability for different frame models in different flood loading conditions in Zone II

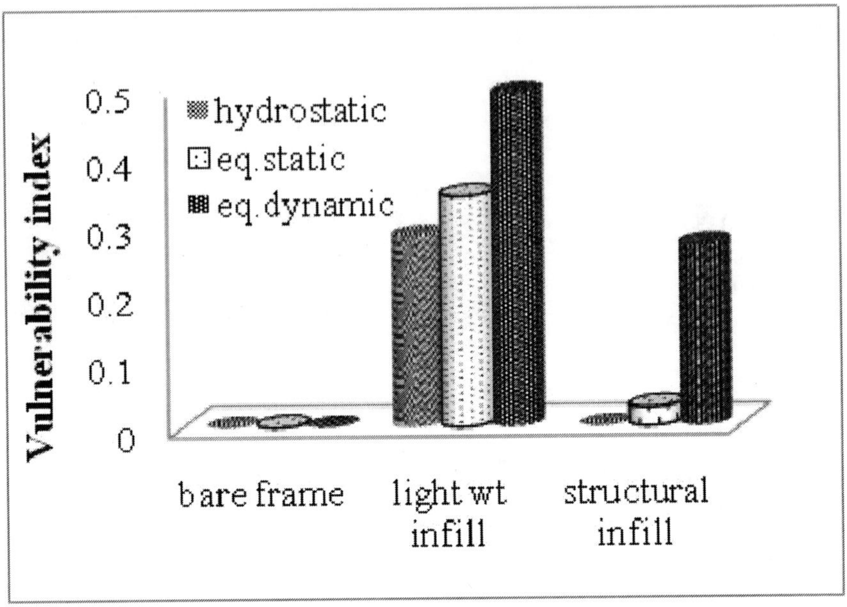

Figure 12. Vulnerability for different frame models in different flood loading conditions in Zone V

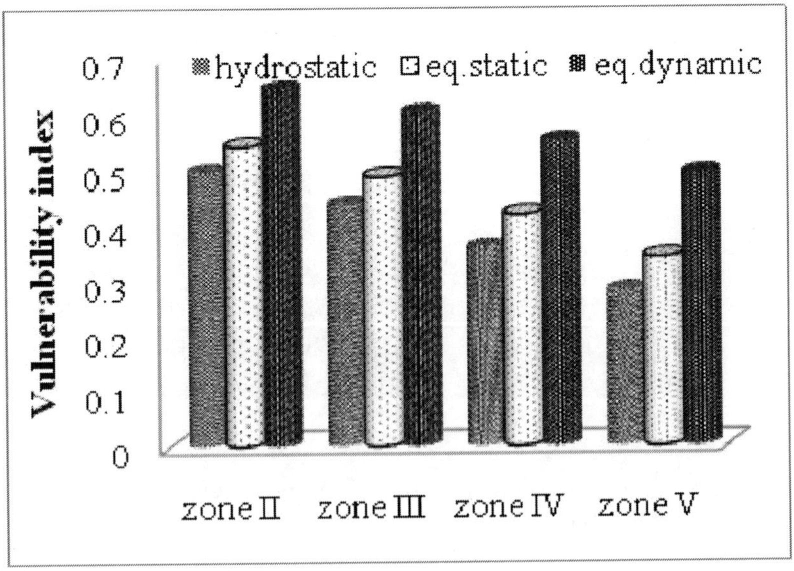

Figure 13. Vulnerability for light weight infill frame under different flood loading conditions in different zones

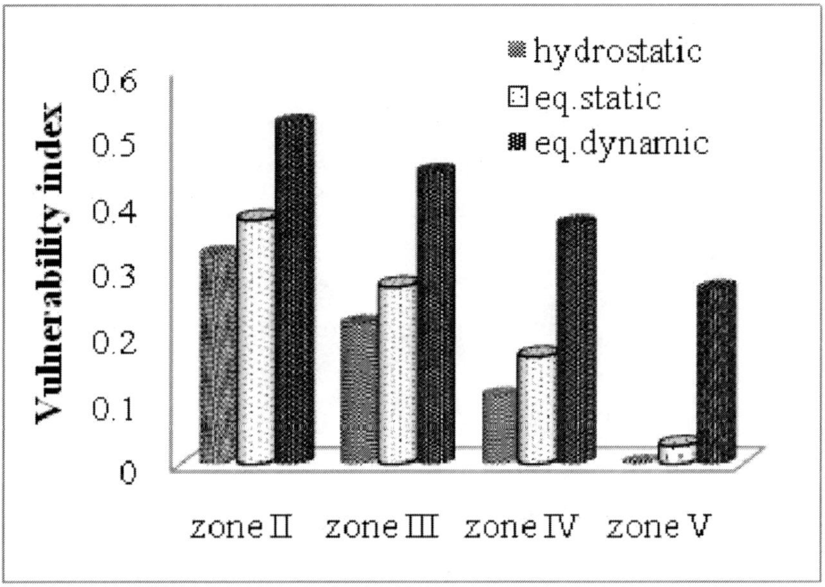

Figure 14. Vulnerability masonry infill frame under different flood loading conditions in different zones

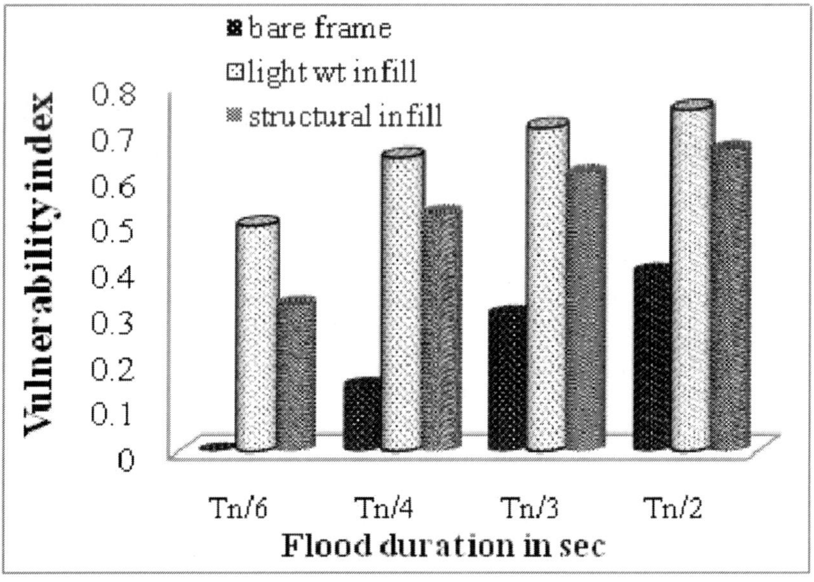

Figure 15. Vulnerability for different frame models under different flood duration in Zone II

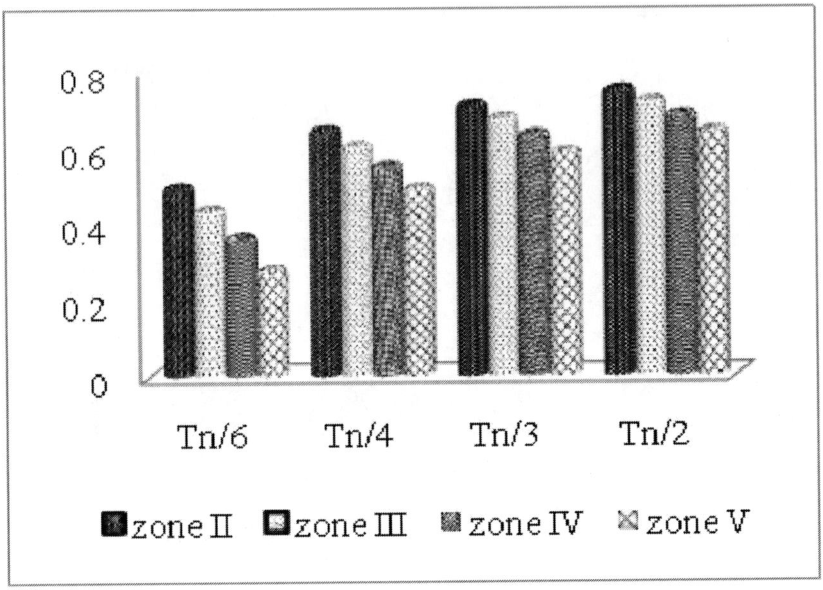

Figure 16. Vulnerability for the frame with light weight infill in different zones

Analysing different frame models under different flood duration in Zone II (Fig.15.), vulnerability increases with the duration of flood, but it is lower for bare frame (39.3%) if compared to frame with partitions (74.7% for light weight infill and 66% for frame with structural infill). It is due to the free movement of flood water between the columns of bare frame. The results of vulnerability for the frame with light weight infill (Fig. 16.), show that a building in zone V with flood duration $T_n/3$ is less vulnerable (58.4%) than a building in zone II with flood duration of $T_n/2$ (64.2%).

The analysis was carried out for all the cases, keeping the support of columns as fixed. The earthquake load calculations were made for all the zones and all the models analysed and designed as per IS 456:2000, for each zone and maximum design moments (Table 17). The maximum moment is lower for the fixed support condition, so the cross sections required is lower when compared to hinge support condition. The sizes of frame sections are given in Table 18. Fig. 18. shows the variation of flood moments for different frame models due to hydrostatic force. The flood moments parabolically increase as flood water height increases.

Table 17. Maximum design moment in kN-m

Zone	Bare Frame	Light weight infill	Structural infill
II	16.1389	16.1389	33.6639
III	25.3319	25.3319	49.3
IV	30.6598	30.6598	69.5349
V	42.6198	42.6198	100.8072

Table 18. Frame cross-sections in mm

Frame model	Column size	Beam size
Bare Frame	250 x 250	250 x 300
Light wt infill	250 x 250	250 x 300
Structural infill	300 x 300	250 x 300

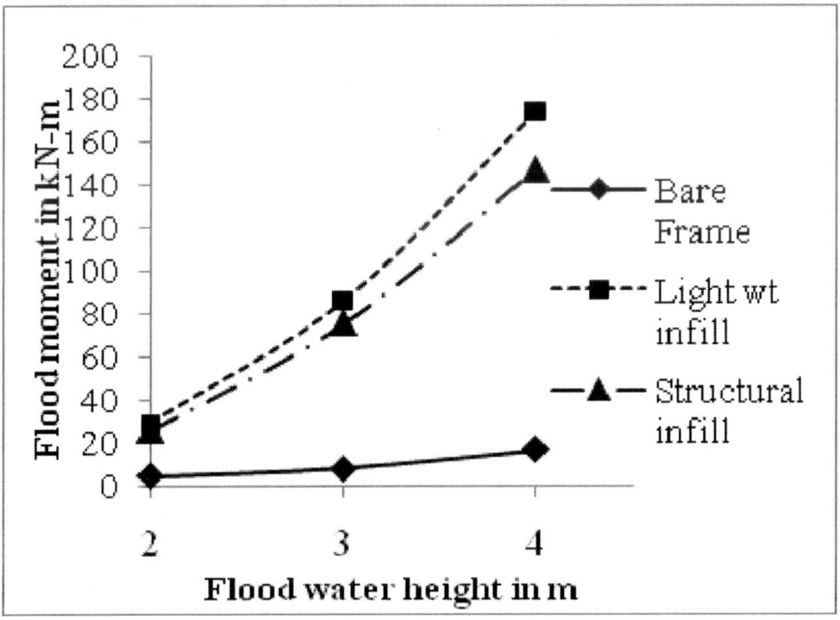

Figure 17. Variation of flood moment to hydrostatic force (without impact factor) with water height

The maximum moments obtained from the analysis for fixed support condition are shown in Table 19. For all frame models, the moments linearly increase as impact load increases. This is because, for the present case, impact force is considered as factor of hydrostatic load. Non-linearity will come only while considering flood duration. The duration of flood load (td) considered for various R values for hinged support condition are shown in Table 20 and the flood moments due to dynamic flood loads in various zones for fixed support condition are shown in Table 21. Vulnerability for the bare frame is zero in all the seismic zones but it is non-zero for frame with partitions (Fig. 17.). This is due to the free movement of water in between the columns of the bare frame, so that the contact area of flood water will be low if compared the other frames. The vulnerability of frame with light weight partition wall is very high (60.3%), while for frame with structural infill it reaches 44.6% and it is not present for bare frame model.

Vulnerability indexes obtained due to hydrostatic and dynamic impact forces for fixed support condition are shown in Table 22 and 23, respectively.

Table 19. Moment due to hydrostatic and impact forces in kN-m

hf (m)	Bare Frame		Light weight infill		Structural infill	
	0.1γhf	0.2γhf	0.1γhf	0.2γhf	0.1γhf	0.2γhf
2	4.4614	4.462	35.7516	42.4092	31.2512	36.9778
3	9.6472	11.2901	102.16	118.589	87.7618	100.1991
4	19.6866	22.5464	202.5541	231.1526	164.8106	182.9759

Table 20. Duration of flood (t_d) in sec

Frame type	R=1	R=1.414	R=1.732	R=2
Bare frame	0.0401	0.0601	0.0801	0.1202
Frame with masonry infill	0.0098	0.0147	0.0196	0.0294

TABLE 21. Flood moment due to dynamic flood forces in kN-m

Zone	Bare frame				Frame with structural infill			
	R=1	R=1.4142	R=1.7321	R=2	R=1	R=1.4142	R=1.7321	R=2
II	16.14	22.82	27.95	32.28	33.66	47.61	58.31	67.33
III	25.33	35.82	43.88	50.66	49.30	69.72	85.39	98.60
IV	30.66	43.36	53.11	61.32	69.53	98.34	120.44	139.07
V	42.62	60.27	73.82	85.24	100.81	142.56	174.61	201.61

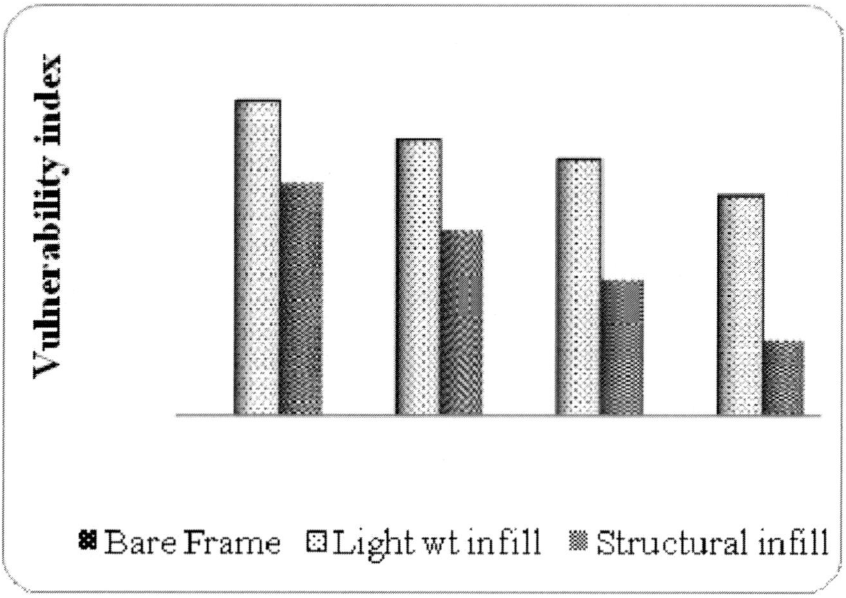

Figure 18. Variation of vulnerability in various zones

Table 22. Vulnerability index due to hydrostatic and impact forces

Zone	Bare Frame		Light weight infill		Structural infill	
	$0.1\gamma h_f$	$0.2\gamma h_f$	$0.1\gamma h_f$	$0.2\gamma h_f$	$0.1\gamma h_f$	$0.2\gamma h_f$
II	0	0	0.638	0.663	0.488	0.523
III	0	0	0.566	0.596	0.400	0.438
IV	0	0	0.531	0.564	0.308	0.348
V	0	0	0.466	0.501	0.193	0.235

The storey drift is lower for fixed support condition and the maximum value concerns the frame with light weight partition walls (Fig. 19.). The frame with structural infill wall show the smallest storey drift: this indicates the significance of infill in resisting lateral storey drift. Storey drift reaches the maximum of 20.188 mm for the frame with light weight partition walls, which is less than that specified for seismic resistant building (Table 24).

For the frame with structural infill wall, even though the initial relative cost is high, the vulnerability is lower if compared to frame with non-structural partition walls (Table 25).

Table 23. Vulnerability index due to dynamic flood forces

R	1	1.414	1.732	2
Bare frame				
Zone II	0.000	0.298	0.427	0.503
Zone III	0.000	0.178	0.329	0.419
Zone IV	0.000	0.128	0.288	0.383
Zone V	0.000	0.040	0.217	0.321
Frame with light weight infill				
Zone II	0.603	0.719	0.771	0.802
Zone III	0.528	0.666	0.727	0.764
Zone IV	0.491	0.640	0.706	0.746
Zone V	0.422	0.592	0.667	0.711
Frame with structural infill				
Zone II	0.445	0.608	0.680	0.723
Zone III	0.355	0.544	0.628	0.678
Zone IV	0.262	0.478	0.574	0.631
Zone V	0.145	0.396	0.507	0.573

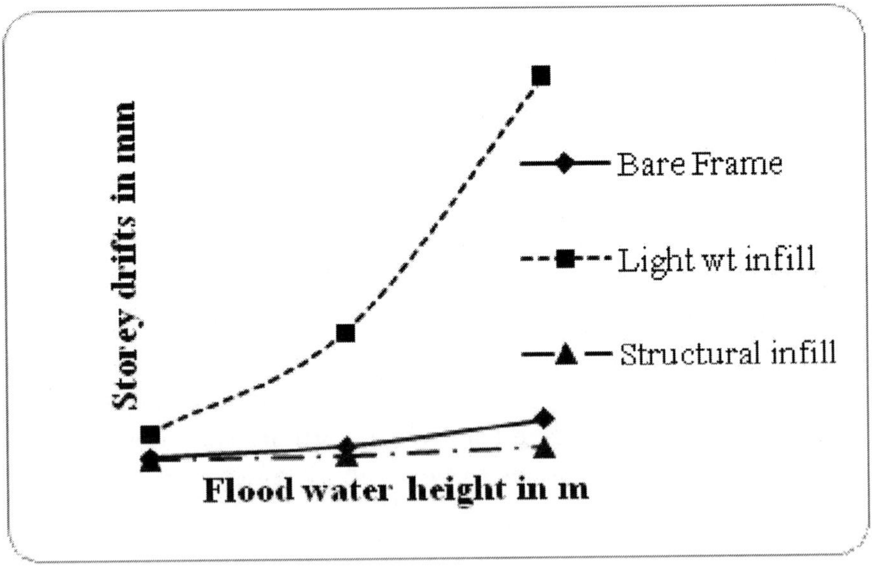

Figure 19. Variation of storey drift with flood water height

Table 24. Storey drifts due to hydrostatic and impact forces in mm

h_f (m)	Bare Frame		Light weight infill		Structural infill	
	$0.1\gamma h_f$	$0.2\gamma h_f$	$0.1\gamma h_f$	$0.2\gamma h_f$	$0.1\gamma h_f$	$0.2\gamma h_f$
2	0.18	0.22	1.24	1.64	0.09	0.103
3	0.605	0.778	5.497	7.229	0.22	0.274
4	1.618	2.074	15.621	20.188	0.534	0.68

Table 25. Relative cost as a factor of design moment for three frame models

Zone	Bare Frame		Light weight infill		Structural infill	
	DM	cost relative	DM	cost relative	DM	cost relative
II	16.139	0.000	16.139	0.000	33.664	1.086
III	25.332	0.570	25.332	0.570	49.300	2.055
IV	30.660	0.900	30.660	0.900	69.535	3.309
V	42.620	1.641	42.620	1.641	100.807	5.246

The vulnerability results obtained for different flood loadings are compared with respect to partitions (Fig. 20. and 21.). The frame with light weight infill wall is more vulnerable and bare frame is least vulnerable. This is due to the free movement of water in between the columns of the bare frame so that the contact area of flood water is lower if compared to the other frames. For the frame with masonry infill, vulnerability is lower if compared to light weight partition, even though the flood moment is the same for both the cases (Fig. 20.). It is due to the structural action of masonry infill against the lateral flood load.

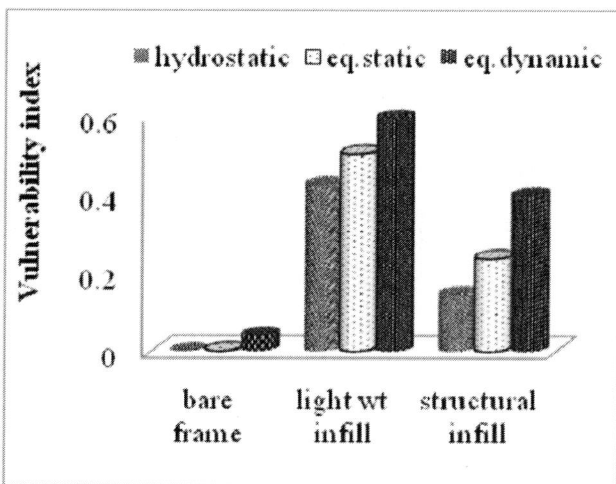

Figure 20. Vulnerability for different frame models under different flood loading conditions in Zone V

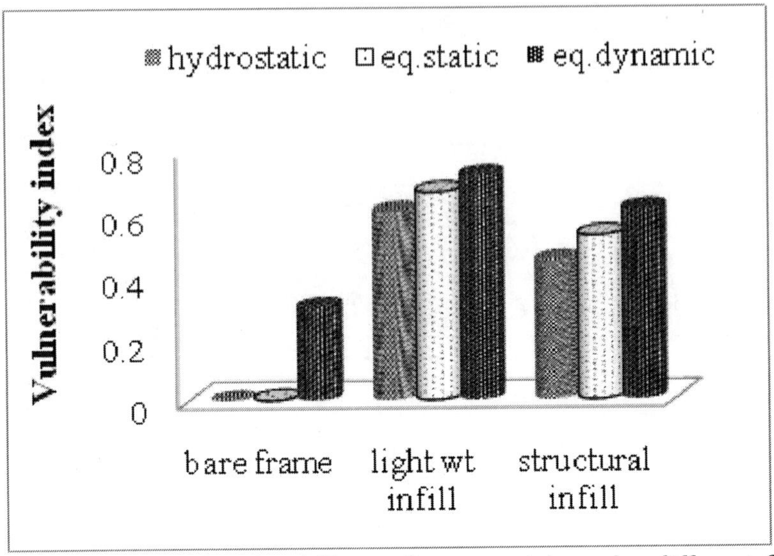

Figure 21. Vulnerability for different frame models under different flood loading conditions in Zone II

The vulnerability reduces from zone II to zone V because the design moment in zone V is higher if compared to zone II and hence the building is more resistive to flood. The variation of vulnerability for the frame with light weight infill and with masonry infill under different flood loading conditions in different zones are shown in Fig. 22. and 23, respectively.

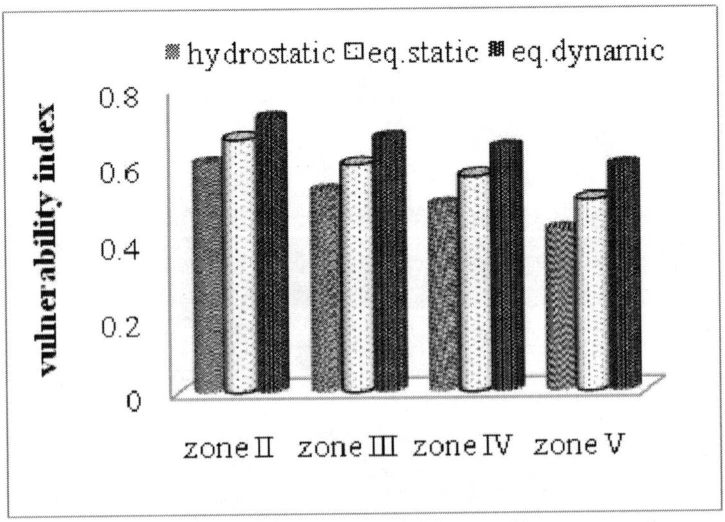

Figure 22. Vulnerability for the frame with light weight infill under different flood loading conditions in different zones

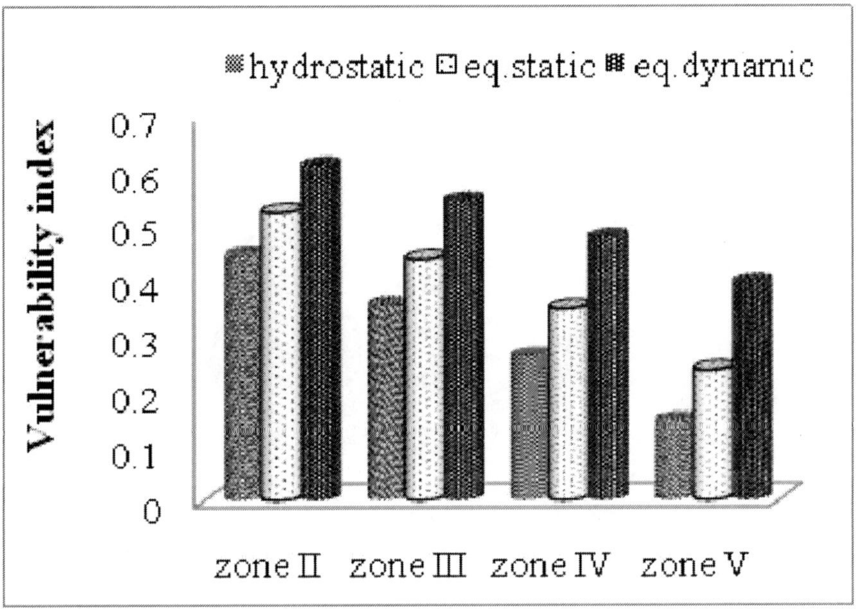

Figure 23. Vulnerability for the frame with masonry infill under different flood loading conditions in different zones

For the frame with light weight infill, vulnerability is higher in Zone II (71.9%) and it reduced as zone increases (Fig. 22.). For frame with masonry infill, vulnerability is higher in Zone II (60.8%) and it decreases as zone increases (Fig. 23.). This is because the design moment of building in zone V is higher if compared to zone II and hence the building in zone V will be more resistant to flood.

The vulnerability results obtained for different flood loadings are compared with respect to seismic zones (Fig. 24. and 25.). As the duration of flood increases, vulnerability increases (Fig. 24.); vulnerability is lower for bare frame than for frame with partitions. A building in zone V with flood duration $T_n/3$ is less vulnerable (66.7%) than a building in zone II with flood duration of $T_n/2$ (71.9%) (Fig. 25.), hence vulnerability is higher for building subjected to longer floods even if it also depends on the seismic zone.

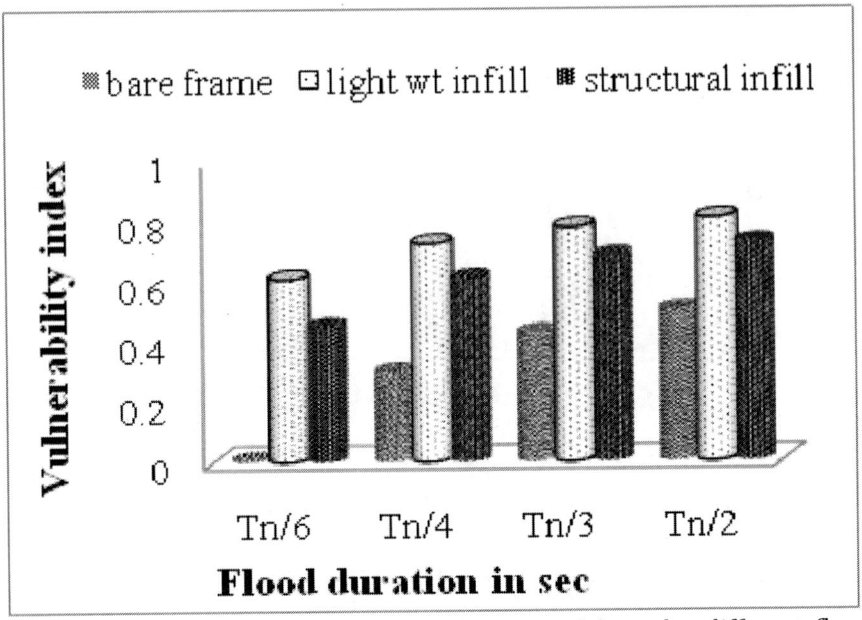

Figure 24. Vulnerability for different frame models under different flood duration in Zone II

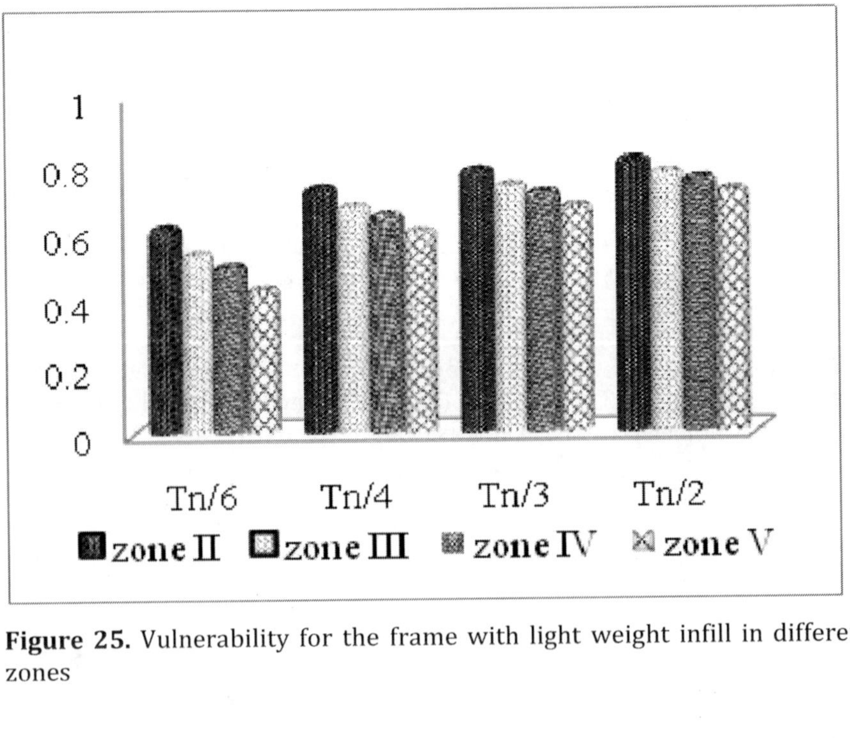

Figure 25. Vulnerability for the frame with light weight infill in different zones

5. Conclusions

Flood physical vulnerability deals with the level of loss that elements at risk or built environment suffer from the occurrence of flooding. This study aims to find out the flood vulnerability limit as a factor of ground floor height under flood forces and to quantify flood load. Three frame models were modelled and the effects of flood forces in each frame were analysed. The significance of infill walls in resisting lateral storey drift during flood is also investigated. The main conclusions of the analysis are:

- The flood moments parabolically increase as flood water height increases and linearly increase as impact load increases.
- The vulnerability of frame with light weight partition wall, for hinged support condition, reaches 64.2% for dynamic flood forces, that is very high if compared to the other frames.
- For frame with light weight partition wall in hinged support condition, storey drift reaches 71.32 mm, which is more than the value specified for seismic resistant building.
- The vulnerability of frame with light weight partition wall, for fixed support condition, is up to 60% in zone II which is very high if compared to the other frames.
- Storey drift for frame with light weight partition wall in fixed support condition is found to be less than hinged condition. The maximum value of storey drift for frame with light weight partition wall is 20.188mm.
- Even though the initial cost is more for frame with structural partitions, its vulnerability is very low if compared to frame with non-structural partitions.
- Buildings in zone II is most vulnerable and the vulnerability is reducing as zone increases. It reaches zero for frame with structural infill ass zone varies from zone II to zone V. This is because the design moment of building in zone V is higher if compared to zone II and hence the building in zone V is more resistive to flood.

Frame with light weight partition wall result as the most vulnerable and bare frame is least vulnerable. Hence frame with non-structural partitions like plywood are not preferred in flood prone areas. The storey drift for the frame with structural infill walls is very low if compared to the other frame models and this indicates the significance of infill in resisting lateral storey drift. Soft storied buildings are less vulnerable compared to ordinary buildings and this depends on the free movement of water in between the columns. Results also indicate the real need of considering the flood loads in the design procedure of reinforced concrete buildings.

References

1. American Society of Civil Engineers (2006), Minimum Design Loads for Buildings and Other Structures, ASCEI/SEI 7-05.
2. Chopra, A.K. Dynamic of Structures - Theory and Applications to Earthquake Engineering, Third Edition, Pearson Education, 2009.
3. Arulselvan, S., Subramanian K., Pillai E. B.P., and Santhakumar A. R.(2007), RC Infilled frames - RC Plane Frame Interactions for Seismic Resistance, Journal of Applied Sciences, 7, 942-950.
4. Federal Emergency Management Agency (2001), Engineering Principles and Practices for Flood Prone Residential Structures, FEMA 259, Edition 2.
5. Haugen E.D., and Kaynia A.M. (2008), Vulnerability of structures impacted by debris flow, Landslides and Engineered Slopes, Taylor & Francis Group, London, ISBN 978-0-415-41196-7, 381-387.
6. Kelman, I. (2002), Physical Flood Vulnerability of Residential Properties in Coastal, Eastern England, Ph.D. Dissertation, University of Cambridge, U.K.
7. Kelman, I., and Spenc, R. (2004), An overview of flood actions on buildings, Journal of Engineering Geology, 73, 297–309.
8. IS 1893 (Part 1): 2002, Indian Standard Criteria for Earthquake Resistant Design of Structures: Part-1 General Provisions and Buildings, 5th Revision, BIS, New Delhi.
9. IS: 456: 2000 Plain and Reinforced Concrete -Code of Practice, 4th Revision, BIS, New Delhi.
10. IS: 875 (Part 1): 1987, Code of practice for design loads (other than earthquake) for buildings and structures: Part-1 Dead loads - unit weights of building materials and stored materials, 2th Revision, BIS, New Delhi.
11. IS: 875 (Part 2): 1987, Code of practice for design loads (other than earthquake) for buildings and structures: Part-2 Imposed Loads, 2th Revision, BIS, New Delhi.
12. Kreibich H., Piroth K., Seifert I., Maiwald H., Kunert U., Schwarz J., Merz B., and Thieken, A. H. (2009), Is flow velocity a significant parameter in flood damage modelling?, Natural Hazards Earth System Sciences, 9, 1679–1692.
13. Messener, F., and Meyer, V. (2005), Flood Damage, Vulnerability and Risk Perception – Challenges for Flood Damage Research, Discussion Papers, Nato Science Series, Springer Publisher.

14. Sagala, S.A.H. (2006), Analysis of flood physical vulnerability in residential areas, M.Sc. Thesis, International Institute of Geo-Information Science and Earth Observation, Netherlands.
15. Scheuren, J. M., de Waroux, O., Below, R., Guha-Saphir, D. and Ponserre, S. (2007), Annual Disaster Statistical Review. CRED Brussels, Belgium.
16. Schwarz, J. and Maiwald, H. (2008), Damage and loss prediction model based on the vulnerability of building types, 4th International Symposium on Flood Defence: Managing Flood Risk, Reliability & Vulnerability, May 6-8.
17. SP-16: 1980, Design aids for reinforced concrete to IS: 456-1978, 11th Edition, BIS, New Delhi.
18. Singh, A. K., and Sharma, A. K. (2009). GIS and a remote sensing based approach for urban floodplain mapping for the Tapi catchment, India. Hydro informatics in Hydrology, Hydrogeology and Water Resources at the Joint IAHS & IAH Convention, Hyderabad, India, September 2009.

CHAPTER 4

Seismic Hazard Analysis for Archaeological Structures — A Case Study for EL Sakakini Palace Cairo, Egypt

Sayed Hemeda[1]

[1] Conservation Department, Faculty of Archaeology, Cairo University, Egypt

1. Introduction

The modern architectural heritage of Egypt is rich, and extensively variable. It covers all kinds of monumental structures from palaces, public buildings, residential and industrial buildings, to bridges, springs, gardens and any other modern structure, which falls within the definition of a monument and belongs to the Egyptian cultural heritage. We present herein a comprhensive geophysical survey and seismic hazard assesment for the rehabilitation and strengthening of Habib Sakakini's Palace in Cairo, which is considered one of the most significant architectural heritage sites in Egypt. The palace located on an ancient water pond at the eastern side of Egyptian gulf close to Sultan Bebris Al-Bondoqdary mosque, a place also called "Prince Qraja al-Turkumany pond". That pond had been filled down by Habib Sakakini at 1892 to construct his famous palace in 1897.

Various survey campaigns have been performed comprising geotechnical and geophysical field and laboratory tests, aiming to define the physical, mechanical and dynamic properties of the building and the soil materials of the site where the palace is founded. All these results together with the seismic hazard analysis will be used for the seismic analysis of the palace response in the framework of the rehabilitation and strengthening works foreseen in a second stage. We present herein the most important results of the field campaign and the definition of the design input motion.

The seismic hazard analysis for El Sakakini Palace has been performed based on historical earthquakes, and maximum intensity.PGA with 10% probability of exceedance in 50 and 100 years is found equal to 0.15g and 0.19g respectively. P-wave and S-wave seismic refraction indicated a rather low velocity soil above the seismic bedrock found at depths higher than 20m. Ambient noise measurements have been used to determine the natural vibration frequency of soil and structure of El-Sakakini Palace. The fundamental frequency of El-Sakakini palace is 3.0Hz very close to the fundamental frequency of the underlying soil, which makes the resonance effect highly prominent.

Some floors are considered dangerous since it show several resonance peaks and high amplification factors (4th and 5th floors) these floors are made of wood so, warnings to decision makers are given for the importance of such valuable structures.

The seismic design and risk assessment of El Sakakini palace is performed in two steps. In the first one we perform all necessary geotechnical and geophysical investigation together with seismic surveys and seismic hazard analysis in order to evaluate the foundation soil properties, the fundamental frequency of the site and the structure, and to determine the design input motion according to Egyptian regulations. The second phase comprises the detailed analysis of the palace and the design of the necessary remediation measures. IN the present pare we present the results of the first phase.

2. Seismic Hazard

2.1. Historical Seismicity

Egypt possesses a rich earthquake catalogue that goes back to the ancient Egyptian times. Some earthquakes are reported almost 4000 years ago. Figure 1 shows the most important historical events affecting ElSakakini palace. We can see that the Faiyum area as well as the Gulf of Suez is the most important earthquake zones affecting the place.

2.2. Maximum Intensity

Historical seismicity and maximum reported intensity is a good preliminary index of the expected severity of a damaging earthquake. Available isoseismal maps in the time period 2200 B.C. to 1995 were digitized and re-contoured to determine the maximum intensity affecting the place. This was done using a cells value of equal area 0.1 lat. × 0.1 long. Figure 2 present the produced IMM intensity showing that a maximum IMM of VII is good design value.

2.3. Probabilistic Hazard Assessment

An improved earthquake catalogue for Egypt and surrounding areas affecting El Sakakini Palace has been prepared for the purposed of this study partially based on recent work of Gamal and Noufal, 2006. The catalogue is using the following sources:

- For the period 2200 B.C to1900: Maamoun,1979; Maamoun et al., 1984 ; Ben-Menahem 1979and Woodward-Clyde consultants, 1985.
- For the period 1900 to 2006: Makropoulos and Burton, 1981; Maamoun et al., 1984 ; Ben-Menahem 1979; Woodward-Clyde consultants, 1985; Riad and Meyers, 1985; Shapira, 1994and NEIC, 2006; Jordan seismological observatory 1998-2000.

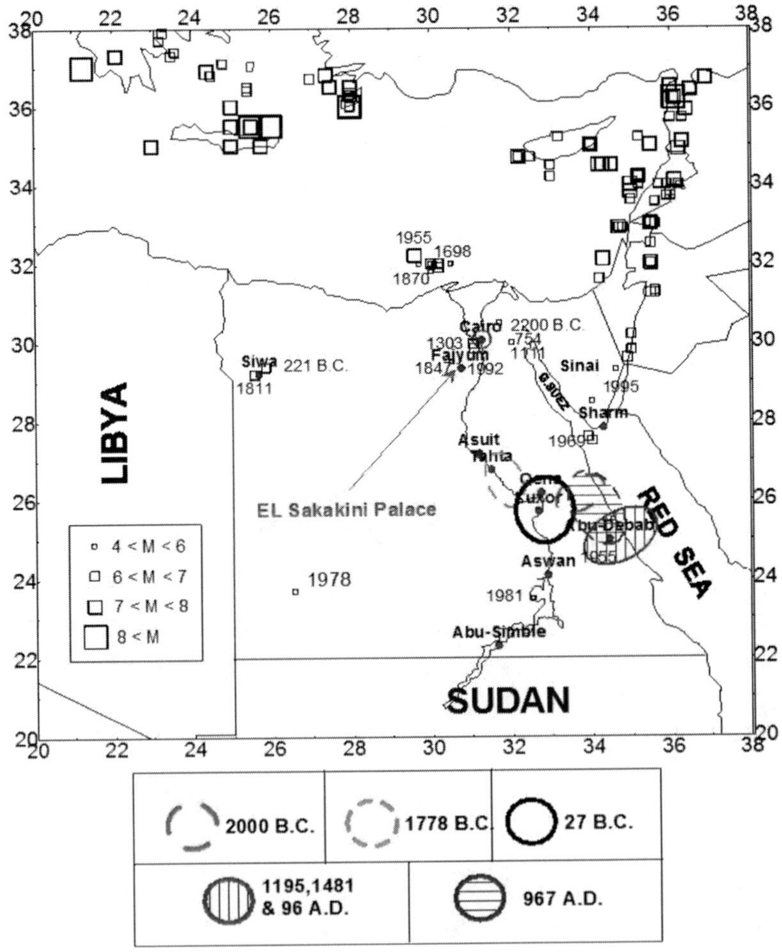

Figure 1. Important and historical earthquakes occurred in and around El Sakakini Palace area in the period 2200 B.C to1995.

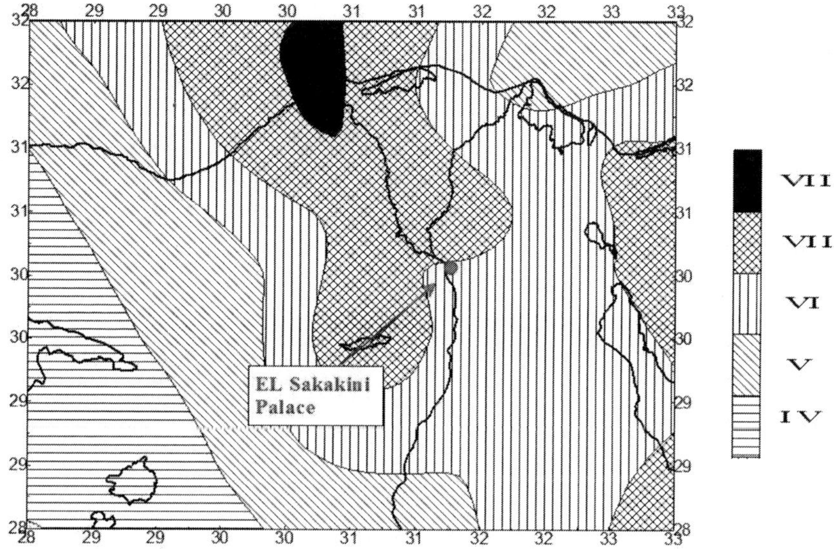

Figure 2. Maximum intensity zonation map based on the historical seismicity reported in the time period 2200 BC to1995

The horizontal peak ground acceleration over the bedrock of El Sakakini area was estimated using Mcguire program 1993. 37 seismic source zones were used to determine the horizontal PGA over the bedrock (Figure 3), while PGA attenuation formula of Joyner and Boore, 1981 was used because of its good fitting to real earthquake data in Egypt. A complete analysis for the input parameters to estimate the PGA values over the bedrock can be found in Gamal and Noufal, 2006.

$$(\mathrm{PGA}) = 2.14\,\mathrm{e}1.13^{M} D^{-1} e^{-0.00590}\ D = \left(R^2 + 4.0^2\right)^{0.5} \qquad (1)$$

The probabilistic analysis provided the following results: The peak horizontal acceleration in gals with 10 % probability of exceedance over 50 years is 144cm/sec^2(or 0.147g) For 10% probability in 100 years the estimated PGA for rock conditions is 186 (cm/sec^2) (or 0.19g)(Figures 3 and 4). These values are quite high and considering the local amplification they may affect seriously the seismic design and stability of El. Sakakini Palace.

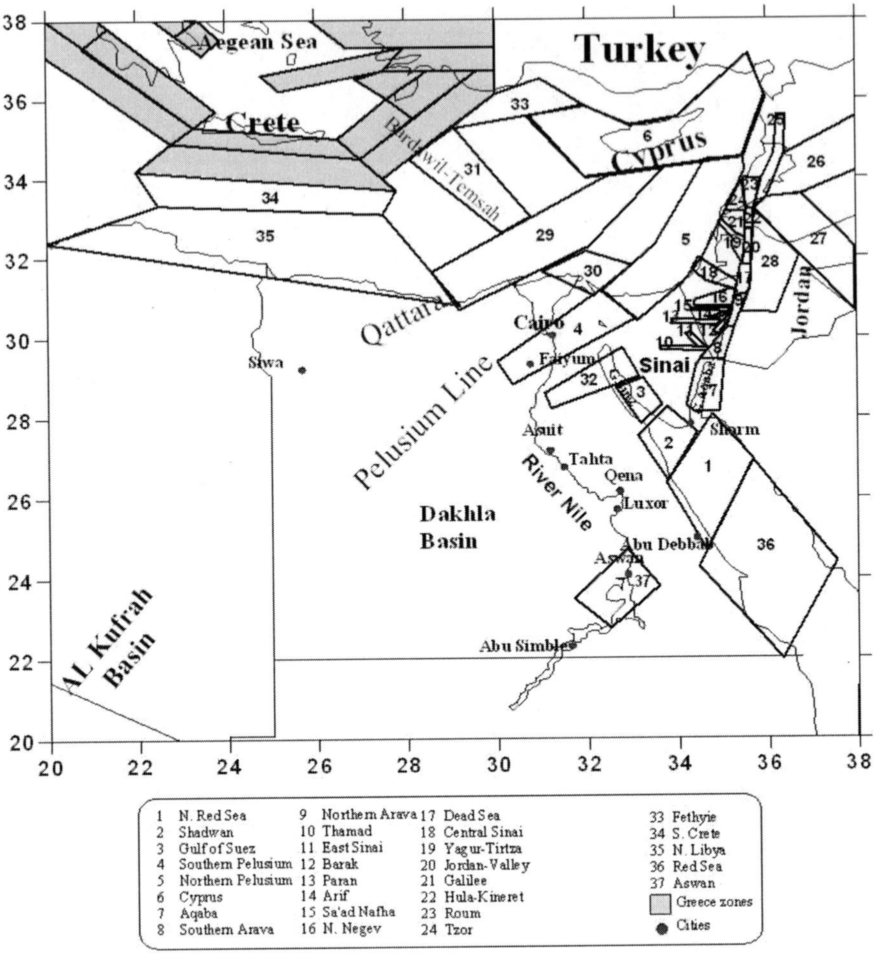

Figure 3. Seismic source regionalization using 37 seismic source zone (except greece zones) adopted for Egypt and surrounding areas (Gamal and Noufal, 2006).

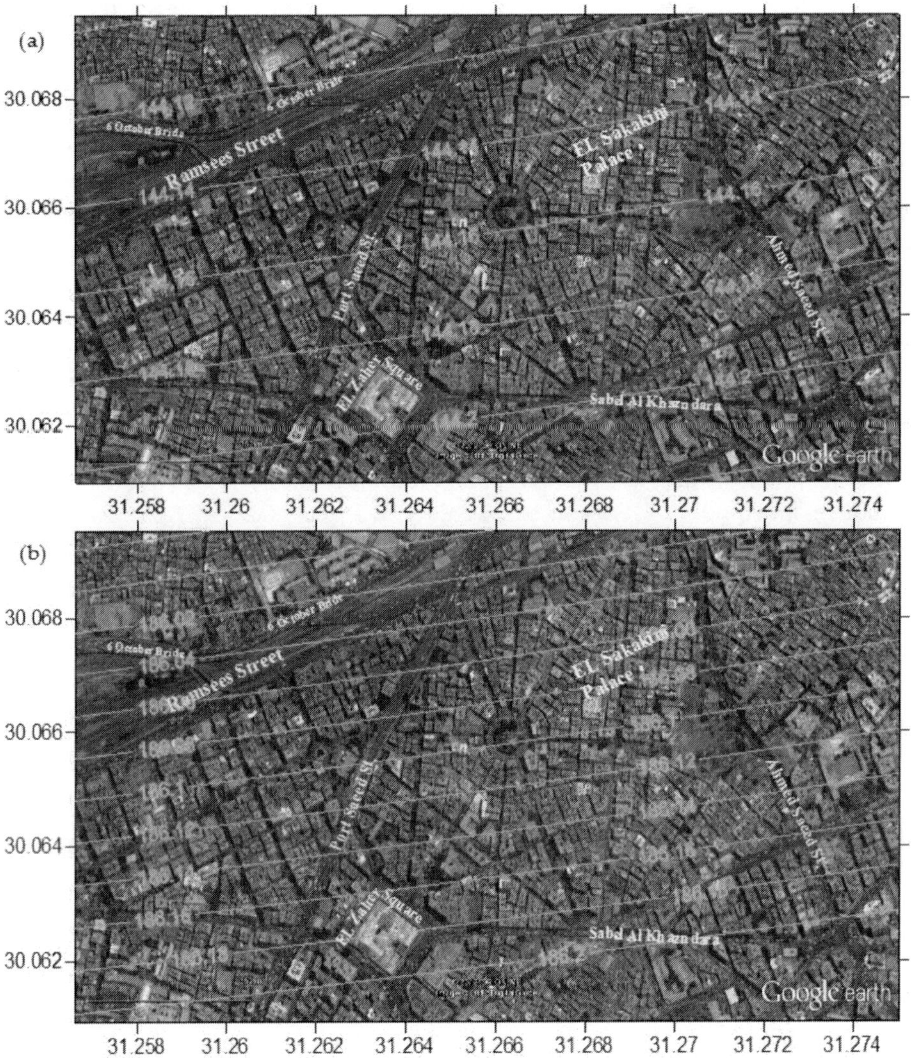

Figure 4. a: Peak Horizontal Acceleration in gals (cm/sec^2) for the seismic bedrock with10 % probability of exceedance in 50 years. b: Peak Horizontal Acceleration in gals (cm/sec^2) for the seismic bedrock with 10 % probability of exceedance over 100years

3. Geotechnical investigation

Core drilling is among the routine methods for subsurface exploration. Most commonly, NX-size core drill is used, representing a hole diameter of 76 mm (3") and a core diameter of 54 mm (2 1/8"). The drilling often has multiplier purposes, of which the following are in most cases the most important:

Verification of the geological interpretation. Detailed engineering geological description of rock strata. To obtain more information on rock type boundaries and degree of weathering. To supplement information on orientation and character of weakness zones. To provide samples for laboratory analyses. Hydro geological and geophysical testing. Input data for engineering classification of rock masses.

The geotechnical investigation, six geotechnical boreholes with Standard Penetration Test (SPT) measurements have been carried out in the archaeological site included the drilling of three geotechnical boreholes with integral sampling to a depth 20 meters, one borehole to depth 15 meters and two boreholes to depth 10 meters at six locations in the site. The geotechnical data also indicated the ground water level at the archaeological site. We did all the boreholes inside the site with hand boring machine.

The results of laboratory tests which have been carried out on the extracted soil samples from the boreholes, which include specific gravity (Gs), water content (Wn), saturated unit weight (γsat), unsaturated unit weight (γunsat), Atterberg limits and uniaxial compressive strength (UCS), in addition to the ground water table (GWT), are shown in the figures (7a,7b).

The shear wave profile obtained by using ReMi compared very well to geotechnical boreholes and geophysical survey data. In addition, the shear wave profile obtained by using ReMi Performed much better than commonly used surface shear-wave velocity measurements.

Geotechnical boreholes (1) through (3) indicated that:

Filling of Fill (silty clay and limestone fragments, calc, dark brown) From ground surface 0.00m to 3.50m depth. Sand Fill (silty clay, medium, traces of limestone& red brick fragments, calc, dark brown) From 3.50m to 5.00m depth. Silty clay, stiff, calc, dark brown From 5.00m to 6.50m depth. Clayey silt, traces of fine sand & mica, yellowish dark brown From 6.50m to 8.50m depth. Silty sand, fine, traces of clay & mica. Dark brown. From 8.50m to 11.00m depth. Sand, fine, some silt, traces of mica, yellowish dark brown. From 11.00m to 14.00m depth. Sand, fine to medium, traces of silt& mica, tracesof fine to medium gravel, traces of marine shells, yellowish dark brown. From 14.00m to 16.00m depth. Sand, fine, traces of silt & mica, yellowish dark brown. From 16.00m to 18.00m depth. Sand & Gravel, medium sand, graded gravel, traces of silt, yellow darkbrown. From 18.00m to 20.00m depth. End of drilling at 20.00m.

Geotechnical boreholes (4) through (6) indicated that:

Fill (silt, clay and fragments of limestone and crushed brick, from ground surface 0.00m to 4m depth. Fill (silty cal with medium pottery and brick fragments, calc dark brown) from 4 to 5 m depth. Brown stiff silty clay and traces of limestone gravels, from 5.00m to 7.50m depth. silt, traces of brown fine sand & traces of clay from 12.00m to 14.00m depth. Dark brown clay silt with traces of fine sand. from 14.00m to 15.00m depth.

Figure 5. El- Sakakini palace and the Geotechnical investigations.

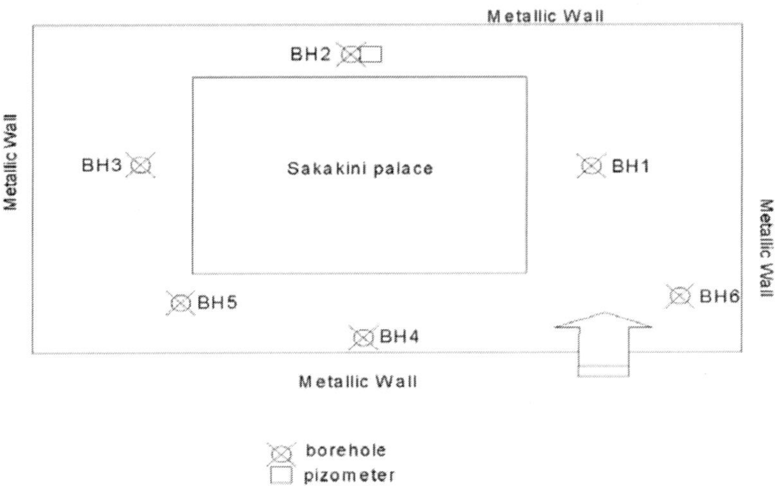

Figure 6. General layout & boreholes locations.

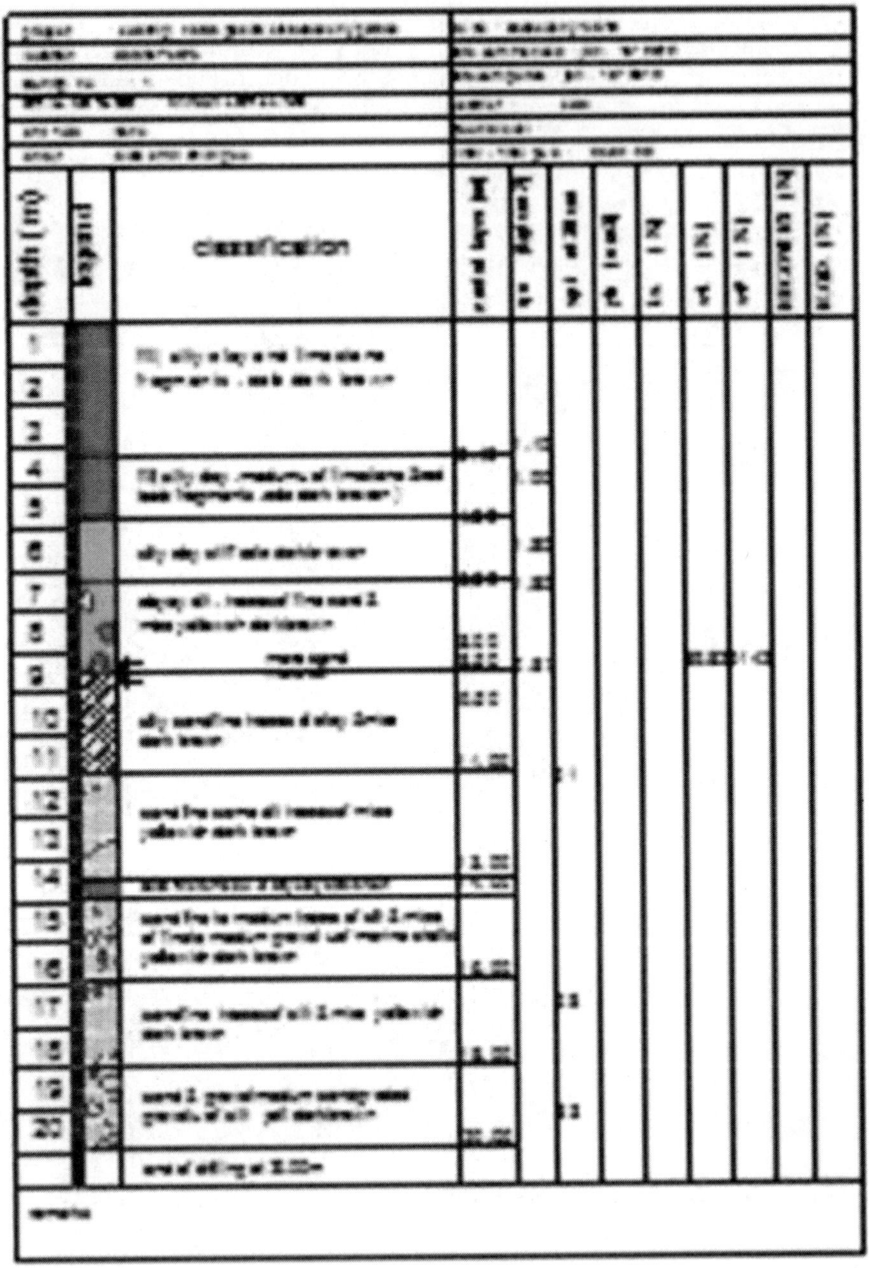
classification

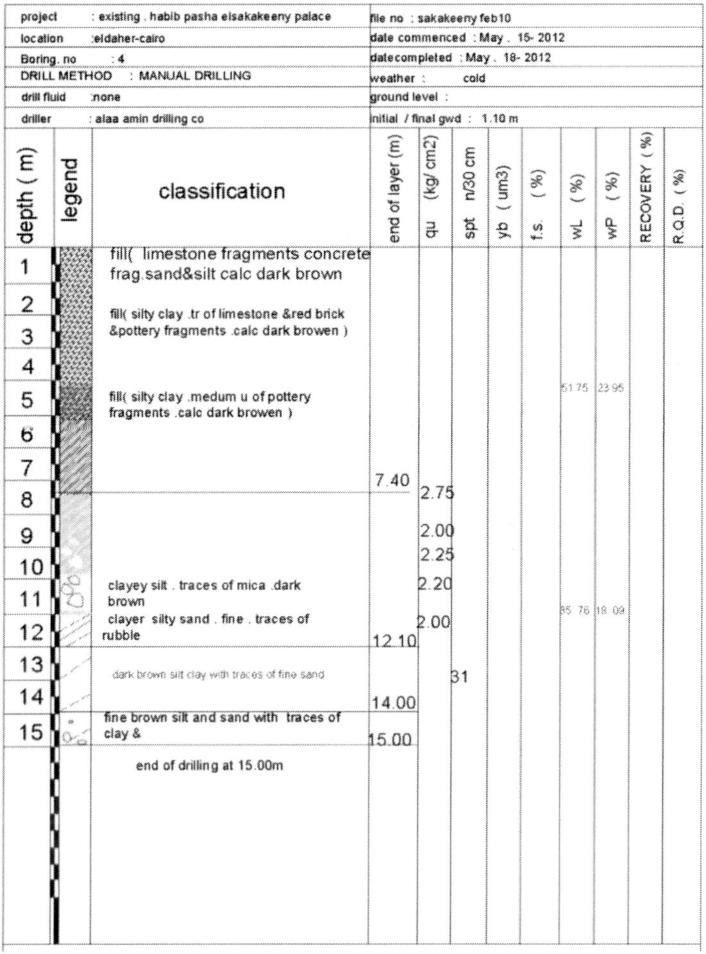

Figure 7. a. Geotechnical Borehole_1, El Sakakini Palace. b. Geotechnical Borehole_4, El Sakakini Palace.

4. Geophysical Campaign

4.1. P-Wave Refraction

A total of 10 seismic profiles are conducted at El Sakakini palace area (Figure 8). All profiles are carried out using 12 receivers, P-type geophones with 5m intervals and 2 shots. The forward and reverse shots were carried at a distance of 1 m at both ends. The seismic shots layouts are described inTable 1.

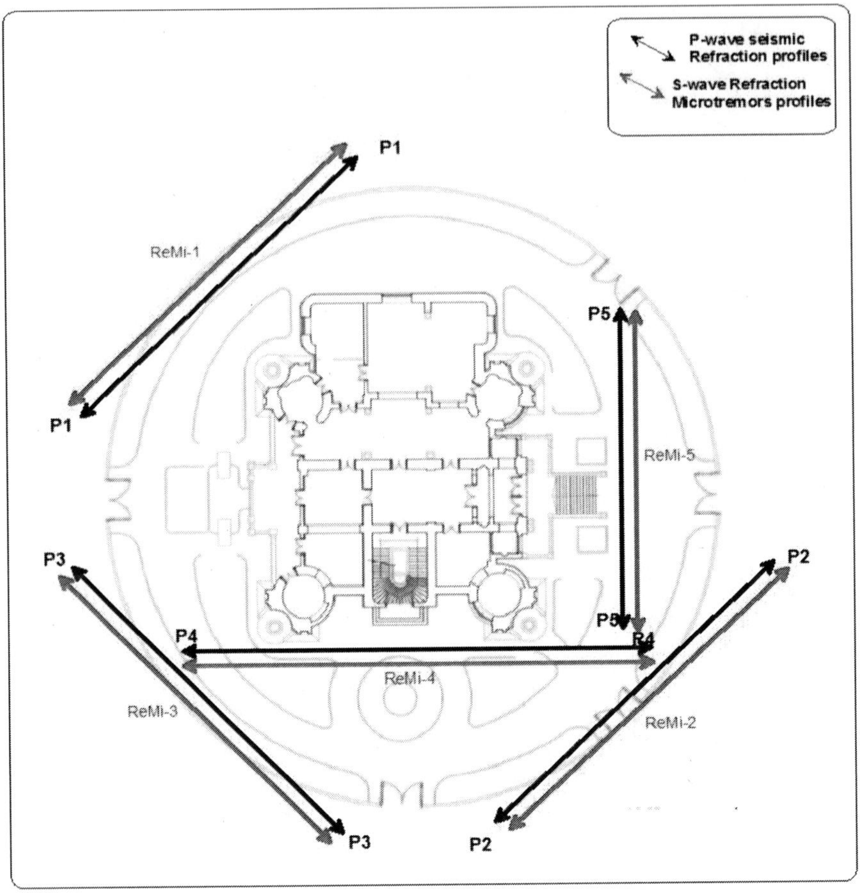

Figure 8. Location of the P-wave seismic refraction, S-wave refraction and ReMiprofiles conducted at ElSakakini Palace.

Table 1. Seismic shots.

Shot #	Name	Offset X (m) (relative to R1)
S2	Forward	-1
S4	Reverse	56

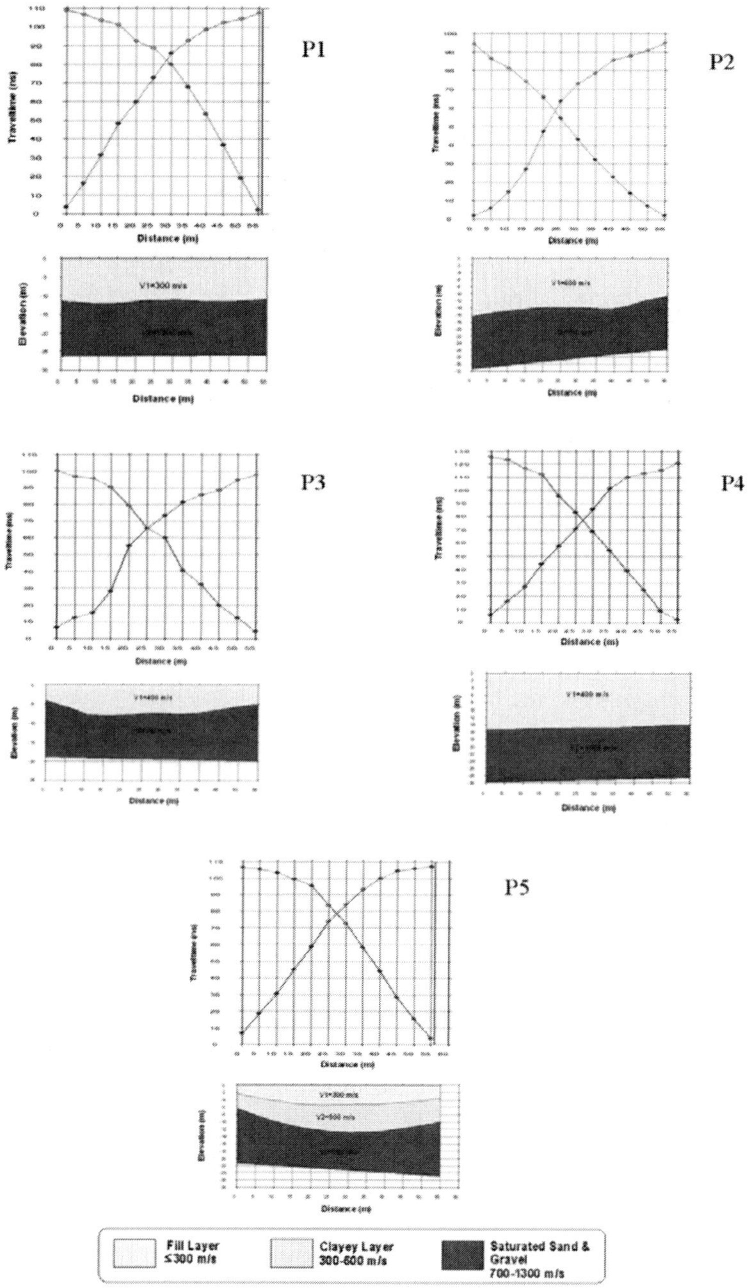

Figure 9. P-wave travel time distance curve and its corresponding geoseismic model for profiles # P1-P5 (Figure 8).

The conducted profiles are interpreted using time-term inversion method; an example of the conducted profiles and corresponding geoseismic model is shown in Figure 9. Table 2 summarizes the measured Vs values and the corresponding soil thicknesses. The soil stratification is not uniform and horizontal, as it should be expected for a filled area. However it is possible to distinguish the following three main layers: the soil layering can be summarized in the following (table 2).

Soil A- Fill (<300 m/s): A surface highly heterogeneous material (mainly man-made fill) with an average thickness of 10 m and an average velocity Vs lower than 300m/s. It is composed of very loose and low strength sediments such as silt, clay and limestone fragments. It is not found in all locations.

Soil B-Clayey soil (400-600 m/s): Below the surface layer (soil A) there is a clayey or silty clay layer with an average thickness of 10 m meters and Vs velocity 400-600 m/s.

Soil C-Saturated Sand & Gravel (700-1300 m/s): Below soil B there is a stiff soil layer with various thicknesses. it shows a considerable increase of Vs seismic velocity reaching sometimes values as high as 1300m/s. The soil is composed of compacted stiff saturated sand and gravel with an average Vs velocity equal or higher than 700m/s. It may be considered as the "seismic bedrock" for the local site amplification analyses.

Table 2. P-wave refraction geophysical campaign conducted at El-Sakakini palace area.

	Layer A	Layer B		Layer C	
Profile N°	Velocity in m/s	Velocity in m/s	Depth (m)	Velocity in m/s	Depth in (m)
1		300	10	1300	25
2		600	16	900	32
3		400	9	700	20
4		400	14	1200	30
5	< 300	300	5	500	10

4.2. Refraction- Microtremor (Remi Method)

We have used the ReMi (refraction microtremors) method to determine the S-wave seismic velocity with depth. The method is based on two fundamental ideas. The first is that common seismic-refraction recording equipment, set out in a way almost identical to shallow P-wave refraction surveys, can effectively record surface waves at frequencies as low as 2 Hz (even lower if low frequency phones are used). The second idea is that a simple, two-dimensional slowness-frequency (P-f) transform of a microtremors record can separate Rayleigh waves from other seismic arrivals, and allow recognition of true phase velocity against apparent velocities. Two essential factors that allow exploration equipment to record surface-wave velocity dispersion, with a minimum of field effort, are the use of a single geophone sensor at each channel, rather than a geophone "group array", and the use of a linear spread of 12 or more geophone sensor channels. Single geophones are the most commonly available type, and are typically used for refraction rather than reflection surveying. There are certain advantages of ReMi method: it requires only standard refraction equipment, widely available, there is no need for a triggering source of energy and it works well in a seismically noisy urban setting. (Louie, 2001,Pullammanappallil et al. 2003).

A 12 channel ES-3000 seismograph was used to measure background 'noise' enhanced at quiet sites by inducing background noise with 14Hz geophones in a straight line spacing 5m Figure 5 shows the map were ReMi measurements were made. Almost all the sites were noisy. In particular big hammer used to break some rocks generated noisy background at El Sakakini Palace.30 files of 30sec records (unfiltered) of 'noise' were collected at each site. Five profiles were taken inside the Palace (Figure 8).Figure 10-11 shows an example of the dispersion curves and its P-F image (Remi Spectral ratio of surface waves) for refraction microtremors profile ReMi-1. The estimated average Vs for all profiles are shown in Figure 12.

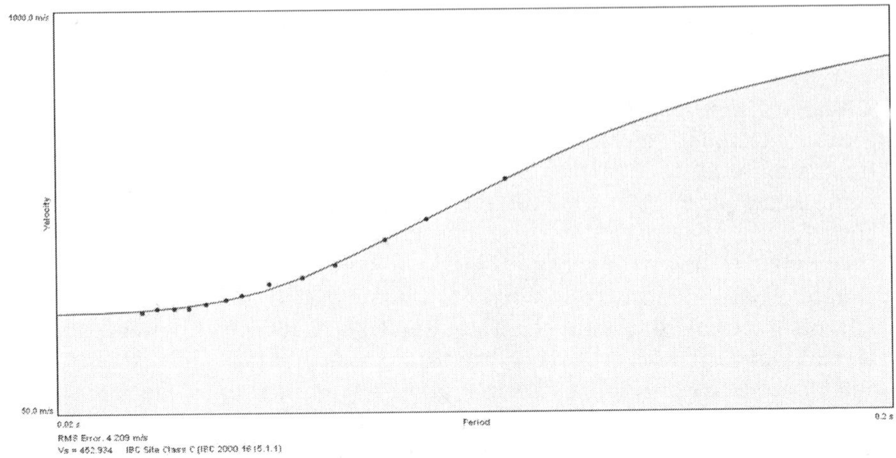

Figure 10. Dispersion curve showing picks and fit for Profile ReMi-1

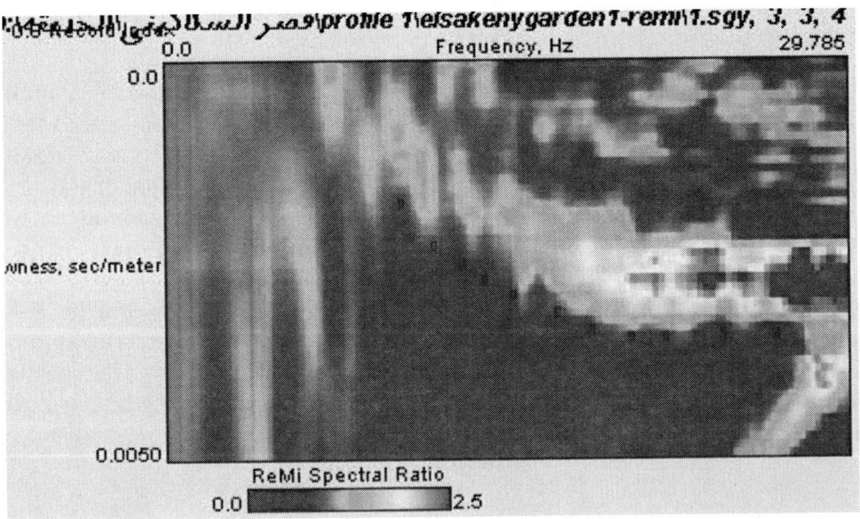

Figure 11. P-F image with dispersion modeling picks for Profile ReMi-1

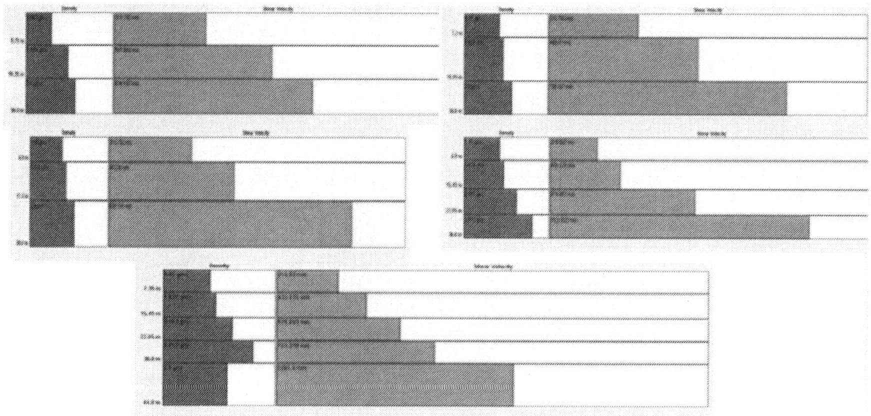

Figure 12. Shear wave velocity model calculated for refraction microtremors profiles ReMi-1 To ReMi-5 (Figure 9).

5. Frequency Charactersitics of the Soil and the Building Using Microtremrs

Microtremors are omnipresent low amplitude oscillations (1-10 microns) that arise predominantly from oceanic, atmospheric, and urban or anthropogenic actions and disturbances. The implicit assumption of early studies was that microtremors spectra are flat and broadband before they enter the region of interest (soil or building). When microtremors enter preferable body it changes and resonate depending on the nature of the material, shape, and any other characteristics of this body.

It may be considered to compose of any of seismic wave types. We have two main types of microtremors, Local ambient noise coming from urban actions and disturbances and long period microtremors originated from distances (e.g. oceanic disturbances). There is still a debateongoing on the characteristics of the ambient noise that should be used for site characterization and ground response. While some are using only the longer period microtremors originated from farther distances (e.g. Field et al, 1990), others considered that traffic and other urban noise sources are producing equally reliable results. In general low amplitude noise measurements comparable results give with strong motion data (Raptakis et al, 2005., Pitilakis, 2011., Apostolidis et al., 2004., Manakou et al, 2010., Mucciarelli, 1998).

Kanai 1957, first introduced the use of microtremors, or ambient seismic noise, to estimate the earthquake site response (soil amplification). After that lots of people followed this work but from the point of soil amplification of earthquake energy for different frequencies (e.g. Kanai and Tanaka 1961and Kanai 1962, Kagami et al, 1982 and 1986; Rogers et al., 1984; Lermo et al., 1988; Celebi et al. 1987).

5.1. Instrumentation and Data Acquisition

A high dynamic range Seismograph (Geometrics ES-3000 see Figure 13) mobile station with triaxial force balance accelerometer (3 channels), orthogonally oriented was used. The station was used with 4Hz sensors to record the horizontal components in longitudinal and transverse directions in addition to the vertical components. For the data acquisition and processing we followed the following steps:

- Recording 10-min of ambient noise data using a mobile station moving among variable soil stations or El Sakakini building floors/
- Zero correction to the total 10-minnoise at time domain
- Subdivision of each 10-minsignal into fifteen 1-min sub-windows,

Figure 13. High dynamic range ES-3000 Geometrics mobile station and triaxial geophone used 4 Hz to drive soil response of El-Sakakini Palace.

Each of these series was tapered with a 3-sec hanning taper and converted to the frequency domain using a Fast Fourier transform,

- Smoothing the amplitude spectrum by convolution with 0.2-Hz boxcar window,
- Site response spectrum for a given soil site (or certain floor) is given by dividing the average spectrum of this site over the spectrum of the reference site. The reference site is choose carefully in the site as deepest and calmest station in the basement floor with least soil response (usually we choose a certain basement floor location with least soil response to be used as reference site).
- Smoothing the final response curves by running average filter for better viewing. A complete description of the methodology can be found in Gamal and Ghoneim, (2004).

5.2. Ground Response

Figure 14 shows the locations of microtremors stations used to determine the ground response at EL Sakakini Palace area. The predominant frequency of the ground at EL Sakakini Palace is about 3 Hz (see Figure 15 & Table 1), a value almost identical to the theoretical estimation according to Kennett and Kerry (1979) (Figure 16 & Table 4). The amplification factor is about 2, which is relatively low.

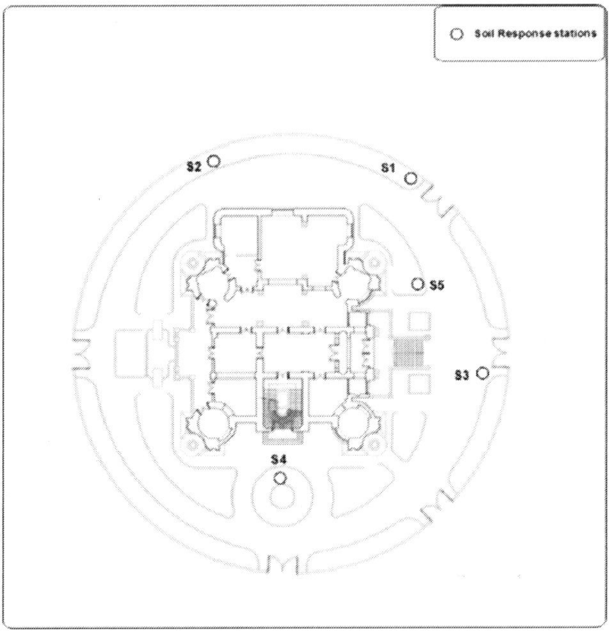

Figure 14. Ambient noise measurement locations

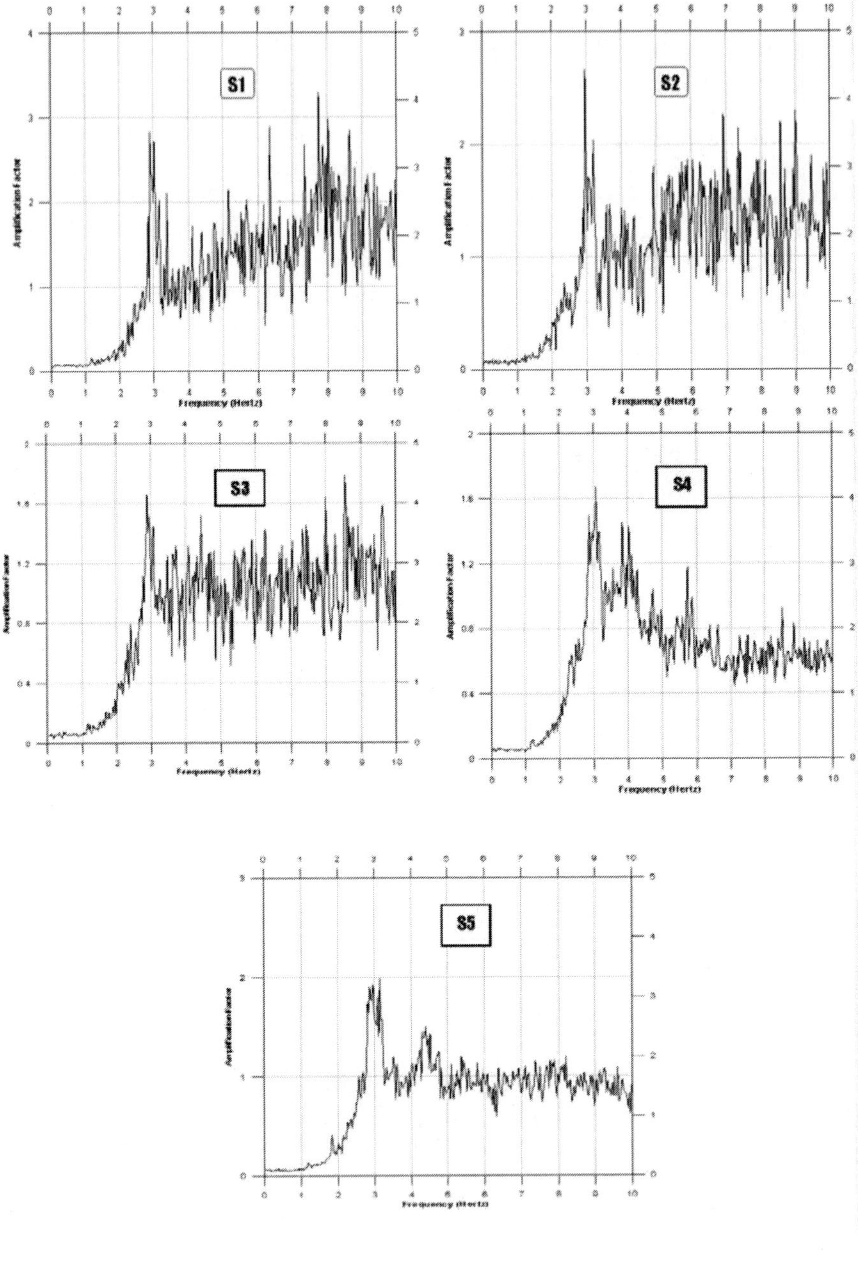

Figure 15. Microtremors soil response for El Sakakini Palace Sites S1 to S5.

Table 3. Fundamental frequencies and amplification factors at five locations

Site	Fundamental frequency (Hz)	Amplification Factor
S1	3	2.5
S2	3.2	2
S3	3	1.6
S4	3	1.6
S5	3	2

Table 4. Parameters used for the Kennett and Kerry method (1979)

Thickness	P-wave velocity (m/s)	S-wave velocity (m/s)	Dry Density (gm/c.c)	Quality factor Qs
10	1300	315 (350?)	1.6	7
25	1300	500	1.7	15
>10.5	2000	700	2	100
	3000	1200	2.5	200

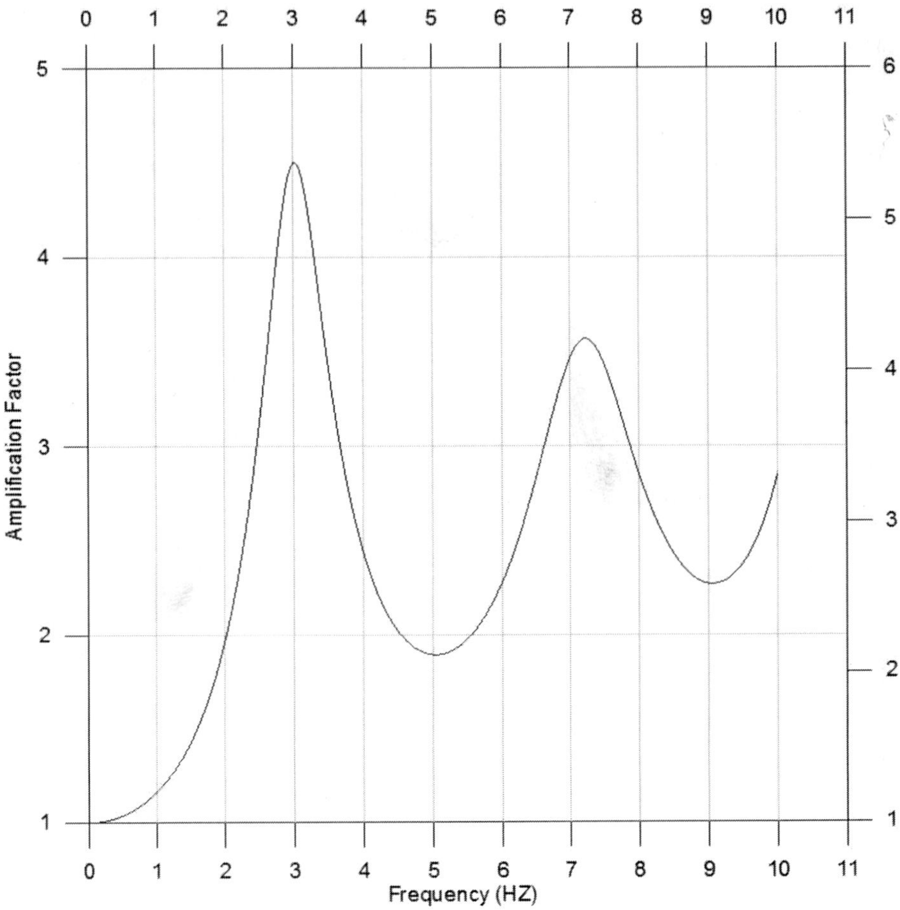

Figure 16. Theoretical ground response analysis at EL Sakakini Palace using Kennett at al. (1979) method.

5.3. Building Response

The El Sakakini building is composed of a basement and five floors the upper two being wooden.Figures 17 to 19 show the locations of recording stations used to drive El Sakakini building response.Figures 21 to 26 and Table 5 show the recorded natural frequency of vibration for each floor. All floors show nearly the same resonance frequency with the soil (3-4 HZ). The wooden floors (Figure 25 & 26) show very high amplification and multi peak as fundamental and other harmonics. The fundamental natural frequency of vibration is always the most important frequency that insert the maximum earthquake vibration energy into structure. However when we find other mode of vibrations with big amplification factors we consider this as a warning that this structure may suffer from vibration. This could be very good warning for its unstable performance during vibration.

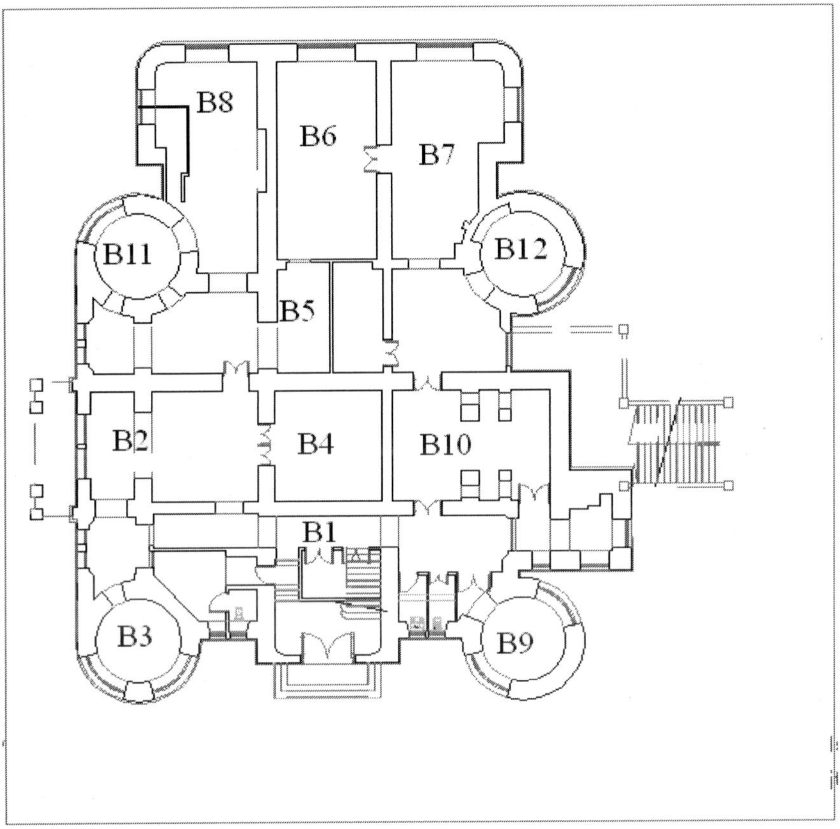

Figure 17. Location of stations at the basement of EL-Sakakini Palace.

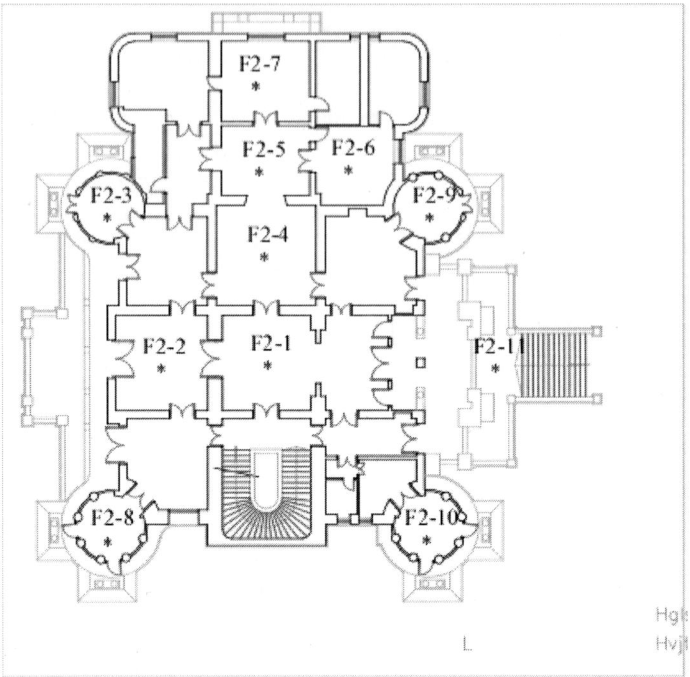

Figure 18. Location of stations at 2nd floor of El-Sakakini Palace.

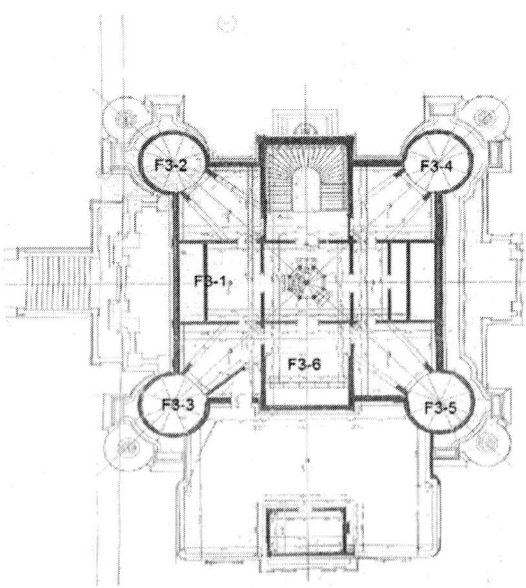

Figure 19. Location of stations at the 3rd floor.

Figure 20. High dynamic range ES-3000 Geometrics mobile station and triaxial geophone used 4 Hz to drive structure response of El-Sakakini Palace.

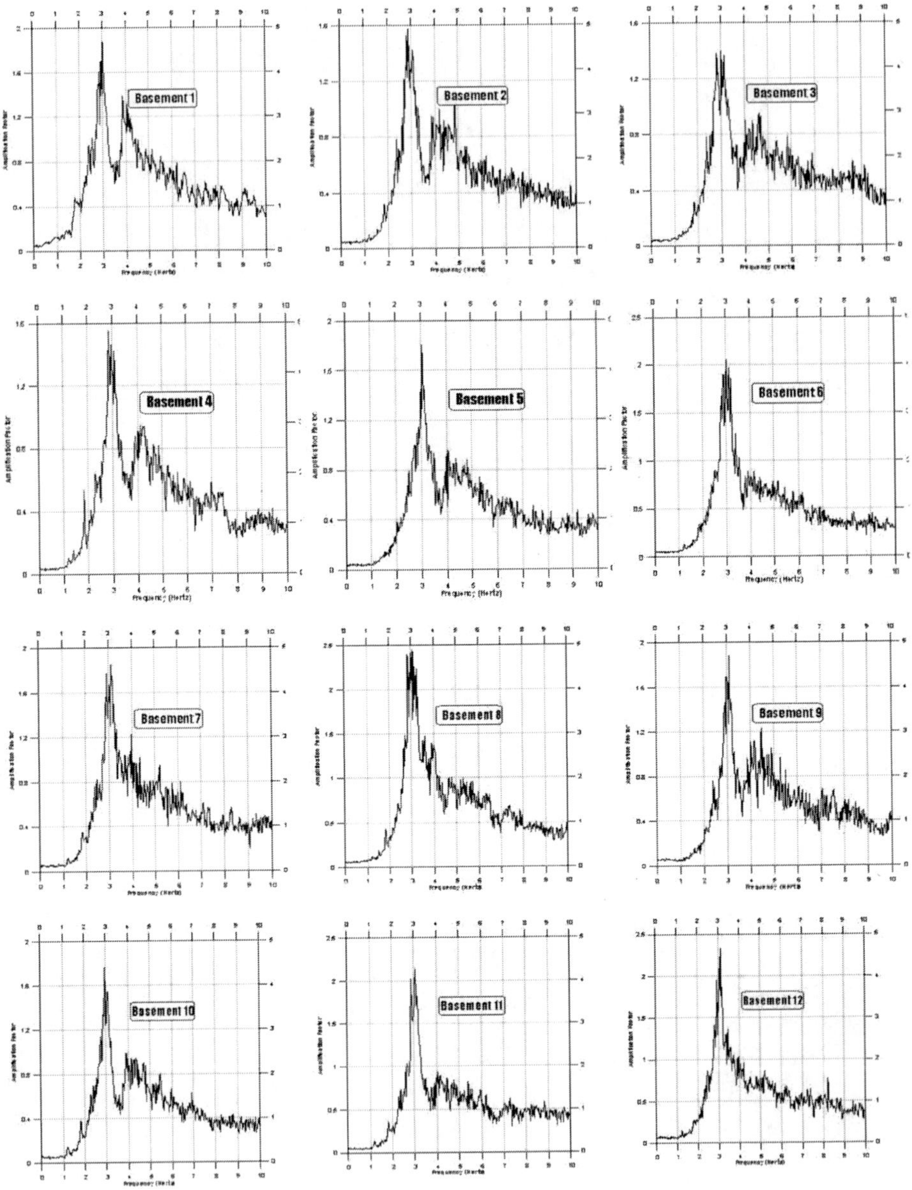

Figure 21. Natural frequency of vibration for basement floor.

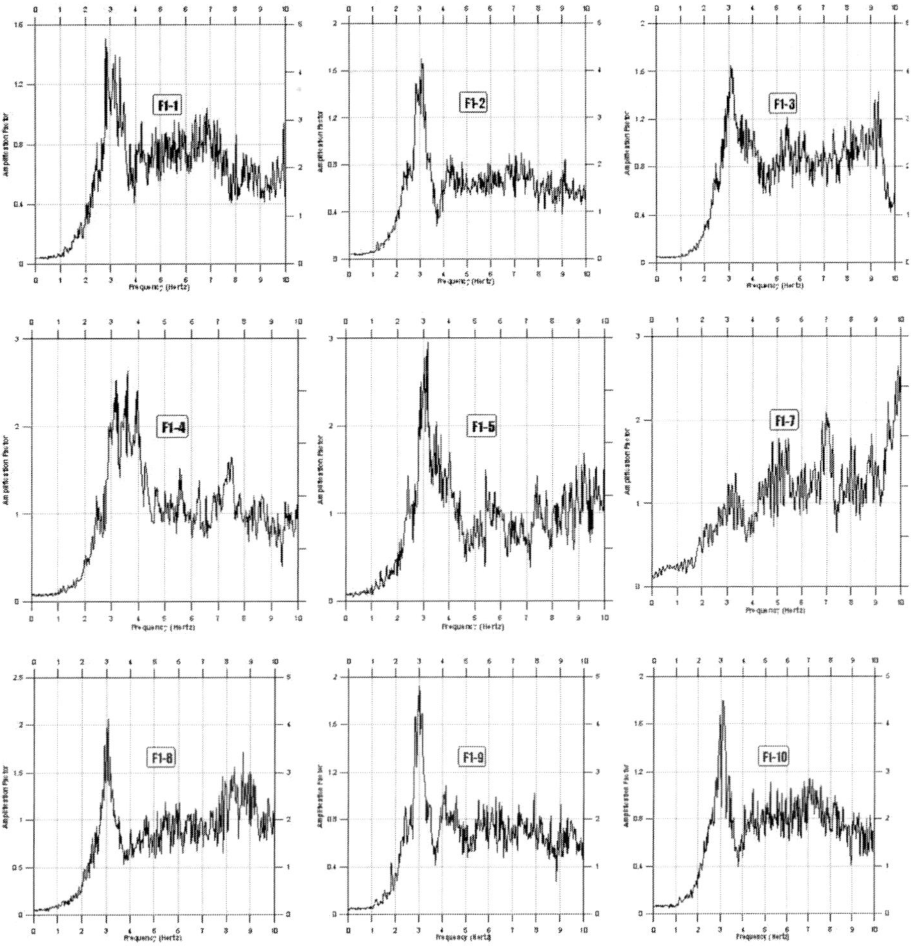

Figure 22. Natural frequency of vibration for the 1st floor.

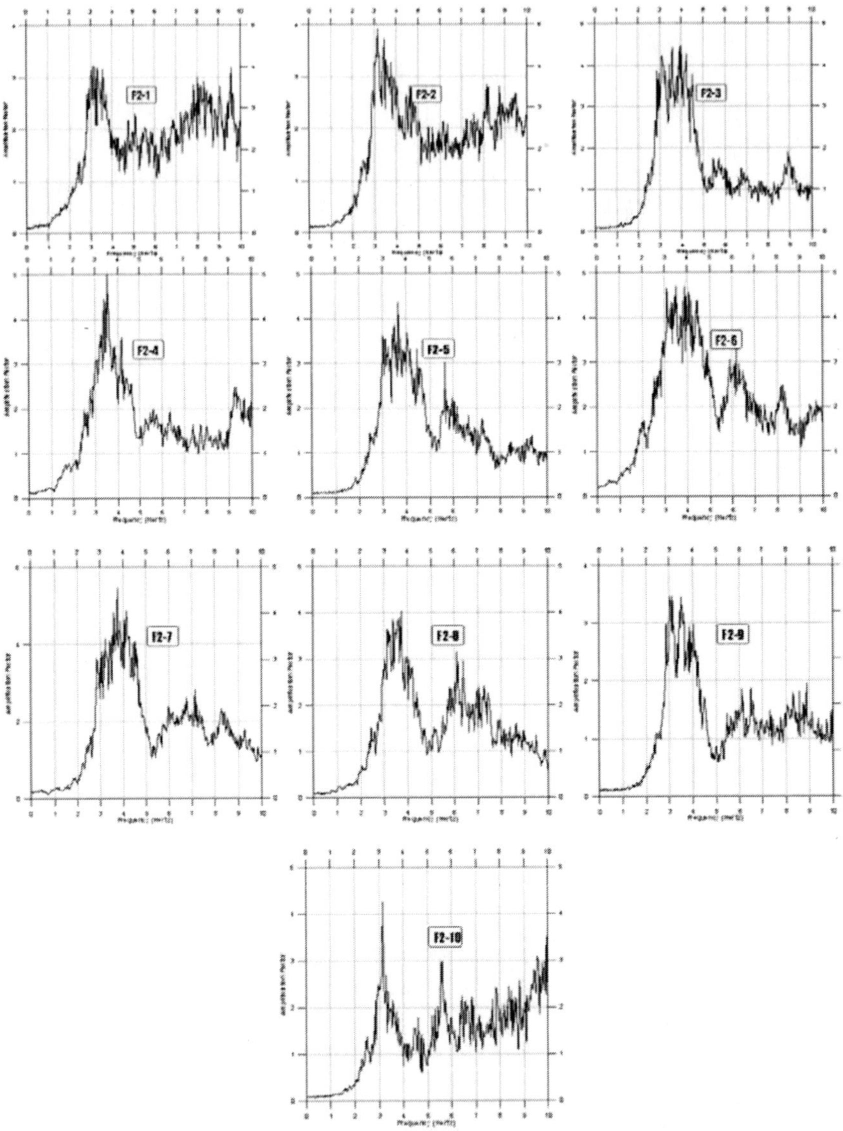

Figure 23. Natural frequency of vibration for the 2nd floor.

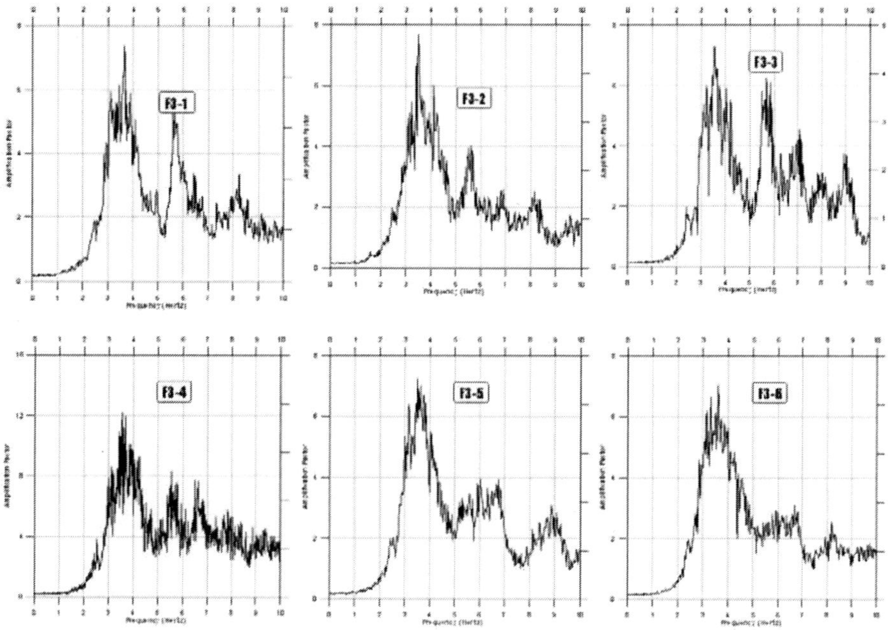

Figure 24. Natural frequency of vibration for the 3rd floor.

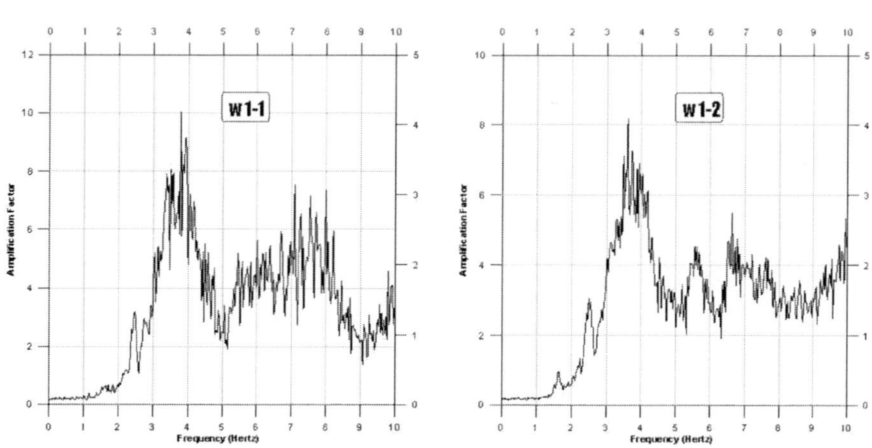

Figure 25. Natural frequency of vibration of the 4thfloor (wooden).

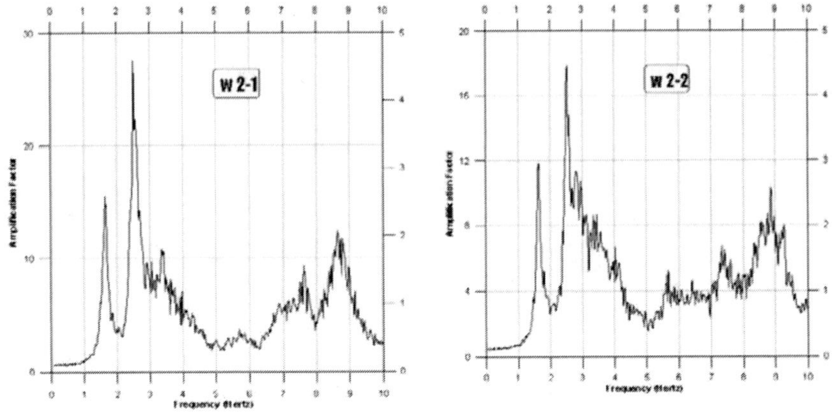

Figure 26. Natural frequency of vibration of the 5thfloor (wooden).

Table 5. Natural frequencies of vibration of El Sakakini Palace.

Floor	Fundamental resonance frequency (HZ)	Amplification Factor
Basement	3	1.3-2
1st Floor	3-3.5	1.4-2.8
2nd Floor	3-4	3-4.5
3rd Floor	3-4.3	5.5-8
4th wood Floor	3-4.5	7.5-8
5th wood Floor	2.5	12-20

6. Conclusions

ElSakakini Palace is an important monument in Egypt. We presented the main results of the seismic hazard analysis and the geophysical campaign to estimate the main characteristics of the ground response and the structure. Based on the available maximum intensity maps for historical earthquakes (>2200BC) the maximum Mercalli Intensity expected at ElSakakini Palace site is VII.

The peak horizontal acceleration at the seismic rock basement found at -35m approximately, and for 10% probability of exceedance in 50 years is 144 cm/sec^2 (0.147g), while for 100 years is 186 (cm/sec^2) (0.19 g). We determined the average soil profile using different geophysical campaigns. It is found that the upper layer has an average shear wave velocity lower than 300m/s and a thickness of 5 to 10meters. It is a man made fill material in rather loose conditions. Below there is a clayey material with average Vs velocity equal to 400-600m/s. At -35m in average we found saturated compacted sand and gravels with Vs velocity exceeding 700m/s. It is considered as the seismic bedrock for the foreseen detailed site-specific analysis of the ground response.

Based on the ambient noise campaign the fundamental frequency of the ground is of the order of 3.0 to 3.5sec very close to the fundamental frequency of the palace. Resonance phenomena should be expected and considered seriously in the detailed analysis of the structure. There are strong evidences that the upper two stories with wooden floors, which are presenting high amplification factors, are subjected to several damages and degradation of their bearing capacity.

References

1. Apostolidis P., Raptakis D., Roumelioti Z., Pitilakis K., (2004), Determination of S-Wave velocity structure using microtremors and SPAC method applied inThessaloniki (Greece). Soil Dynamics & Earthquake Engineering, 24, 1, 49-67.
2. Ben-Menahem, A., (1979). Earthquake catalogue for the Middle East (92 B. C. - 1980 A. D.): Bollettinodi Geofisica Teoricaed Applicata, 21, No. 84, 245 - 310
3. Castro, R. R., F. Pacor, A. Sala, and C. Petrungaro (1996). S Wave attenuation and site effects in the region of Friull, Italy, J. Geophysics. Res. 101, 22355-22369.

4. Celebi, M., C. Dietel, J. Prince, M. Onate, and G. Chavez (1987). Site amplification in Mexico City (determined from 19 September 1985 strong-motion records and from records of weak motion), in Ground Motion and Engineering Seismology, A. S. Cakmak (Editor), Alsevier, Amsterdam, 141-152.
5. Chopra, A. K. (1981). Dynamics of structures, a primer Earthquake engineering research institute.
6. Chopra, A. K., (1995). "Dynamic of structures – Theory and Application to Earthquake Engineering, Englewood Cliffs, New Jersey.
7. Clayton, R. W., and McMechan, G. A., 1981, Inversion of refraction data by wavefield continuation: Geophysics, v. 46, p. 860-868.
8. Egyptian Code. Egyptian code of Practice (ECP-1993) for estimating loads and forces, 1993.
9. F. Nateghi-A, (1994). Seismic strengthening of a ten story steel framed hospital: Earthquake resistant Construction & design, Savidis (ed.), Balkema, Rotterdam, ISBN 90 5410 392 2, P 849-856.
10. Field E. H., S. E. Hough, and. H. Jacob (1990). Using microtremors to assess potential earthquake site response, a case study in Flushing Meadows, New york city. Bulletin of the Society of America, Vol. 80. No. 6, pp. 1456-1480, December, 1990.
11. Gamal M. A. and Noufal A., A Neotectonics model AND seismic hazard assessment for Egypt, 8 th international conference on geology of the Arab world, Cairo university Egypt 13-16 2006.
12. Jordan Sesimoloical observatory (1998-2000), national resources authority geophysics and technical services, bulletin (1998-2000).
13. Joyner, W. B., and D. M. Boore, 1981, Peak horizontal acceleration and velocity from strong motion record including records from the 1979 Imperial Valley, California, earthquake, Bull. Seism. Soc. Am., 17, 2011-2038.
14. Kagami, H., C. M. Duke, G. C. Liang, and Y. Ohta (1982). Observation of 1-5 second microtremors and their application to earthquake engineering. Part II. Evaluation of site effect uppon seismic wave amplification due to extremely deep soils., Bull. Seism. Soc Am. 72, 987-998.
15. Kagami, H., S. Okada, K. Shiono, M. Oner, M. Dravinski, and A.K. Mal (1986). Observation of 1-5 seconds microtremors and their application to earthquake engineering. Part III. A two-dimensional of the site effect in San Fernando valley, Bull. Seism. Soc. Am. 76, 1801-1812.
16. Kanai, K. (1957). The requisite conditions for predominant vibration of ground, Bull. Earthq. Res. Inst., Tokyo University, 31, 457.

17. Kanai, K. (1962). On the spectrum of strong earthquake motions, Primeras J. ArgentinasIng.Antisismica 24, 1.
18. Kanai, K. and Tanaka (1961). On microtremors. VII, Bull. Earthq. Res. Inst., Tokyo Univ., 39, 97-115.
19. Langston, C. A. (1977). Corvallis, Oregon, Crustal and upper mantle receiver structure from teleseismic P and S waves, Bull. Seism. Soc. Am. 67, 713-724.
20. Lermo, J. and F. J. Chavez-Garcia, (1993). Site effect evaluation using spectral ratios with only one station, Bull. Seism. Soc. Am. 83, 1574-1594.
21. Lermo, J., M. Rodriguez, and S. K. Singh (1988). The Mexico Earthquake of September 19, 1985: Natural periods of sites in the valley of mexico from microtremorsmeasurments and from strong motion data, Earthquake Spectra 4, 805-814.
22. Louie, J, N., 2001, Faster, Better: Shear-wave velocity to 100 meters depth from refraction microtremors arrays: Bulletin of the Seismological Society of America, v. 91, p. 347-364.
23. Maamoun, M., 1979:Macroseismic observations of principal earthquake in Egypt, HelwanInstitue of Astronomy and Geophysics, Bulletin 183, 120.
24. Maamoun, M., A. Megehed, and A. Allam, (1984).Seismicity of Egypt, Institute of Astronomy and Geophysics 4.
25. Makropoulos, K. C., and P. W. Burton, (1981).A catalogue of Seismicity in Greece and adjacent areas.Geophys. J. R. Astr. Soc. 65, 741 - 762.
26. Manakou M., Raptakis D., Apostolidis P., Chávez-García F. J., Pitilakis K., (2010), 3d soil structure of the Mygdonian basin for site response analysis, Soil Dynamics and Earthquake Engineering, vol. 30, pp. 1198-1211
27. McGuire, R.K., (1993), FORTRAN computer program for seismic risk analysis. U.S. Geol. Surv.
28. Miller, R. D., Park, C. B., Ivanov, J. M., Xia, J., Laflen, D. R., and Gratton, C., 2000, MASW to investigate anomalous near-surface materials at the Indian Refinery in Lawrenceville, Illinois: Kansas Geol. Surv. Open-File Rept. 2000-4, Lawrence, Kansas, 48 pp.
29. Mohamed A. Gamal, A. Noufal and E. Ghoneim, ANeotectonics model and seismic hazard assessment for Egypt (2006): (Accepted for publication) 8th international conference of geology of the Arab world, 13-16 Feb. 2006 Cairo university Giza, Egypt.

30. Mucciarelli M., (1998). Reliability and applicability of Nakamura's technique using microtremors: an experimental approach, Journal of Earthquake Engineering, Vol.2, No.4 625-638 Imperial college press.
31. Nakamura, Y. (1989). A method for dynamic characteristic estimation of subsurface using microtremors on the ground surface. Q. R. Railway tech. Res. Inst. Rept. 30, 25-33.
32. NEIC, (1973-2006). Prelimenary determination of Epicenters, monthly listing, U.S. Department of the interior/ Geological survey, National Earthquake information center:.
33. Nogoshi, M. and T. Igarashi (1970).On the propagation characteristics of microtremors, J. Seism. Soc. Jpn. 23, 264-280.
34. Pitilakis K., Anastasiadis A., Kakderi K., Manakou M., Manou D., Alexoudi M., Fotopoulou S., Argyroudis S., Senetakis K., 2011, "Development of comprehensive earthquake loss scenarios for a Greek and a Turkish city: Seismic hazard, Geotechnical and Lifeline Aspects", Earthquakes and Structures,Vol. 2, No 3, September 2011.
35. Pullammanappallil, S., Honjas, B., Louie, J., Siemens, J. A., and Miura, H., 2003, Comparative study of the refraction microtremors method: Using seismic noise and standard P-wave refraction equipment for deriving 1D shear-wave profiles, Proceedings of the 6th SEGJ International Symposium (January 2003, Tokyo), 192-197.
36. Raptakis D., Manakou M., Chavez-Garcia F., Makra K., Pitilakis K., (2005), 3D configuration of Mygdonian basin and preliminary estimate of its site response,Soil Dynamics and Earthquake Engineering, 25, 871-887.
37. Riad, S., and H. Meyers, (1985).Earthquake catalog for the Middle East countries 1900 - 1983. World Data Center.
38. Rogers, A. M., R. D. Borcherdt, P. A. Covington, and D. M. Perkins (1984). A comparative ground response study near Los Angeles using recordings of Nevada nuclear tests and the 1971 San Fernando earthquake, Bull. Seism. Soc. Am. 74, 1925-1949.
39. Shapira, A., (1994).Seimological Bulletin of Israel. 1900 - 1994.
40. Thorson, J. R., and Claerbout, J. F., 1985, Velocity-stack and slant-stack stochastic inversion: Geophysics, v. 50, p. 2727-2741.
41. Uniform Building Code (UBC,1985). International Conference of building officials whittier, California.
42. Wilson, E.L, and Habiballah, A (1990)."SAP2000, Structural Analysis Programe" computer and structures Inc., Berkeley, California.

43. Woodword Clyde Consultant WWCC, (1985).Earthquake activity and dam stability for the Aswan High Dam, Egypt.High and Aswan Dames Ministry of Trregation, Cairo 2.

CHAPTER 5

Uncertainty in Seismic Capacity of Masonry Buildings

Fulvio Parisi * and Nicola Augenti

Department of Structural Engineering, University of Naples Federico II, Via Claudio 21, Naples 80125, Italy

Abstract

Seismic assessment of masonry structures is plagued by both inherent randomness and model uncertainty. The former is referred to as aleatory uncertainty, the latter as epistemic uncertainty because it depends on the knowledge level. Pioneering studies on reinforced concrete buildings have revealed a significant influence of modeling parameters on seismic vulnerability. However, confidence in mechanical properties of existing masonry buildings is much lower than in the case of reinforcing steel and concrete. This paper is aimed at assessing whether and how uncertainty propagates from material properties to seismic capacity of an entire masonry structure. A typical two-story unreinforced masonry building is analyzed. Based on previous statistical characterization of mechanical properties of existing masonry types, the following random variables have been considered in this study: unit weight, uniaxial compressive strength, shear strength at zero confining stress, Young's modulus, shear modulus, and available ductility in shear. Probability density functions were implemented to generate a significant number of realizations and static pushover analysis of the case-study building was performed for each vector of realizations, load combination and lateral load pattern. Analysis results show a large dispersion in displacement capacity and lower dispersion in spectral acceleration capacity. This can directly affect decision-making because both design and retrofit solutions depend on seismic capacity predictions. Therefore, engineering judgment should always be used when assessing structural safety of existing masonry constructions against design earthquakes, based on a series of seismic analyses under uncertain parameters.

Keywords: masonry buildings; material properties; uncertainties; seismic capacity

1. Introduction

Modeling uncertainties have a crucial role in performance-based earthquake engineering (PBEE) because they propagate throughout each step of seismic loss estimation according to the Pacific Earthquake Engineering Research (PEER) Center framework [1,2]. Some approaches to characterize uncertainties in ground motion hazard, building response, damage to building components, and losses (*i.e.*, casualties, financial losses, and business interruption) have been proposed by Baker and Cornell [3]. They assessed uncertainty propagation with the PEER framework up to the estimation of the expected value and variance of mean annual loss, mean rate of collapse, and mean rate of exceeding a given cost. Past research on reinforced concrete (RC) buildings has indicated that modeling uncertainties associated with material strengths have a relatively small effect on seismic pre-collapse performance predictions [4,5]. On the contrary, Ibarra and Krawinkler [6] have shown that uncertainties associated with deformation capacity and post-peak softening branch of force–displacement diagrams of individual structural components can significantly affect the predicted collapse performance. Liel *et al.* [7] have assessed seismic collapse risk of both standard (*i.e.*, ductile) and substandard (*i.e.*, non-ductile) RC framed buildings, incorporating structural modeling parameter uncertainties. Based on the use of Monte Carlo methods and response surface analysis [8,9], a response surface has been derived to describe each median collapse capacity parameter as a multivariate function of modeling random variables. The analysis of fragility curves has shown that modeling uncertainties related to both the structure and earthquake ground motion have greater impact on seismic capacity estimation when modeling parameters are more uncertain and their relationship with structural response is highly nonlinear. In this case, modeling uncertainties cause a dispersion increase and a shift in the median of the structural response fragility, which may decrease by as much as 20%.

In the case of masonry structures high dispersion in material properties and highly nonlinear response under earthquake actions are to be considered. The intrinsic randomness of material mechanics and the limited knowledge provided by experimental tests on material properties can significantly affect the confidence in structural analysis results. The assignment of physical properties is a difficult task even in the presence of regular assemblages, such as running and American bond masonries. In fact, several factors of uncertainty arise from workmanship techniques, including partial or complete filling of joints with mortar, constant or variable size ratio between joints and bricks, *etc*. Another problem is related to the acquisition of experimental data from past investigations: the same masonry property can be characterized through different laboratory or *in situ* experimental tests, test setups, and specimens. In many cases nonstandard procedures have been employed in the past for masonry characterization. These factors clearly induce a significant dispersion in results, so engineering judgment is needed to select reliable values from an experimental data set, or ranges suggested by building codes and handbooks. Comprehensive studies on ranges of masonry properties and empirical/analytical formulations available for their estimation have been carried out by Augenti [10] and Tassios [11]. Experimental data on sliding shear properties of different masonry types are summarized and discussed in [12]. The Italian Building Code Commentary (IBCC) [13] provides typical ranges of a number of masonry types to be used for structural assessment of existing masonry constructions. In view of an IBCC revision, Augenti and Parisi [14] have performed a statistical analysis of experimental data on tuff masonry selected from the former release of *MADA*, which is an online database of masonry properties collected in more than 400 files [15]. More recently, Oropeza [16] has collected relevant data on the variability of masonry properties. However, anyone could find high variance in both experimental data and empirical/analytical estimates of properties.

Figure 1 shows the typical stress-strain behavior of masonry under uniaxial compression, which is highly nonlinear even at small strain levels. Mechanical nonlinearity can also be detected under shear deformations. Microcracking in the elastic range and material instability caused by fracture in the inelastic range call for a probabilistic characterization of masonry mechanics. Uncertainties affect both limit strains and the corresponding strengths of masonry. Cracking strain ε_p associated with peak strength σ_p is less uncertain than ultimate strain ε_u and residual strength σ_u. Strain softening of masonry induces lower confidence in the simulation of plastic behavior given that the fracture process is noticeably sensitive to local material discontinuities. For instance, pumice inclusions and voids can be detected in tuff stones, which are widely spread in many earthquake-prone regions. Compressive tests on tuff masonry in the direction orthogonal to bed joints [17] have shown a coefficient of variation (*CoV*) equal to 12.5% and 9.1% for σ_p and ε_p, respectively; such values increase even to 23.4% and 73.1% for strength and limit strain at collapse. Young's and shear moduli at $\sigma_p/3$ have been found to have *CoV* equal to 16.7% and 17.7%, respectively, whereas *CoV* = 26.8% has been computed for the relevant Poisson's ratio. Different *CoV*-values have been detected under compression parallel to bed joints. The representation of confidence interval under varying axial strain has confirmed that uncertainty increases with inelasticity. It is emphasized that such an outcome depends on the randomness of fracture process and the ability of the test setup in capturing cracks. Figure 1 also shows probability density functions (PDFs) of peak and ultimate strengths, as well as conditional PDFs of cracking and ultimate strains. This is a typical representation for strains' uncertainties because they are usually estimated at given strength levels. Ultimate strain is associated with a predefined strength drop.

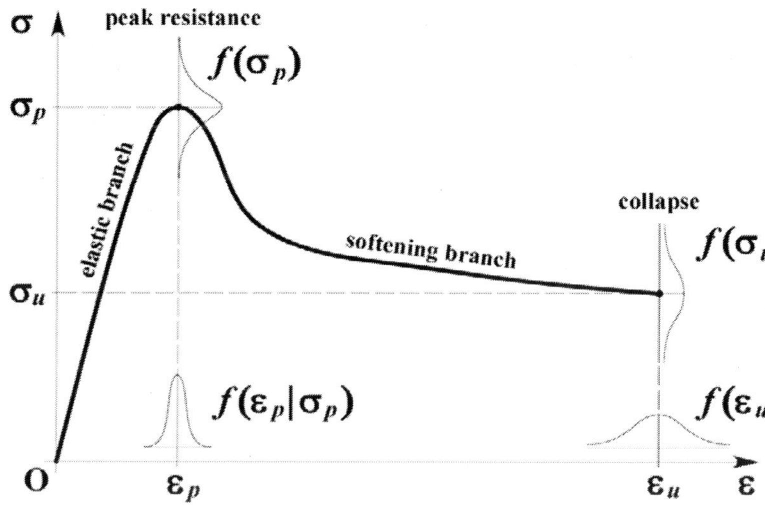

Figure 1. Typical stress–strain diagram of masonry in compression and uncertainties.

The aforementioned studies motivated this paper, which investigates the uncertainty propagation within nonlinear seismic analysis of unreinforced masonry (URM) buildings. A case-study building was analyzed to estimate mean and variance of seismic capacity parameters under uncertain material properties, regardless of ground motion uncertainty whose influence on seismic vulnerability of URM buildings has already been assessed by Pasticier *et al.* [18]. This investigation attempts to provide an answer to the following question of practical interest: are numerical results sufficiently reliable to assess structural safety of URM buildings, or does engineering judgment still play a key role in decision-making?

2. Accounting for Uncertainty in Seismic Capacity Estimation of a Case-Study Masonry Building

The effects of masonry uncertainties were assessed in the case of an ideal two-story URM building, which was composed by seven load-bearing walls with openings (Figure 2). Each wall was made of tuff masonry. The case-study building was selected from typical residential buildings of southern Europe [19,20]. The fundamental period of the masonry buildings was estimated as $T_1 = 0.24$ s through a simplified code formula [21]. A simulated design procedure was employed in accordance to Eurocode 8 (EC8)—Part 3 [22] on the basis of practice rules, without considering specific seismic design provisions.

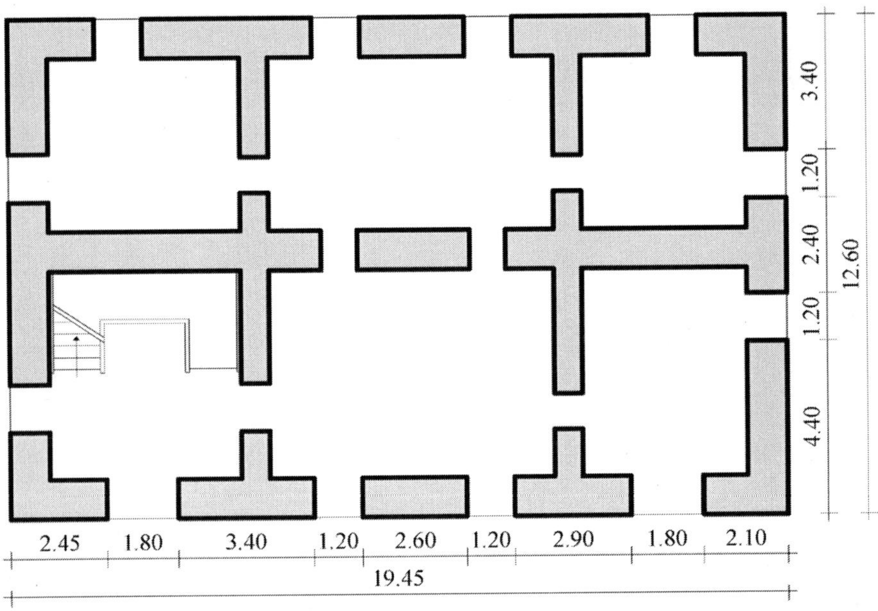

Figure 2. Typical floor plan of case-study building (dimensions in meters).

Actually, no seismic design codes for masonry buildings have been released in many European countries prior to EC8—Part 1 [21]. Assuming parapets to be nonstructural components, wall openings were considered to have the same size and to be both vertically and horizontally aligned in elevation. The staircase system was assumed to be made of RC slabs with 15 mm thickness, which were supported by three walls belonging to a corner masonry cell of the building. Perimeter and internal load-bearing walls had a thickness of 1.00 and 0.70 m at the ground floor, and 0.90 and 0.60 m at the first floor, respectively. RC bond beams connected all walls at each floor level and one-way RC joist slabs with thickness of 250 mm were introduced to distribute earthquake actions among walls. Finally, wooden lintels were placed above openings and were supposed to be well anchored within surrounding masonry. Orthogonal walls were assumed to be well connected through masonry interlocking and sufficient flange lengths. The building was 19.45 × 12.60 m^2 in plan and 8.16 m high, given that interstory height was set to 4.08 m.

2.1. Material Properties and Related Uncertainty

Based on statistics derived for tuff masonry [14], a uniform PDF was assumed for unit weight (γ) and a Gaussian PDF was assumed for the following parameters: uniaxial compressive strength (f_m); shear strength at zero confining stress (τ_0); Young's modulus (E_m); shear modulus (G_m); and available ductility in shear (μ_s). Lower and upper bounds, mean and standard deviation of such modeling variables are listed in Table 1, where one can note that *CoV* was equal to: 15% in the case of f_m, E_m, and G_m; 12.19% in the case of τ_0; and 23.90% in the case of μ_s. Young's modulus was about 1000 times compressive strength, whereas shear modulus was 0.4 times Young's modulus. According to EC6 [23], friction coefficient of masonry was assumed to be 0.4 at any confining stress level. The spatial distribution of material properties was assumed to be uniform throughout the building and wall sizes were assumed to be deterministically known. Realizations of the random modeling vector $\Theta = [\gamma, f_m, \tau_0, E_m, G_m, \mu_s]$ were generated through the full Monte Carlo method.

Table 1. Random properties and their statistics.

Parameter	γ [kN/m^3]	f_m [MPa]	τ_0 [MPa]	E_m [MPa]	G_m [MPa]	μ_s
Lower bound	14.00	1.40	0.038	1746	698	2.64
Upper bound	18.00	2.40	0.057	2444	978	3.71
Mean	–	1.90	0.047	2090	836	3.18
Standard deviation	–	0.28	0.006	314	125	0.76

2.2. Capacity Modeling

Simplified macroelement models are widely used to assess seismic performance of standard masonry buildings, like that investigated in this study. Such models are based on an equivalent frame idealization of masonry walls with openings and their three-dimensional (3D) assemblage through floor diaphragms and, if any, RC bond beams or other tying elements (e.g., steel or wooden ties). Continuous masonry strips along the height and length of the wall are referred to as *piers* and*spandrels*. Their intersection allows the definition of: (1) rigid macroelements named *joint panels*; and (2) flexible macroelements named *pier panels* or *spandrel panels*, depending on whether they are vertical or horizontal structural components, respectively (Figure 3).

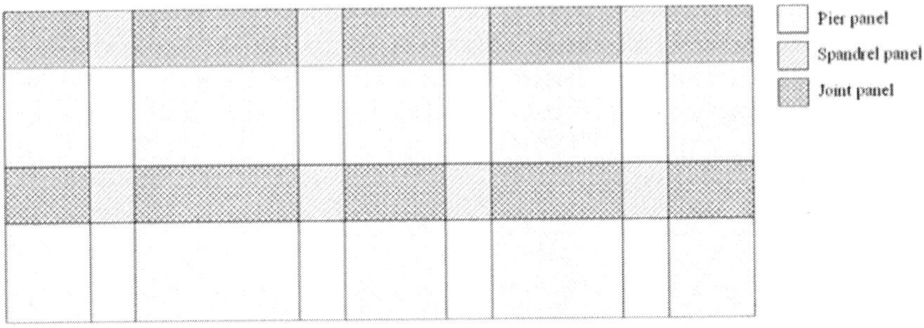

Figure 3. Macroelement model of masonry wall with openings.

Figure 4 shows the 3D macroelement model of the case-study building where shadows indicate floor slab spanning. The 3D model was developed in RAN Code, a computer program which performs nonlinear macroelement analysis of masonry buildings according to RAN method [10]. Pier and spandrel panels had axial, flexural and shear flexibilities and their effective resistant volume Ω changed with the given lateral drift as a result of tensile cracking and masonry crushing at compressed corners (see [24] for details).

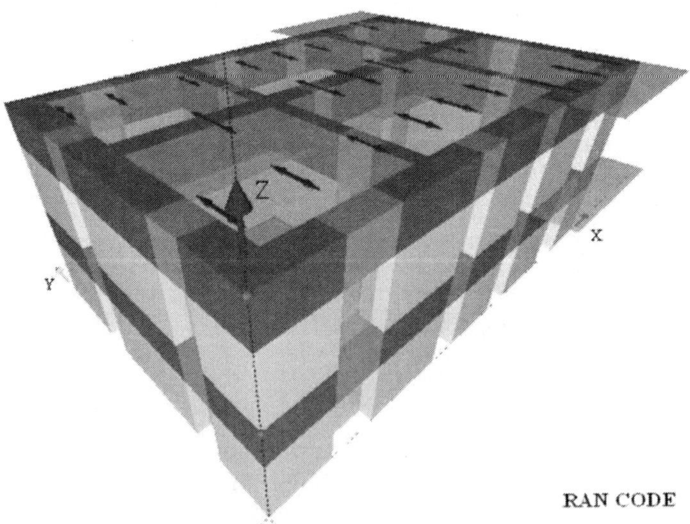

Figure 4. 3D macroelement model of case-study masonry building.

Nonlinear behavior of each URM macroelement under lateral loading was simulated by a force-displacement diagram where the lateral resisting force is derived as the maximum axial strain on the end sections is increased. Assuming a plane axial strain distribution within the macroelement, the flexural component of lateral displacement is derived through double integration of axial strains over Ω and the corresponding lateral resisting force is computed according to closed-form balance equations of end sections under eccentric compression. A constitutive model with strain softening [25] was assumed, considering deterministic limit strains at cracking and collapse equal to 0.20% and 0.35%, respectively (note that the statistical characterization of such mechanical properties have not yet been carried out in the case of tuff masonry). Consequently, the available ductility in compression was 1.75. Furthermore, masonry was considered as equivalent homogeneous material with zero tensile strength. The shear component of lateral displacement was then computed at any lateral resisting force, accounting for Ω evolution [10]. In this computation the available ductility μ_s was used to amplify the total displacement of the macroelement in the case of shear failure.

3. Static Pushover Analysis

Four hundred response-controlled static pushover (SPO) analyses of the case-study building were carried out. Out-of-plane mechanisms were not included in the SPO analysis because they were inhibited by the presence of rigid floor diaphragms and RC bond beams connecting walls at each floor. SPO analysis included a step-by-step updating of (1) lateral strength of macroelements on the basis of current axial force demands and limit strength domains; and (2) stiffness center at each floor level to account for inelastic torsion. RAN Code uses the procedure presented in [24] and extended to 3D macroelement models of masonry buildings. Accidental eccentricities of mass centers were set to 5% of the building plan dimensions in both directions according to EC8—Part 1 [21], so four load combinations for each direction of the building plan were accounted for the SPO analysis. Two lateral load patterns were considered: uniform load pattern (ULP) and inverse triangular load pattern (ITLP), which are a mass- and a first mode-proportional distribution of lateral actions. Figure 5 shows a typical outcome for regular building structures according to EC8 criteria, where SPO curves related to ULP provide an upper bound for earthquake resistance. Each curve is represented as first-mode spectral acceleration S_a *versus*roof lateral displacement d_c up to the collapse of the building, which was assumed to occur when all piers minus one at the same story reached their displacement capacity. It is emphasized that displacement capacity of the whole building is different from that of a macroelement, opposed to the case of other structural systems (e.g., RC frames) where the attainment of the limit chord rotation in a member is also assumed to be a global limit state. In the case of masonry constructions, the building displacement capacity at life safety (d_u) is typically defined as the lateral displacement of a control point on the roof corresponding to 20% resistance drop on the post-peak descending branch of the SPO curve. Therefore, one can estimate spectral acceleration capacity $S_a|d_u$ as the intensity measure causing the life safety limit state.

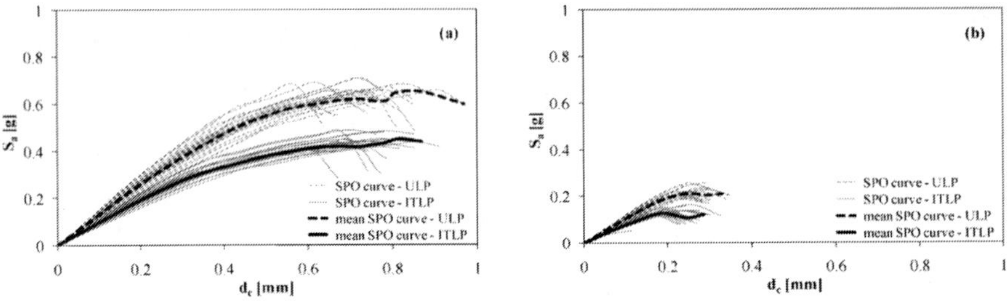

Figure 5. Static pushover (SPO) curves and their means related to the analysis cases: (**a**) 1X; (**b**) 1Y.

The curves in Figure 5 is related to specific analysis cases, but it could be shown that the same occurred in other cases. Spectral acceleration capacity of the case-study building in the X-direction is more than three times that related to the Y-direction. This can be motivated on one hand by the fact that gravity loads on floors were mainly distributed to walls parallel to the X-axis, and on the other hand by the presence of more piers and spandrel panels in those walls. Figure 6shows all SPO curves of the case-study building.

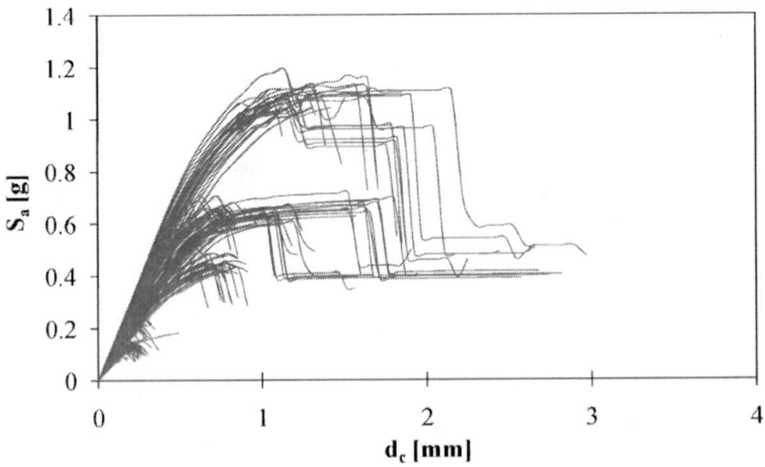

Figure 6. SPO curves related to all analysis cases.

The case-study building has an actual elastic first-mode period ranging between 0.06 and 0.08 s (corresponding to a frequency between 12.50 and 16.67 Hz). Such values can be estimated for each SPO curve as $T_e = 2\pi\sqrt{(d_e/S_{ae}\Gamma_1)}$, where d_e is the limit elastic displacement, S_{ae} is the spectral acceleration associated with d_e, and Γ_1 is the first-mode participation factor of the structure. Therefore, T_e is about one-fourth of the fundamental period T_1 estimated in accordance with EC8 [21] (see Section 2). This produces a low deformation capacity, but also a limited displacement demand on the structure. Note that the mean limit elastic displacement of the building is 0.6 and 0.2 mm in the X- and Y-direction, respectively.

4. Estimation of Seismic Capacity Uncertainty

Based on SPO analysis results, mean and standard deviation of spectral acceleration conditioned on d_c, denoted by $\mu_{Sa|dc}$ and $\sigma_{Sa|dc}$ respectively, were investigated for each loading direction (Figure 7 and Figure 8). It is confirmed that the earthquake resistance of the case-study building is not the same for both loading directions and lateral load patterns. In addition, different SPO curves were obtained for positive and negative orientations of seismic actions, especially in the case of the X-direction. This was caused by the lack of symmetry in plan which induced higher seismic response sensitivity to axial forces in piers generated by overturning moments at each floor level. Variations in axial forces produce a change in lateral capacity of pier panels, as analytically shown by Augenti [10]. It is worth noting that dispersion of conditional spectral acceleration increases until collapse of some macroelements is attained, Figure 8.

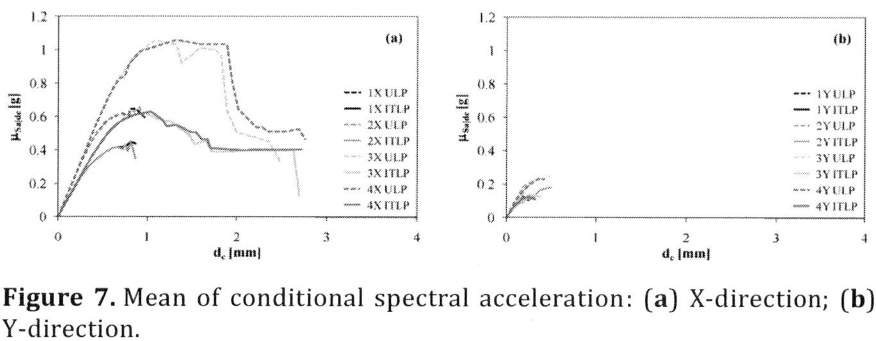

Figure 7. Mean of conditional spectral acceleration: **(a)** X-direction; **(b)** Y-direction.

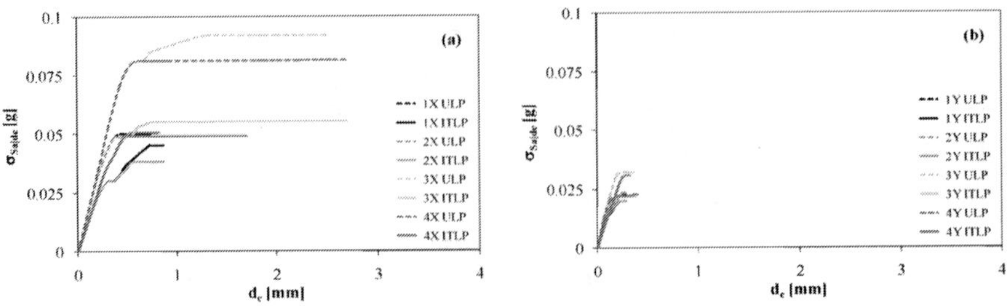

Figure 8. Dispersion of conditional spectral acceleration: (**a**) X-direction; (**b**) Y-direction.

At larger displacement demands, $\sigma_{Sa|dc}$ reaches a constant value falling in the ranges [0.04 g, 0.09 g] and [0.02 g, 0.03 g] for seismic actions acting in the X- and Y-direction, respectively. This also indicates a reliable estimation of seismic capacity uncertainty, which is quantified in Table 2. In particular, one can note that CoV is 9%–26% in the X-direction and 12%–55% in the Y-direction, in the case of d_u. Uncertainty reduces in the case of spectral acceleration capacity $S_a|d_u$ for which CoV is 8%–11% in the X-direction and 11%–24% in the Y-direction. For each code-based load combination, uncertainty in the Y-direction is higher than that in the other direction of the building plan.

Table 2. Statistics of displacement capacity and conditional spectral acceleration.

Analysis Case	Load Pattern	μ_d [mm]	CoV_d	$\mu_{Sa\|du}$ [g]	$CoV_{Sa\|du}$
1X	ULP	0.77	12%	0.61	8%
	ITLP	0.76	11%	0.42	11%
2X	ULP	0.77	11%	0.61	8%
	ITLP	0.75	9%	0.41	9%
3X	ULP	1.51	22%	1.02	9%
	ITLP	1.28	19%	0.57	10%
4X	ULP	1.30	26%	1.06	8%
	ITLP	1.32	21%	0.55	9%
1Y	ULP	0.29	12%	0.20	11%
	ITLP	0.21	25%	0.11	16%
2Y	ULP	0.28	13%	0.20	11%
	ITLP	0.24	29%	0.11	18%
3Y	ULP	0.20	55%	0.16	19%
	ITLP	0.18	51%	0.09	24%
4Y	ULP	0.26	51%	0.20	11%
	ITLP	0.18	37%	0.11	20%

Uncertainty levels provided by SPO analysis were estimated by using PDFs of masonry properties, each of them expressed as $f_\theta(\theta,\Theta_f)$ where the vector Θ_f included the mean μ_θ and standard deviation σ_θ estimated over experimental data available in the literature. Actually, such properties depend on mechanical parameters and relative size of masonry constituents (*i.e.*, masonry units and mortar joints), as well as masonry bond and boundary conditions [10,11]. This means that, in line of principle, the overall uncertainty could be reduced assuming that it is partly aleatory and partly epistemic [26]. Aleatory uncertainty cannot be reduced by increasing the size of the data set of modeling parameters or modeling rules, because it is assumed to be dependent on the randomness of the phenomenon to be modeled. Conversely, epistemic uncertainty is treated as a variable which decreases as the knowledge level increases [27]. In the case of existing building, material and geometric properties are epistemic variables, whereas loads and their associated demands on the structure, be it existing or future, have always future realizations so their uncertainty is typically classified as aleatory. An important source of epistemic uncertainty is also due to the use of a physical structural model which is a simplification of the real construction. On the material side, experimental data sets on parameters of both masonry assemblages and their constituents are still lacking to be reliably used, especially in the case of tuff masonry. This causes the following missing submodels: (1) physical submodels $\theta = g(\beta,\Theta_g) + \varepsilon$ describing each tuff masonry property θ as a function of the vector β of geometric and mechanical parameters of masonry constituents, which is characterized by the vector of parameters Θ_g fitted to experimental data and statistical uncertainty (residual) ε; and (2) probabilistic submodels of masonry constituent properties $f_\beta(\beta,\Theta_f)$. When such an information will be available, the use of Monte Carlo-type or other methods to obtain realizations of masonry constituent properties will enable to reduce uncertainty in tuff masonry properties, and hence in seismic capacity of tuff masonry structures. Nevertheless, at this state of knowledge, the identification of epistemic variables ensures transparency in the quantitative analysis presented above, because one can know which reducible uncertainties have been left unreduced in the modeling phase. The present paper thus provides a first estimation of seismic capacity of a typical existing URM building which could be improved in the future on the basis of more observed data on mechanical properties.

5. Conclusions

This paper has dealt with the issue of seismic capacity uncertainty. After uncertain properties of masonry have been identified and discussed, their influence on seismic capacity estimation has been assessed for an ideal case-study URM building which falls in the class of residential buildings located in southern Europe. The case-study building was designed assuming a low-seismicity area, according to EC8 and past rules of practice. Based on a statistical characterization of six fundamental masonry properties, realizations of a random modeling vector were implemented in a macroelement capacity model of the structure and SPO analysis was performed for each code-based load combination and lateral load pattern.

Analysis results demonstrate that seismic capacity estimation of masonry buildings can be affected by a large dispersion, as shown for other building structures in the literature. Dispersion in SPO curves increases up to the failure of one or more macroelements, after which standard deviation of spectral acceleration remains almost constant up to collapse. Most part of the uncertainty lies in (1) the definition of the building displacement capacity rather than spectral acceleration capacity; and (2) seismic capacity estimation related to lateral actions applied in the plan direction corresponding to the minimum earthquake resistance. Such observations confirm that, at the present state of knowledge, engineering judgment still plays a key role when assessing structural safety of existing masonry buildings against design earthquakes. This has a direct consequence in decision-making because both design and retrofit solutions depend on seismic capacity predictions. It is recommended to use a set of values for each of the most uncertain parameters and to perform seismic analysis for each vector of parameter realizations, in order to get a reliable idea of the actual safety of a masonry building. Future research could explore the influence of an uncertain location of material properties in the building and uncertain size of load-bearing walls. Nevertheless, uncertainty in seismic capacity could be reduced by increasing the amount of experimental data on masonry units and mortar, provided that correlation models to estimate their relevant masonry properties will be developed and/or validated.

Finally, a key issue is how the information on seismic capacity uncertainty provided in this study could be used to estimate the actual seismic intensity measure (IM) at collapse (e.g., the peak ground acceleration PGA_C or elastic spectral acceleration $S_{ae,C}$), as well as the corresponding probability of failure P_f in a given time interval, which are crucial for decision-making in PBEE. This computation should include the uncertainty in seismic demand, which has not been estimated in this study. To this end, in the context of SPO-based approaches, two alternative procedures are available to determine the relationship between IM and a given engineering demand parameter (EDP), e.g., the roof displacement d_c, without using incremental dynamic analysis (IDA) [28] for a large suite of earthquake ground motion records: SPO2IDA tool [29,30] and incremental N2 (IN2) method [31]. Both methods could provide approximate IDA curves, but different procedures should be employed to estimate central values (*i.e.*, mean or median) and dispersion in seismic performance, accounting for aleatory uncertainty due to record-to-record variability. Then, the inclusion of seismic hazard uncertainties could allow to fully estimate P_f under varying IM, deriving the fragility curve at collapse. Nevertheless, further research is needed to assess the applicability of SPO2IDA tool and IN2 method to URM buildings. Furthermore, extensive sensitivity studies will have to be performed in order to define reliable default values for dispersion measures associated with displacement demand.

Acknowledgments

This research was carried out in the framework of the ReLUIS-DPC 2010-2013 project (Line AT1-1.1—"Evaluation of the Vulnerability of Masonry Buildings, Historical Centers and Cultural Heritage") funded by the Italian Department of Civil Protection.

References

1. Porter, K.A. An overview of PEER's performance-based earthquake engineering methodology. In Proceedings of the 9th International Conference on Applications of Statistics and Probability in Civil Engineering, San Francisco, CA, USA, 6–9 July 2003.
2. Deierlein, G.G. Overview of a comprehensive framework for earthquake performance assessment. In Proceedings of the International Workshop on Performance-BasedSeismic Design Concepts and Implementation, Bled, Slovenia, 28 June–1 July 2004.
3. Baker, J.W.; Cornell, C.A. Uncertainty propagation in probabilistic seismic loss estimation. *Struct. Saf.* **2008**, *30*, 236–252.
4. Porter, K.A.; Beck, J.L.; Shaikhutdinov, R.V. Sensitivity of building loss estimates to major uncertain variables. *Earthq. Spectr.* **2002**, *18*, 719–743.
5. Lee, T.H.; Mosalam, K.M. Seismic demand sensitivity of reinforced concrete shear-wall building using FOSM method.*Earthq. Eng. Struct. Dyn.* **2005**, *34*, 1719–1736.
6. Ibarra, L.; Krawinkler, H. *Global Collapse of Frame Structures under Seismic Excitations; Report No. 152*; John A. Blume Earthquake Engineering Center: Stanford, CA, USA, 2005.
7. Liel, A.B.; Haselton, C.B.; Deierlein, G.G.; Baker, J.W. Incorporating modeling uncertainties in the assessment of seismic collapse risk of buildings. *Struct. Saf.* **2009**, *31*, 197–211.
8. Rubinstein, R.Y. *Simulation and the Monte Carlo Method*; John Wiley & Sons: New York, NY, USA, 1981.
9. Pinto, P.E.; Giannini, R.; Franchin, P. *Seismic Reliability of Structures*, 2nd ed; IUSS Press: Pavia, Italy, 2007.
10. Augenti, N. *Il calcolo sismico degli edifici in muratura (in Italian)*, 2nd ed; Unione Tipografico-Editrice Torinese: Turin, Italy, 2004.

11. Tassios, T.P. Seismic engineering of monuments. In *Earthquake Engineering in Europe*; Garevski, M., Ansal, A., Eds.; Springer: Berlin, Germany, 2010; pp. 1–42.
12. Augenti, N.; Parisi, F. Constitutive modelling of tuff masonry in direct shear. *Construct. Build. Mater.* **2011**, *25*, 1612–1620.
13. Italian Ministry of Infrastructures and Transportation, *Circ. 02.02.2009, No. 617: Istruzioni Per l'applicazione delle «Nuove Norme Tecniche per le Costruzioni» di Cui Al Decreto Ministeriale 14 Gennaio 2008 (in Italian)*; Italian Ministry of Infrastructures and Transportation: Rome, Italy, 2009.
14. Augenti, N.; Parisi, F. Experimental data analysis on mechanical parameters of tuff masonry. In Proceedings ofWorkshop on Design for Rehabilitation of Masonry Structures (Wondermasonry 2009), Lacco Ameno, Italy, 8–10 October 2009.
15. Augenti, N.; Parisi, F.; Acconcia, E. MADA: Online experimental database for mechanical modelling of existing masonry assemblages. In Proceedings of the 15th World Conference on Earthquake Engineering, Lisbon, Portugal, 24–28 September 2012.
16. Oropeza, M. Fragility Functions for Seismic Risk in Regions with Moderate Seismicity. Ph.D. Dissertation, École Polytechnique Fédérale de Lausanne, Lausanne, Switzerland, 2011..
17. Augenti, N.; Parisi, F. Constitutive models for tuff masonry under uniaxial compression. *J. Mater. Civ. Eng.* **2010**, *22*, 1102–1111.
18. Pasticier, L.; Amadio, C.; Fragiacomo, M. Non-linear seismic analysis and vulnerability evaluation of a masonry building by means of the SAP2000 V.10 code. *Earthq. Eng. Struct. Dyn.* **2008**, *37*, 467–485.
19. Kappos, A.J. Seismic vulnerability assessment of existing buildings in Southern Europe. In Proceedings of the 10th Italian conference on earthquake engineering, Potenza-Matera, Italy, 9–13 September 2001.
20. Bramerini, F.; Di Pasquale, G. Aggiornamento delle mappe di rischio sismico (in Italian). *Ing. Sismica* **2008**, *2*, 5–23.
21. Comité Européen de Normalisation, *Eurocode8: Design of Structures for Earthquake Resistance—Part 1: General Rules, Seismic Actions and Rules for Buildings*; Comité Européen de Normalisation: Brussels, Belgium, 2004.
22. Comité Européen de Normalisation, *Eurocode8: Design of Structures for Earthquake Resistance—Part 3: Assessment and Retrofitting of Buildings*; Comité Européen de Normalisation: Brussels, Belgium, 2005.

23. Comité Européen de Normalisation, *Eurocode 6: Design of Masonry Structures—Part 1-1: General Rules for Reinforced and Unreinforced Masonry Structures*; Comité Européen de Normalisation: Brussels, Belgium, 2005.
24. Parisi, F.; Augenti, N. Seismic capacity of irregular unreinforced masonry walls with openings. *Earthq. Eng. Struct. Dyn.* **2012**. in press..
25. Turnšek, V.; Čačovič, F. Some experimental results on the strength of brick masonry walls. In Proceedings of the 2nd International Brick and Block Masonry Conference, Stoke-on-Trent, UK, 12–15 April 1970.
26. Der Kiureghian, A.; Ditlevsen, O. Aleatory or epistemic? Does it matter? *Struct. Saf.* **2009**, *31*, 105–112.
27. Faber, M.H. On the treatment of uncertainties and probabilities in engineering decision analysis. *J. Offshore Mech. Arct. Eng.* **2005**, *127*, 243–248.
28. Vamvatsikos, D.; Cornell, C.A. Incremental dynamic analysis. *Earthq. Eng. Struct. Dyn.* **2002**, *31*, 491–514.
29. Vamvatsikos, D.; Cornell, C.A. Direct estimation of the seismic demand and capacity of oscillators with multi-linear static pushovers through IDA. *Earthq. Eng. Struct. Dyn.* **2006**, *35*, 1097–1117.
30. Fragiadakis, M.; Vamvatsikos, D. Fast performance uncertainty estimation via pushover and approximate IDA.*Earthq. Eng. Struct. Dyn.* **2010**, *39*, 683–703.
31. Fajfar, P.; Dolšek, M. A practice-oriented estimation of the failure probability of building structures. *Earthq. Eng. Struct. Dyn.* **2012**, *41*, 531–547.

CHAPTER 6

Assessment of Seismic Vulnerability of a Historical Masonry Building

Francesca Ceroni *, Marisa Pecce, Stefania Sica and Angelo Garofano

Department of Engineering, University of Sannio, Piazza Roma, 21-82100 Benevento, Italy

Abstract

A multidisciplinary approach for assessing the seismic vulnerability of heritage masonry buildings is described throughout the paper. The procedure is applied to a specific case study that represents a very common typology of masonry building in Italy. The seismic vulnerability of the examined building was assessed after the following: (a) historical investigation about the building and the surrounding area, (b) detailed geometrical relieves, (c) identification of materials by means of surveys and literature indications, (d) dynamic *in-situ* tests, (e) foundation soil characterization, (f) dynamic identification of the structure by means of a refined Finite Element (FE) model. After these steps, the FE model was used to assess the safety level of the building by means of non-linear static analyses according to the provisions of Eurocode 8 and estimate of the q-factor. Some parametric studies were also carried out by means of both linear dynamic and non-linear static analyses.

Keywords: masonry buildings; knowledge level; multidisciplinary *in-situ* inquiries; FE models; dynamic behavior; non-linear analyses

1. Introduction

The assessment of safety and, in general, of the structural behavior of existing masonry buildings is often a complex issue due to the inevitable uncertainness regarding geometry, typologies and mechanical properties of materials, effects of age and past loading history referring both to service and accidental loads (earthquake, fire, explosion, *etc.*), presence of architectonical and structural interventions with different techniques or materials, interaction with surrounding buildings and foundations, knowledge of foundation soil properties. Thus, for an existing masonry building, a complete knowledge is needed before starting any study of its structural safety or designing an eventual retrofitting intervention.

A complete knowledge needs detailed *in situ* inquiries according to a multidisciplinary approach providing both traditional and innovative investigation techniques. A detailed knowledge of the structure, even if it could appear expensive and time-consuming in the initial stage, in the later stages allows optimizing an eventual repairing design; such an approach reduces the costs of intervention and is in agreement with the 'minimum intervention' philosophy for heritage buildings [1,2]. Indeed, the knowledge level of the geometrical and structural characteristics of the examined building strictly influences the definition of the numerical models and, thus, the reliability of the building response prediction. The Italian Guidelines for cultural heritage [2] suggest to use methods characterized by different levels of knowledge and detail for the assessment of the seismic safety of historical buildings in function of the aim of such studies. The final level of knowledge of the structure depends on quality, quantity and deepening of *in situ* relieves, historical investigations, and experimental tests. The knowledge phases are aimed to define a model able to predict qualitatively the structural behavior or to carry out more refined structural analyses. Clearly, the reliability of each type of model, qualitative or quantitative, depends on the deepening level of the inquiries and on the amount of available data.

Simple approaches, based on indexes related both to a simplified estimation of the seismic resistance of the structure and the expected seismic action [2], can be useful instruments to develop seismic vulnerability maps functional to define a priority scale for further detailed analysis.

More refined analyses request a larger and more detailed database of information about the examined structure (geometry, mechanical properties in the linear and non-linear field, variability of parameters and anomalies in the constituent elements, *etc.*...) that can be collected by several fonts. Furthermore, for historical buildings the simplified methods can give useful indications particularly when detailed experimental *in situ* investigations cannot be carried out, *i.e.*, in presence of strong preservation requirements, so that detailed approaches become unreliable. Note that the simplified methods are in general reliable only for regular and symmetrical buildings, stiff floors and predominant shear failure modes in the walls.

Historical investigations can be useful to characterize and analyze the origin of the structure and its vicissitudes. In particular, the following topics can be assessed by means of this type of investigation:

(a) the reasons for earlier structural interventions: severe damages or collapse due to catastrophic events (earthquake, explosion, fire) or normal use of the building, change of use destination, change of owners, change of architectonic features related to the current architectonic style;

(b) the overall evolution of the examined building as well as of the surrounding buildings related to the urban evolution of the area;

(c) the presence of not contemporaneous construction phases characterized by different materials and techniques;

(d) the individuation in coeval buildings placed in the same area of typical construction materials and techniques. Indeed, the choice of the construction materials, especially if natural carved stones are used, was often related to the geographical location of the building and the local availability of quarries in the neighborhood.

Most of this information can be collected only by means of archival and bibliographical documentations and their importance should be not neglected.

Successively, detailed surveys of geometry and materials make it possible to detect characteristic features such as type and configuration of floors, disposition of resistant masonry walls, nature and dimensions of masonry component elements (*i.e.*, type and dimensions of masonry blocks), wall texture (*i.e.*, disposition of blocks and mortar joints, thickness of joints, presence of a single or double panel with chopped materials in the central cavity), construction details (connections between the floors and the vertical resistant elements, local connections between parallel walls as steel chains or between orthogonal walls in the corners along their height for improving the global behavior of the building, *etc....*). These aspects can be better assessed and/or integrated by means of information collected by historical investigations and the knowledge of the construction techniques typical of that area. Indeed, thanks to the comparison with similar buildings, the amount of the specific investigations to be realized *in situ* can be reduced.

The assessment of the mechanical properties of the various constituent materials (*i.e.*, masonry blocks, mortar) and of their assembling has to be investigated by means of detailed *in situ* testing [3]. The original mechanical properties of the construction materials often have been modified by the age of the building and by loading and environmental effects over time. Thus, the detection of crack patterns and/or of decay conditions of the materials is a first check for individuating potential troubles in structural elements. The mechanical properties of masonry blocks and mortar joints are usually assessed by means of destructive tests *in situ*, even if in the last years, a significant effort has been spent to use non-destructive testing techniques [4,5,6,7,8] in order to limit destructive tests to only a few points.

Investigation techniques based on *in situ* dynamic tests carried out through ambient or forced vibrations [9,10,11,12,13,14,15] represent an innovative and useful instrument to observe different aspects of historical buildings thanks to the study of their dynamic response in terms of frequencies and modal shapes [2].

Firstly, based on the comparisons of experimental evidence (mainly consisting of the principal modal shapes and the related frequencies) and predictions given by linear dynamic models, the values of elastic constants and unit weight of masonry can be easily assessed, taking into account the overall assembly of stones/bricks and mortar.

Moreover, dynamic *in situ* tests can be aimed at studying the response of the structure under seismic actions and at investigating the interaction of the examined building with adjacent constructions or with the foundation soil. This latter topic in particular is rarely accounted for in common structural modeling because in most cases, type and dimensions of the foundations or properties of the subsoil are unknown. When the foundation soil is soft, the hypothesis of a fixed-base condition for the superstructure is no longer valid and the effect of Soil-Structure-Interaction (SSI) should be accounted for. If dynamic *in situ* monitoring is carried out, the role of SSI should be properly accounted for, because measures acquired in different parts of a building may contain the effect of coupling between soil and structure.

Finally, a periodical control of the main dynamic parameters of the building by *in situ* dynamic tests is useful to verify if some changes in the building have occurred, especially in the case of historical ones [2].

The information collected in these preliminary steps are functional to the structural model; it is clear that, if the amount of information is huge, the modeling can be more refined and the results more reliable. For example, models by Finite Elements (FE) allow understanding the structural behavior for a fixed action, identifying the original causes of the current damage patterns, individuating the resisting mechanisms, and predicting the behavior for exceptional actions (*i.e.*, seismic action).

After the building has been modeled, linear analyses under gravity loads are useful instruments to verify the accuracy of the geometrical layout and the reliability of the values of unit weight by comparing the theoretical stress distributions with the experimental data obtained *in situ* in some significant points (*i.e.*, by means of flat jacks). Moreover, a

3D model allows identifying local problems related to stress concentration by comparing them with actual situations and taking into account a rehabilitation design. As previously cited, FE modeling techniques are also a useful support to develop linear dynamic analyses and to assess the mechanical parameters of materials by means of comparison with *in situ* dynamic tests.

Modern code provisions advise to assess the safety of masonry buildings taking into account the non-linear behavior for masonry, by means of 'push-over analysis' carried out thanks to more or less refined models (methods based on storey mechanisms, equivalent frame approaches, two- or three-dimensional continuum approaches [16,17,18,19,20,21,22]. Moreover, pushover analyses are frequently used to assess the seismic vulnerability of historical masonry constructions [23] because the effort spent for the detailed modeling is compensated by the relevance of the structure.

However, the post-elastic behavior of masonry materials is quite uncertain to define, but it is a key aspect to simulate the non-linear behavior of the structure and, thus, needs to be assessed especially for seismic verifications. However, because affordable non-linear FE analyses applied to a continuum require high expertise and should be used when the knowledge of the structure is very detailed, several methods based on simplified element meshing have been developed. The reliability of these approaches for the prediction of the actual behavior of each structure strictly depends on its geometrical configuration. Moreover, in structures endowed of particular architectures (churches, arches, domes, *etc.*...) it is often possible to identify sub-structures characterized by well-defined resistant mechanisms, so that single parts of the building are verified separately [22,24,25].

In this paper a complete approach for the study of an existing masonry building with the final aim to evaluate its safety conditions is followed by analyzing a typical case study: a historical building (Palazzo Bosco Lucarelli) located in Benevento (Italy), a zone of high seismic hazard in the core of the Apennine.

2. The Case Study of a Historical Building

2.1. History and Description of the Structure

For the examined building, 'Palazzo Bosco Lucarelli', an old noble palace located in Benevento (Italy), research in the archives of City Hall and in public libraries allowed collecting information about the different construction phases of the building thanks to historical reports, maps, urban plans and, for the XX century, photos.

The original nucleus of the current building probably goes back to the XI century; successively, some documents of XIII and XIV century confirm the presence of a construction with an adjacent church in the same site of the current palace. This building was severely damaged during the strong earthquake of 1702, after that it was rebuilt and amplified assuming the present 'palace' structure with an internal closed court. Some maps of Benevento of XVIII century (Figure 1a) show a central hallway, the entrance on the current back prospect of the palace (South-West prospect), and the adjacent church. In the years 1768–1774 the building was retrofitted and part of the church was incorporated in the construction, such as indicated in some maps of the beginning of XIX century and in the urban plan of the town dating back to 1880 (Figure 1b). At the end of XIX century, another floor was added and in 1926, when the church was completely demolished, the west wing of the building was doubled and the current geometrical configuration was assumed (Figure 1c, Figure 2a). Nowadays the court is not central anymore, but is located asymmetrically with respect to the west wing that is double the size compared to the east one (see the plan of the ground floor in Figure 2b).

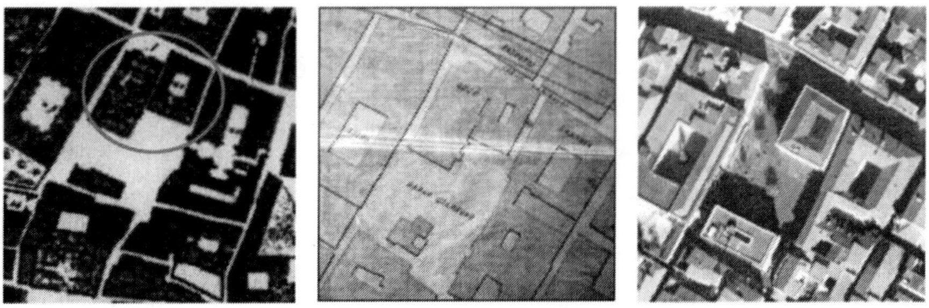

(a) (b) (c)

Figure 1. (**a**) Map of Benevento at the end of XVIII century; (**b**) regulator plan of Benevento in 1880; (**c**) aerial view of historical center of Benevento near Palazzo Bosco Lucarelli in 2012 from a satellite [26].

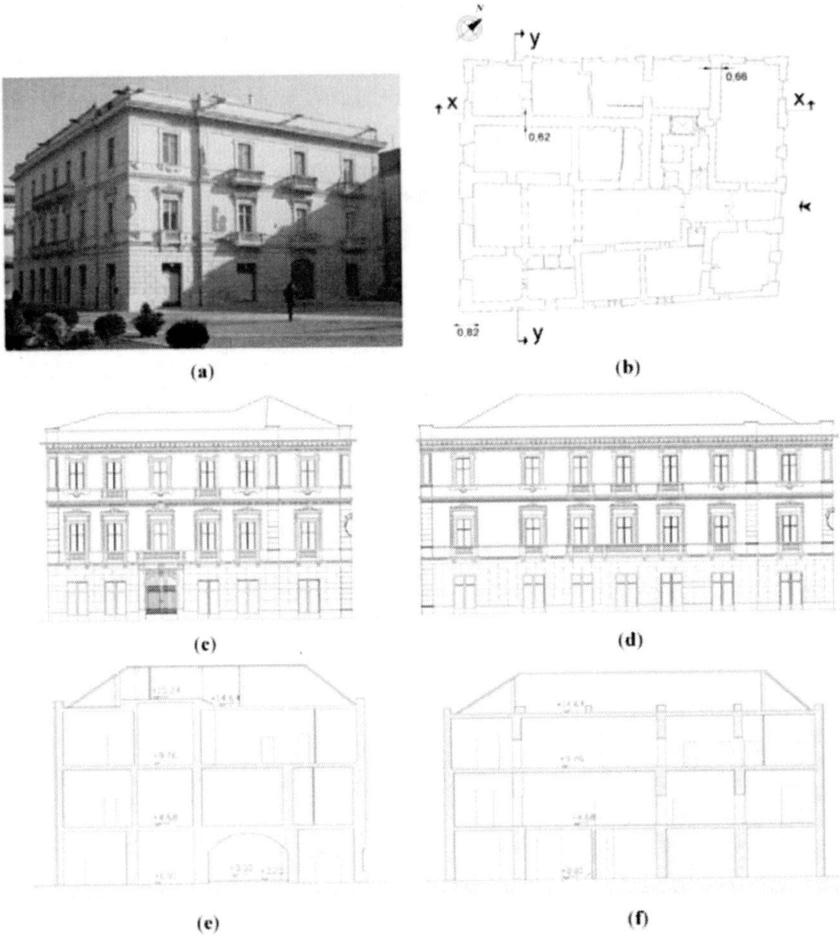

Figure 2. Palazzo Bosco Lucarelli: (a) Current photo of the Northwest and Southwest wings of the building; (b) plan of the ground floor; (c) survey of the Northeast façade (main entrance); (d) survey of the Northwest façade; (e) longitudinal section Y-Y; (f) longitudinal section X-X.

The last interventions were carried out at the end of the 90s when the building became the headquarters of the Engineering Faculty of the University of Sannio. No substantial modifications were carried out in the structural or external architectural layout, only a functional redistribution of the rooms was realized.

Nowadays the structure has a rectangular holed floor plan (Figure 2b), consisting of an underground floor, which is not completely accessible, a ground floor, two upper levels, and an attic under the pitched roof. The largest dimensions in the plan are 20 m and 16 m, the total height is 18.2 m.

The thickness of the walls varies in the range of 0.3 m–1.0 m with the greatest values attained at the ground floor (Figure 2b). A complete geometrical and structural stwo main façades (northeast and northwest prospects, Figure 2c,d) and two orthogonal sections (Figure 2e,f) are reported to illustrate the geometry of the building.

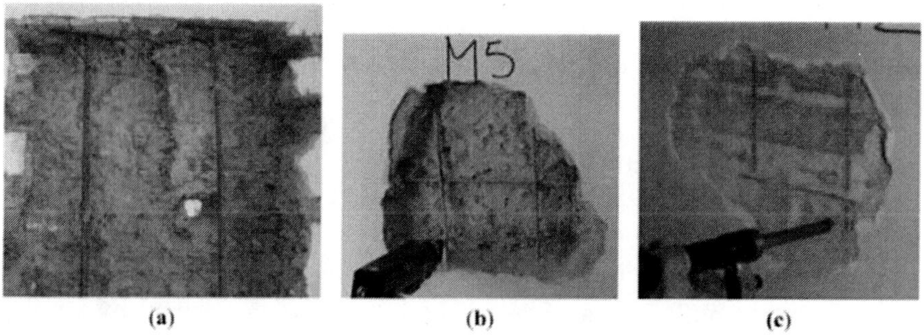

Figure 3. (**a**) Mixed limestone and tuff masonry at the ground floor level; (**b**) tuff panel at the ground floor level; (**c**) Clay bricks at first and second floor levels.

The original floors in most cases have been substituted by double standard 'T' shaped steel profiles with a height of 200 mm (standard type IPE 200) and spaced at a distance of 0.85 m interspersed with bricks and covered by a concrete deck 100 mm thick. Only the ceilings of the staircase and of some rooms in the entrance hall on the ground floor are still made of the original masonry vaults. The floors can be considered as rigid diaphragms and some local tests evidenced a good connection with the vertical masonry walls (reinforced concrete joists are present along the upper perimeter), so that out-of-plane mechanisms can be neglected. The connections were probably improved when the old original floors were substituted by the current steel-concrete ones. Moreover, some steel chains are present below the first floor between the outside walls and the internal walls of the closed court; these chains were positioned at the beginning of the XX century when the last significant structural and geometrical changes of the building were performed (doubling of the west wing and construction of a new level) in order to improve the global behavior of the masonry walls between the ground and the first floor. Nowadays no sign of out-of-plane mechanism can be observed in the structure.

The roof is made of a steel truss that is supported on a perimetric reinforced concrete joist. A profiled steel sheeting, externally covered with brick tiles, closes the structure.

2.2. Assessment of Masonry Properties

The *in situ* surveys allowed detecting a masonry structure made of different materials and textures at the three levels, due to the several construction phases of the building.

At the underground and ground levels, the walls are made of irregular blocks of limestone and conglomerates (Figure 3a) with some inclusions of tuff (Figure 3b) or bricks, probably used in the past to close existing openings. At the underground level, blocks are greater and more irregular suggesting that the underlying foundations were realized as further enlargement of the masonry walls. The first two levels probably date back the construction age of the original nucleus in the IX century, since the masonry texture is typical of coeval palaces present in Benevento and dating to the same period.

At the higher levels, the walls are made of clay bricks (Figure 3c) and were probably built at the end of XIX century, as confirmed by similar masonry textures evidenced in other coeval fabrics and by the information collected about the history of the building.

At each level the walls are covered by a reinforced cement plaster (with a thickness of 50 mm and with a grid of steel bars with a diameter of 4 mm and spaced at a distance of 150 mm in two orthogonal directions, see Figure 3b,c) which can be dated back to the XX century.

Few coring examples of masonry were extracted from the walls and endoscope tests were carried out in the corresponding drilling holes; the images evidence that the walls are made of blocks along the entire thickness, without spacing between the external panels.

Compressive tests were carried out on some single blocks extracted from the masonry. The compressive test on a calcareous stone sample (cube with a side length of 50 mm) extracted at the ground floor level provided a strength of 11.6 MPa, while tests on cubes made of bricks (side length of 60 mm) extracted from the walls at the first and second levels provided strengths of 3.7 MPa and 2.7 MPa (mean values of two tests), respectively. Penetrometer tests on the mortar of the joints gave a compressive strength of about 0.65 MPa for the ground floor and 0.40 MPa for the superior levels; these values allow to classify the mortar as being of a poor typology [27,28].

However, because these data refer to single stones and, thus, are not sufficient for a complete mechanical characterization of the whole masonry assembling, the main mechanical properties (compressive strength, f_m, shear strength without vertical loads, $\tau_{o,m}$, and tensile strength, f_{tm}) were assessed based on the indications given by some recent national guidelines [28] (Table 11.D1, Appendix 11.D) because they provide lower values compared with the ones indicated in the applicative document [29] joined to the current Italian code [27]. The mechanical properties are provided as function of the masonry typology identified *in situ* by the surveys of materials (Table 1). In particular, the masonry of the ground and underground floors has been identified in the category "*Cut stone masonry with good bonding*", while for the higher levels of the building, the masonry has been identified as being made of "*clay bricks with lime mortar*". Considering an intermediate level of knowledge [27,28,30], achieved thanks to the *in situ* surveys, the central value of the interval reported in [28] was adopted for the compressive and shear strength. Then, an amplifying factor of 1.5 due to the beneficial effect of the reinforced plaster [27,29] and a reduction factor of 1.2 (confidence factor for taking into account the knowledge level) were considered to calculate the final values reported in Table 1.

Table 1. Mechanical properties of masonry estimated for 'Palazzo Bosco'.

Floor	Masonry Typology	f_m [MPa]	$\tau_{o,m}$ [MPa]	f_{tm}=1.5 $\tau_{o,m}$ [MPa]	W [kN/m³]	E [MPa]	E·1.5 [MPa]
Ground	*Cut stone masonry with good bonding*	2.19	0.081	0.122	20	1500–1980 (1740)	2250–2970 (2610)
1st–2nd	*Clay bricks with lime mortar*	2.88	0.095	0.143	24	1200–1800 (1500)	1800–2700 (2250)

For the estimation of the unit weight, the samples extracted from masonry provided values of 26 kN/m³, 17 kN/m³ and 19 kN/m³ for the calcareous stone, the bricks and the mortar, respectively; moreover, for the reinforced plaster a unit weight of about 24 kN/m³ can be assumed. Based on these values and on the observed brick texture, for the masonry of the first and second floors the unit weight has been estimated to be about 17.5 kN/m³, which is similar to the value of 18 kN/m³suggested in both the Italian guidelines [28] and code [29]. If the layer of reinforced plaster is considered too, the unit weight increases up to about 20 kN/m³.

Similarly, for the irregular masonry of the ground and underground floors, the unit weight was estimated to be 22 kN/m³, which is comparable with the value of 21 kN/m³ suggested in both the Italian guidelines [28] code [29]. If the layer of reinforced plaster is considered too, the unit weight increases up to about 24 kN/m³.

On the contrary, it was not possible to develop an experimental evaluation of the Young's modulus, E, of the masonry. Literature studies [31] show that the Young's modulus of masonry assembling is affected by a large variability even within the same type of stone (values of CoV also greater than 40%). Some attempts have been made to furnish empirical correlations between the Young's modulus and the compressive strength, but these correlations are a function of the stone type [32,33]. Both national guidelines and code [28,29] were considered for the assessment of Young's modulus. The same range of values (1500–1980 MPa) is given for the *"Cut stone masonry with good bonding"* of the ground floor. Conversely, for the *"clay bricks with lime mortar"*, [28] and [29] suggest the ranges of 1800–2400 MPa and 1200–1800 MPa, respectively. The lower range suggested by [29] was considered for the *"clay bricks with lime mortar"*. Moreover, both codes specify that in case of historical masonry, the suggested values refer to masonry with poor characteristics, thick mortar joints, lack of regular horizontal layers and tidy texture, lack of transversal connecting elements between the internal and external panels of the same wall. Thus, the provided values should be reduced or increased, depending on the current conditions of the masonry. In particular, for the examined masonry types, the presence of the reinforced plaster hence also allows the use of an amplifying factor of 1.5 for the elastic modulus, as is done for the strength. In Table 1 the ranges of E for both types of masonry are reported together with the corresponding central values, assuming an intermediate level of knowledge (2610 MPa and 2250 MPa for the masonry of the ground and the upper floors, respectively). Note that these values represent the starting points for the numerical linear dynamic analyses, aimed at back-calculating the values of E by comparison between experimental and predicted values of frequencies. The above-mentioned values refer to un-cracked condition for the masonry and, thus, should be reduced considering the effective building loading history.

2.3. Characterization of the Foundation Soils

To characterize the foundation soil, a few *in situ* tests carried out near the building were interpreted. In particular, the following tests were considered: (i) a borehole carried out by the local administration very close to the building (SG-003,Figure 4a); (ii) one standard penetration test (SPT) and (iii) one down-hole test (DH). The borehole SG-003 is very deep and reaches a maximum depth of 70 m below ground level (Figure 4b). After the first six to seven meters of man-made ground and historical ruins, the subsoil is made of almost nine meters of fine-grained materials (silt and clay) followed by an alternation of coarser soils, *i.e.*, conglomerate and sandstone up to 60 meters below ground level. The succession ends with grey-azure clay (Pliocenic Age), which is typical of the Benevento area [34] and represents the bedrock formation from the viewpoint of seismic wave propagation. With the velocity of the S-waves in the first 30 meters of the subsoil, V_{S30}, was estimated around 700 m/s. This value allows classifying the site where Palazzo Bosco is located into the category B [27,30].

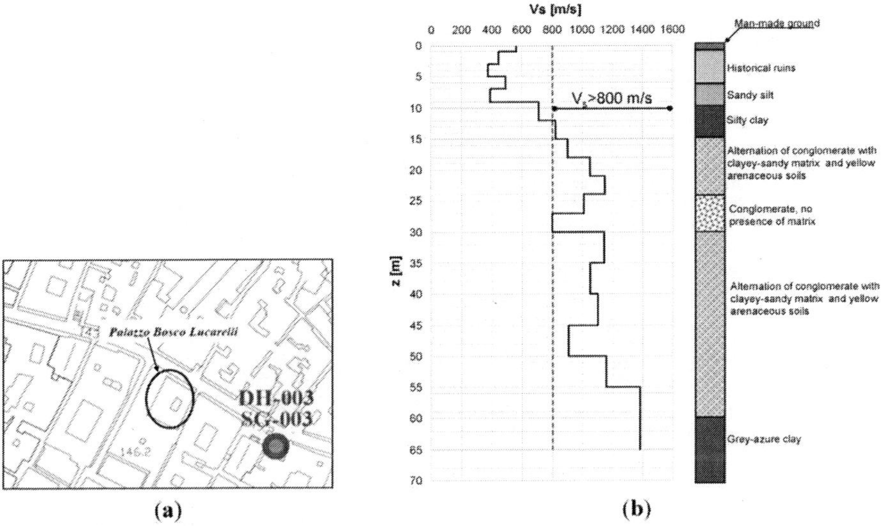

Figure 4. (**a**) Location of the boreholes and down-hole (DH) test; (**b**) shear wave velocity *vs.* depth measured by DH test.

2.4. *In Situ* Dynamic Monitoring

A continuous monitoring system made of nine couples of mono-axial accelerometers was placed on the three floors of the building and a three-axial accelerometer was located on the ground floor. Each couple of sensors is able to measure separately along the X and Y direction. The position of the instruments is reported in Figure 5. The sensors were aligned as accurately as possible along three vertical lines: one corresponding approximately to the center of the building and the other two to the external edges (see Figure 5). On the first floor, the sensors were placed at about 100 mm from the first floor intrados, on the second floor at about 100 mm from the second floor extrados, and on the third floor at about 100 mm from the third floor extrados.

Different sets of accelerometric recordings are available caused by ambient noise and impulsive sources (*i.e.*, the fall of 1 m^3 concrete block on a truck placed close to the building, the rapid braking of the truck along the Y direction of the building, some knocks of an instrumented hammer).

Figure 5. Position of accelerometers in the building.

In particular, the fall of the concrete block originated vibrations in the building with well-defined peaks that were used to have a first rough estimation of the building's dynamic response. The frequency domain representation of such recordings (acquired with a sampling frequency of 500 Hz) was carried out by performing the Fast Fourier Transform (FFT) during a signal-window of eight seconds, centered on the peak value. The energy content of the recorded signals was low at frequencies higher than 50 Hz; for this reason, a low-pass FIR (Finite Impulse Response) filter was adopted in correspondence of 50 Hz. A standard baseline correction was also carried out on the original accelerometric data.

Table 2. Values of frequencies identified by impulsive measurements.

Sensors	Eigenfrequencies [Hz]	
	X	*Y*
A1	4.8	4.2
A2	4.8	4.2
A3	4.5	4.3
A4	4.8	4.2
A5	4.8	4.2
A6	4.5	4.3
A7	4.8	4.2
A8	4.7	4.2
A9	4.5	4.4
Mean	4.69	4.24
St. Dev.	0.15	0.07
CoV	3.1%	1.7%

The values of the first two frequencies of the building (in X and Y direction) identified by each couple of sensors are listed in Table 2 together with the mean value, the standard deviation and the Coefficient of Variation (CoV) of the measured data for each direction.

The frequencies are low scattered among the measures of the nine sensors in both X and Y direction with very small CoVs (3.1% and 1.7% for the X and Y direction, respectively). The frequency identified for the X direction is slightly greater (4.69 Hz *vs.* 4.24 Hz, +10%) than the one associated with the Y direction.

3. 3d fe Model and Numerical Analyses

Linear dynamic analyses can be a useful instrument for identifying the first vibration modes of a building in order to better evaluate the distribution of equivalent horizontal seismic forces in case of the design of a new structure or the assessment of existing ones based on linear static analyses.

Moreover, for historical buildings, the choice of a reliable behavior factor [27,30] for linear static analysis can be very difficult and aleatory, so that non-linear static analysis can prove useful in numerically estimating the actual behavior factor.

In the following sections, both linear dynamic analyses, aimed at assessing the dynamic behavior of the building, and non-linear static analysis (pushover analysis), aimed at estimating the capacity displacement of the building and its behavior factor, will be illustrated.

3.1. Model Description of and Definition of the Loading Conditions

Based on the geometrical survey of the building, a three-dimensional Finite Element Model was implemented with the software DIANA TNO (release 9.4 [35]). The model is made of three-dimensional 'brick' elements, about square shaped in the X-Y plane. The side dimension in the X-Y plane is about 0.3 m which corresponds to half, the third or the whole thickness of each masonry wall, with the masonry thickness variable being in the range of 0.3–1.0 m. The third dimension along the Z-axis is at maximum equal to 1.0 m and allows having a maximum shape ratio between the sides of about three.

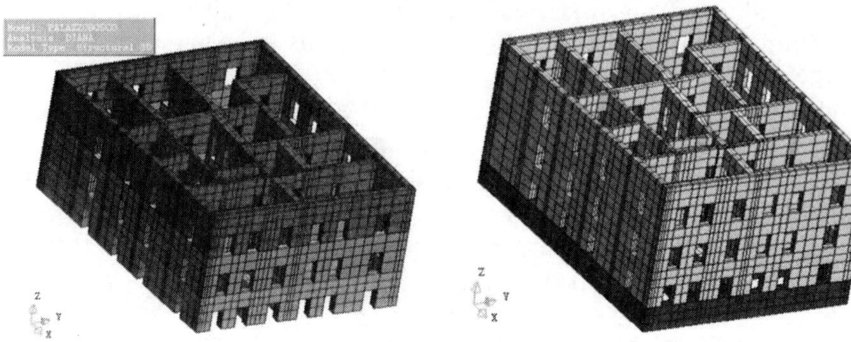

(a) (b)

Figure 6. 3D Finite Element (FE) Model of Palazzo Bosco Lucarelli: (**a**) without the underground floor (Model A); (**b**) with the underground floor (Model B).

The 3D model is made of about 20'000 elements (Figure 6); the three floors were modeled by means of a stiff constraint in each plane and the vertical loads acting on the floors were applied directly on the corresponding masonry walls. The building was modeled both with (Model B, Figure 6b) and without (Model A, Figure 6a) the underground floor according to the type of analysis, as will be explained in the following.

As gravity loads the self-weight of masonry walls was computed using the unit weight reported in Table 1; the dead load of the floors was estimated as $G_k = 5.40$ kN/m^2, while for the accidental loads, at the ground level, the first and the second floor, the values $Q_k = 3.00$ kN/m^2 for eventually crowded ambient [27] were considered and at the level under the roof the lower value of $Q_k = 0.50$ kN/m^2 [27] was assumed. The weight of the steel roof was estimated to be 1.75 kN/m^2, while the weight of the masonry stairs was neglected. The loads of the balconies were applied directly on the external walls ($G_k = 3.60$ kN/m^2, $Q_k = 4.00$ kN/m^2).

Assuming a reduction factor for the accidental loads of $\psi_2 = 0.6$ under the quasi-permanent loading condition [27,36], the total weight is about 68700 kN for Model A and 88900 kN for Model B.

3.2. Linear Dynamic Analysis

Linear dynamic analyses were developed through the 3D FE Model considering all the vibration modes characterized by a participating mass ratio greater than 5%. The aims of such an analysis and its comparison with the dynamic *in situ* tests are multiple, even if they are correlated with each other: (1) to identify the dynamic behavior of the building in terms of main vibration shapes and frequencies; (2) to confirm and/or assess the reliability of the unit weight and the Young's modulus of masonry previously estimated; (3) to have indications about the interaction with the foundation soil. These topics are strictly correlated because the values of the main frequencies depend both on the mechanical properties of masonry (unit weight and Young's modulus) and on the soil-structure interaction effects. Assuming that the geometrical survey of the building is very detailed and conforms to reality and that the estimation of the unit weight of masonry is reliable (this means the global mass of the building is well-defined), the dynamic analysis will be used to assess only the Young's modulus and estimate the effect of the soil-structure interaction, if present. This latter point has been investigated thanks to some information available about the mechanical properties of the foundation soil (see Section 3.3). In order to achieve these results, (1) the mean values of the first two frequencies estimated by the dynamic measurements *in situ* (see Table 2) were used as target for the numerical values of frequencies given by the dynamic analyses; (2) the Young's modulus of masonry were varied in reliable ranges.

Firstly, the effect of the Young's modulus was investigated and, in order to take into account the cracking phenomena as suggested by the codes, the central values of the ranges reported in Table 1 (Young's modulus for the ground floor $E_G = 2610$ MPa, Young's modulus for the 1st and 2nd floor and $E_{1-2} = 2250$ MPa) were reduced. For reinforced concrete structures it is in general suggested to reduce the inertia of the element by 50%, taking into account the cracking phenomena [27]; this hypothesis can be simply introduced into the modeling by reducing the Young's modulus of concrete. For masonry structures it is expected that the reduction of the inertia due to cracking phenomena is lower because of the high compressive stresses usually acting on the masonry walls. Thus, three reductions, lower than 50%, of the Young's modulus were assumed in order to: (1) verify the variation with E of the theoretical values of frequencies given by the FE model, and (2) compare them with the experimental ones:

Case 1: - 30% ($E_G = 1850$ MPa and $E_{1-2} = 1550$ MPa);

Case 2: - 45% ($E_G = 1450$ MPa and $E_{1-2} = 1250$ MPa);

Case 3: - 15% ($E_G = 2200$ MPa and $E_{1-2} = 1850$ MPa).

For these analyses in both Models A and B, the basement was considered completely restrained. Moreover, for the model with the underground level (Model B, Figure 7b), the pressure of the lateral soil was considered as acting on the underground walls by means of an earth pressure coefficient k_o at rest equal to 0.7 (no kinematism was hypothesized).

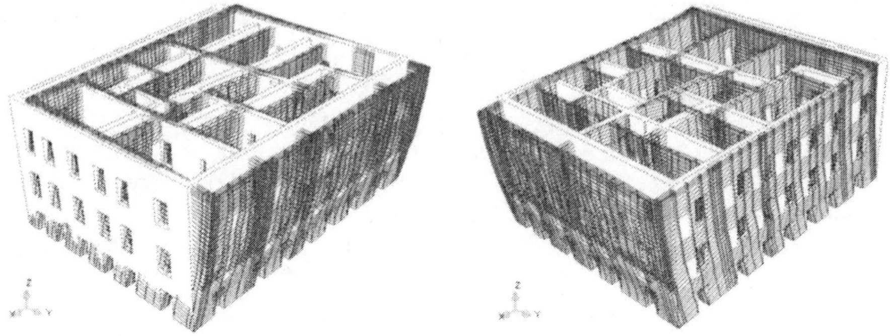

(a) (b)

Figure 7. Modal shapes given by the 3D model for the building without the underground floor (Model A): (**a**) Mode I (Y direction); (**b**) Mode II (X direction).

The numerical values of the first four frequencies are listed in Table 3 for the three cases examined, while the shapes of the first two vibration modes are reported in Figure 7 and Figure 8 for Models A and B of the only Case 1, since they are identical for Cases 2 and 3.

Table 3. Experimental and theoretical main frequencies of Palazzo Bosco for different values of Young's modulus of masonry, E, in the case of fixed basement.

	$f_{th,fix}$ [Hz]						f_{exp} [Hz]
	Case 1 E_G = 1850 MPa E_{1-2} = 1550 MPa		Case 2 E_G = 1450 MPa E_{1-2} = 1250 MPa		Case 3 E_G = 2200 MPa E_{1-2} = 1850 MPa		
Mode	Model A	Model B	Model A	Model B	Model A	Model B	Exp.
1 (Transl. Y)	4.88	4.21	4.36	3.75	5.33	4.60	4.24
2 (Transl. X)	5.45	4.71	4.87	4.19	5.96	5.14	4.69
3 (Local)	11.61	10.81	10.38	9.64	12.71	11.81	-
4 (Local)	13.08	11.58	11.67	10.32	14.29	12.64	-

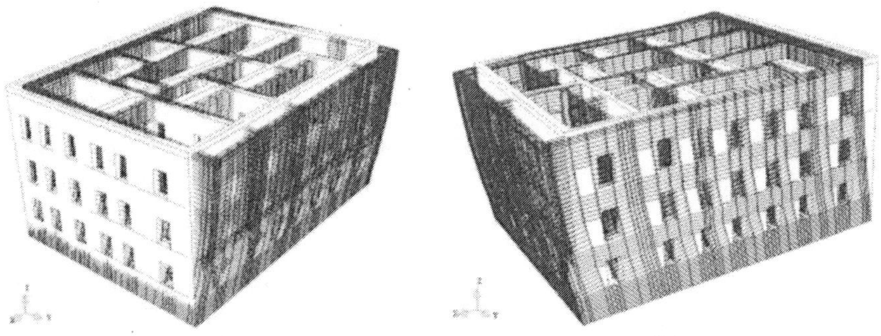

(a) (b)

Figure 8. Modal shapes given by the 3D model for the building with the underground floor (Model B): (**a**) Mode I (Y direction); (**b**) Mode II (X direction).

In both models, the first and the second modal shapes are completely translational in direction Y (Figure 7a and Figure 8a) and X (Figure 7b and Figure 8b), respectively, with a participating mass ratio of about 75% for each direction. In agreement with the experimental results, the first frequency (direction Y) is always about 10% lower than the second one (direction X); this evidences a tendency of the building to be more deformable in direction Y that is parallel to the shorter side (East-West direction, see Figure 2b).

In the model with the underground level (Model B), the first two frequencies are reduced by about 16% with respect to Model A, clearly indicating a greater deformability of the building when the underground floor is modeled too.

The third modal shape, having a share of participant mass equal to 10 % only, is local and regards a wall of the Northwest façade (Figure 2d) on the second floor. This wall is particularly deformable due to the presence of many openings in comparison with the other façades (the two parallel façades along the X direction are not perfectly symmetrical, see plan of the building in Figure 2b). The fourth mode is local too and characterized by mass participant ratios lower than 5%.

When the effect of the Young's modulus is reduced by about 20% (*i.e.*, Case 2 *vs.* Case 1) both first and second frequencies reduce by 10%. Adopting Young's moduli lower than those assumed in Case 2 is unrealistic, because very small numerical frequencies are obtained, and, thus, they are not reliable with the experimental evidence. Moreover, any interaction with the foundation soil has not been considered yet in these analyses, and it is known that the SSI can induce a further reduction of the main frequencies. Similarly, when the Young's modulus increases by about 20% (*i.e.*, Case 3 *vs.* Case 1), the theoretical frequencies increase by about 10%. In conclusion, as expected, a relevant sensitivity of the frequencies to the variation in Young's modulus has been observed: It can be concluded that, for the examined building, a variation of 50% in E leads to a variation of about 25% in terms of frequencies.

In the last column of Table 3, the mean values of the first two frequencies experimentally indentified are listed too. As confirmed by the numerical FE analyses, higher frequencies have not been experimentally identified, probably due to the low rate of participating mass corresponding to these modes and the low intensity energy of the dynamic source. The experimental frequency of the first mode is the mean value of the values corresponding to the first peak of the FFT for each sensor measuring in the Y direction; the same applies analogously for the X direction.

Considering Model A (without the underground level), the results of Case 2 are closer to the experimental evidence (+2.8% and +3.8%). By contrast, considering Model B (with the underground level) that furnishes lower frequencies than Model A, the experimental results are best fitted by the Case 1 (-0.7% and +0.4%).

It is clear that a correct assessment of the Young's modulus of masonry should not ignore the effect of the foundation soil deformability that, in these analyses, has been completely neglected in both Models A and B because the structure has been considered fixed at the basement.

3.3. Effect of Soil Stiffness

In Section 3.2 it has been observed that for Model B, where the underground floor is modeled and lateral pressures of the subsoil are applied on the perimetric underground walls, the first two frequencies decrease by about 16% with respect to Model A (see Table 3), independently of the Young's modulus of masonry.

A first simple approach to consider also the deformability of the foundation soil consists in substituting the fixed constraint at the basement level with vertical elastic springs identified by a coefficient of sub-grade reaction, *k* [37]. This approach, even if it is widely adopted by designers to take into account soil deformability in the global modeling of the structures, is strongly influenced by the values assigned to the parameter *k*. Note that an uncorrected estimation of *k* could lead to an excessive and unrealistic deformability of the structure.

For the case at hand, several values of *k* have been assigned to the vertical springs placed under the basement walls, and it was found that for $k = 0.10$ N/mm^3 the same results of the completely fixed-base schemes are obtained. This led to consider $k = 0.10$ N/mm^3 as an upper limit of the variability range of this parameter. Moreover, varying *k* in the range of 0–0.10 N/mm³, it was found that the main frequencies reduce as k decreases with a quite linear trend: *i.e.*, for k = 0.05 N/mm³, that corresponds to 50% of the upper limit, the frequencies reduce by about 25% compared to the fixed-base schemes. Notwithstanding the simplicity of this approach, the parametric study enhances, thus, that the effect of k is similar to that given by the variation of the masonry Young's modulus discussed in Section 3.2. This aspect highlights how soil deformability could be important in modeling the structure response. However, different and more refined approaches for taking into account Soil-Structure Interaction exist and their use for the case at hand has been illustrated in [38].

3.4. Analysis under Gravity Loads

By means of the 3D FE model of the building, the stress distribution under gravity loads was estimated in order to check the eventual presence of anomalous stress concentrations. The gravity loads were defined in Section 3.1 and were amplified by the coefficients γ_g = 1.3 and γ_q = 1.5 for permanent and accidental loads, respectively, in order to carry out verifications at the ultimate limit state [27,39].

In Figure 9 the distributions of the vertical stresses (σ_{zz}) in the entire building (Figure 9a) and at the ground floor (Figure 9b) are reported for Model A. The highest values of the compressive stresses are 0.85 MPa for the ground floor, 0.70 MPa for the first, and 0.45 MPa for the second one. These values, referring to the strength reported in Table 1 (2.20 MPa for the ground floor and 2.70 MPa for the superior ones), correspond to safety factors of 2.6, 3.9 and 6.0, respectively. However, these values are singular peaks relieved at corners or openings, while the stresses are on average about 0.55 MPa at the ground, 0.45 MPa at the first and 0.35 MPa at the second floor.

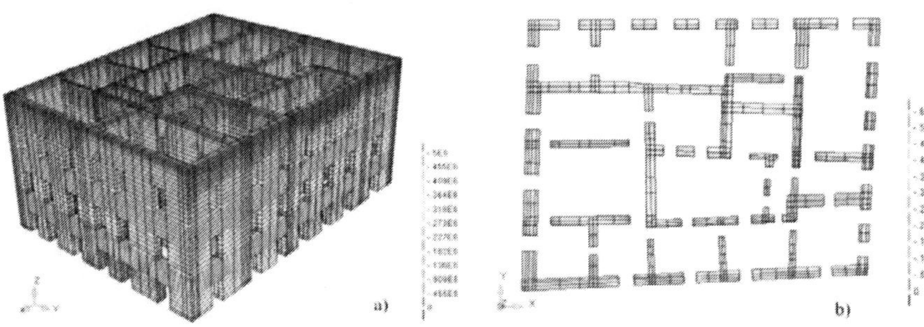

Figure 9. Vertical stress distribution, σ_{zz} (MPa, negative in compression) for Model A: (**a**) whole building; (**b**) plan of the ground floor.

When the underground level is modeled too (Model B) and the pressure of the foundation soil is taken into account on the lateral underground walls by means of the coefficient k_o = 0.7, a contrasting action of the soil on the transversal dilatation of the underground walls is observed. The compressive stresses are a little higher than Model A and are on average equal to 0.90 MPa at the underground level, that is, however, significantly lower that the masonry strength.

3.5. Non Linear Static Analysis under Horizontal Actions

3.5.1. Choice of Parameters for Non-Linear Analysis

Several non-linear static analyses of the structure under horizontal forces (Push Over) were carried out through the 3D FE model. The first purpose of these analyses was the estimation of the sensitivity of the global structural behavior to changes of some parameters typical of masonry constitutive laws, especially in terms of stiffness in the elastic field and ductility in the post-elastic field. Then, the performances of the building under horizontal actions were evaluated.

The masonry was modeled as an equivalent continuous and homogenized material (Poisson coefficient 0.3) adopting the 'total strain' model [40] with 'fixed cracking' for the tension behavior. The 'fixed crack' concept considers that the coordinate system along which the constitutive law is kept in a fixed position, defined upon cracking. Thus, the reference system to evaluate the principal stresses is defined by the first crack plane and, after cracking, a reduction of the shear modulus of the material, G, can be considered by means of a shear retention factor, β [41]. The hypothesis of β = 1 corresponds to not consider any reduction of G after cracking and, in general, leads to overestimation of the global strength of the investigated element or structure. By contrast, when the shear retention factor reduces, the strength reduces too, and the influence of the other parameters of the constitutive laws can become less significant due to the higher decay of G. For the examined building, the effect of three values of β was investigated: 1, 0.5, and 0.01.

Regarding the constitutive laws of masonry, the same Young's modulus were considered in the linear field both for compression and tension behaviors.

For compression, a linear behavior is assumed until reaching 50% of the strength, and then the stiffness reduces according to a parabolic law up to the strength. The values of the compressive strength, f_m, are the ones estimated in Table 1 for the two types of masonry. After the peak, a linear softening is considered until a residual stress of 0.8 MPa (that is 30–35% of f_m[33]) and a maximum strain of 15‰ (see Figure 10a).

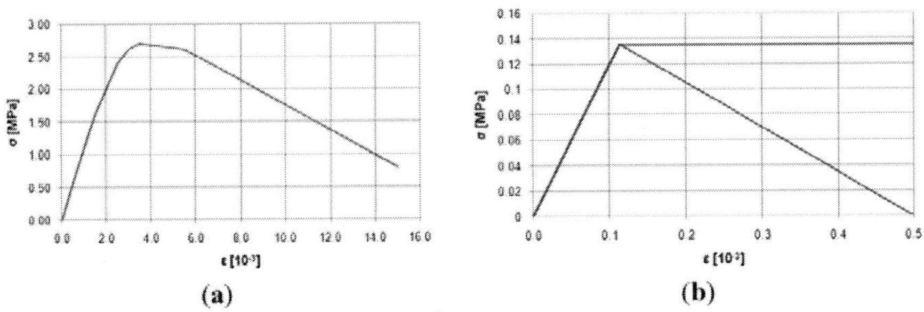

Figure 10. Constitutive relationships of masonry for Floors 1–2: (**a**) compression; (**b**) tension.

For tension, a linear behavior was assumed until the values of strength, f_{tm}, listed in Table 1 and corresponding to about the 5–6% of the compressive one, f_m, [33] were reached. After the peak, two trends were considered in order to analyze the influence of the post-elastic behavior under tension on the global ductility of the building: A linear softening of the stress until a maximum strain of 0.05% or a plastic behavior characterized by an indefinite strain with constant stress (see Figure 10b). The hypothesis of plastic behavior after the peak can be considered as an equivalent and simplified way to take into account the modeling of the homogenized material with the large slips that usually develop in the post-elastic field between masonry blocks and mortar joints.

The shape of the constitutive law in compression was kept unchanged in all the analyses, because it was expected that for a masonry building under horizontal actions the criticism was represented by the shear performances that were related to the modeling of masonry behavior in tension. In particular, it was expected that the post-peak behavior of the compressive law would not be significantly attained, so that the parametric analyses had been carried out only varying β and the post-elastic behavior under tension.

Furthermore, in order to focus the attention on this topic and remove other causes of variability in the structural behavior, the building was modeled without the underground level and under the assumption of a fixed basement (Model A). Thus, no effect of SSI was directly taken into account. However, in order to consider, even if in a simplified way, the additional deformability of the structure due to the neglected effects (underground level and SSI), the lower values of Young's modulus, E_G = 1450 MPa and $E_{1\text{-}2}$ = 1250 MPa corresponding to Case 2, were used in the push-over analyses. Indeed, using these couple of values for E, the numerical frequencies approximate to the experimental ones (see Table 3). This choice for the Young's moduli was also determined by means of the consideration that their values influence the slope of the elastic part of the capacity curve of the structure and, thus, they must be assessed in order to replicate as well as possible the behavior of the structure in the linear field.

In Table 4 a synthesis of the main mechanical parameters assumed in the analyses for the two masonry typologies is reported.

Table 4. Mechanical parameters of masonry for non-linear analysis.

Parameters/Floors	**Ground**	**1st–2nd**
Young's Modulus	1450	1250
Compressive strength [MPa]	2.2	2.7
Tensile strength [MPa]	0.120	0.135
Ultimate strain in compression	0.015	0.015
Compressive stress at the ultimate strain [MPa]	0.80	0.80
Ultimate strain in tension—softening	0.0005	0.0005
Ultimate strain in tension—plastic	indefinite	Indefinite

3.5.2. Assessment of the Capacity Curve

The non-linear behavior of the structure can be synthesized by the relationship (capacity curve) between the base shear, that is the resultant of the horizontal applied forces, and the displacement of a control point of the building. The non-linear static analysis was carried out for the examined building for several reasons: (a) to estimate the displacement capacity of the building under seismic actions, (b) to assess the behavior structure factor *q* [27,30,36] for verifications based on linear-elastic analysis, (c) to check the sensitivity of the building response to some parameters characteristic of the masonry modeling (shear retention factor, β, constitutive law in tension). Because the first two modal shapes of the buildings are the main ones (mass participating ratio of about 75%) and are completely translational in Y and X directions with no torsional effects (seeSection 3.2), the horizontal forces were applied according to a distribution proportional to the two first vibration modes of the building, separately in directions ±X and ±Y ('main distribution' as defined in the current Italian code [27]). The displacement of the centroid of the last floor of the building (about 15 m from the ground level) was assumed as control point for the displacement [27,36]. Note that both the assumption on the force distribution and on the control point may not reproduce correctly the actual dynamic behavior of the building; for example, an adaptive pushover procedure should furnish more reliable results [42]. The only force distributions proportional to the first two vibration modes were considered and no secondary distribution of forces, *i.e.*, an uniform pattern of loads as suggested in both national and European codes [27,30,36], since the attention was mainly focused on having a global view of the building's behavior.

In Figure 11 the capacity curves for the direction +X under the assumption of plastic behavior under tension are graphed for three values of β: 1, 0.5, and 0.01. The analyses were stopped after the base shear had attained a reduction of about 20% of the maximum value [30]. As expected, the lower β is, the lower are both the maximum horizontal load and the maximum displacement. Considering the uncertainness of the effective behavior of masonry, the lowest values for β can be reasonably assumed in order to have the safest predictions of the overall behavior of the building.

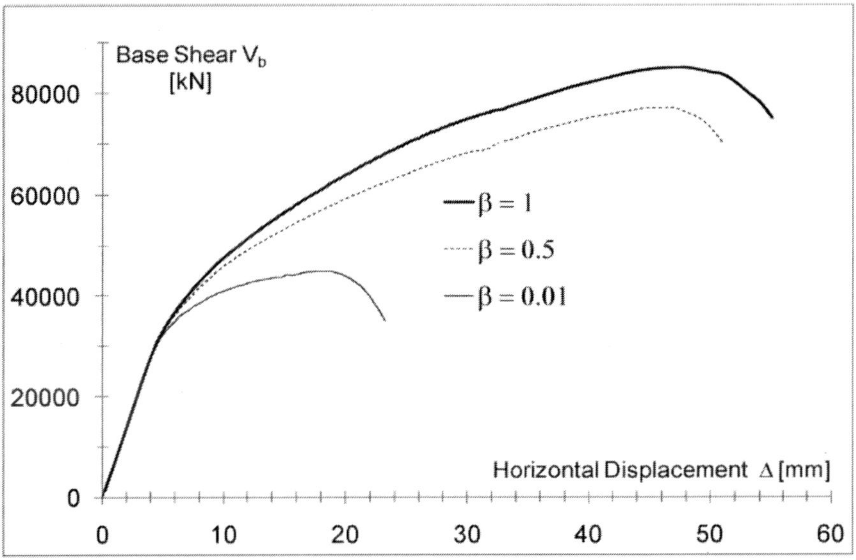

Figure 11. Capacity curves in direction +X for the three different values of β.

In Figure 12 the capacity curves in direction +X and +Y are directly compared under the hypothesis of softening or plastic behavior under tension (see Figure 10b) and assuming β = 0.01. The softening behavior does not influence the stiffness of the building in the elastic phase, but the beginning and the extension of the plastic field. Clearly, under the hypothesis of plastic behavior, the ultimate displacements increase significantly with values that are 1.6–1.7 times larger for both directions. Moreover, the strength is higher too. Because of the asymmetry of the building, the behavior in direction X and Y is not the same, but a larger deformability in the elastic phases is evidenced for direction Y, which is the shorter side and is characterized by a lower number of resistant walls as well (see Figure 2b).

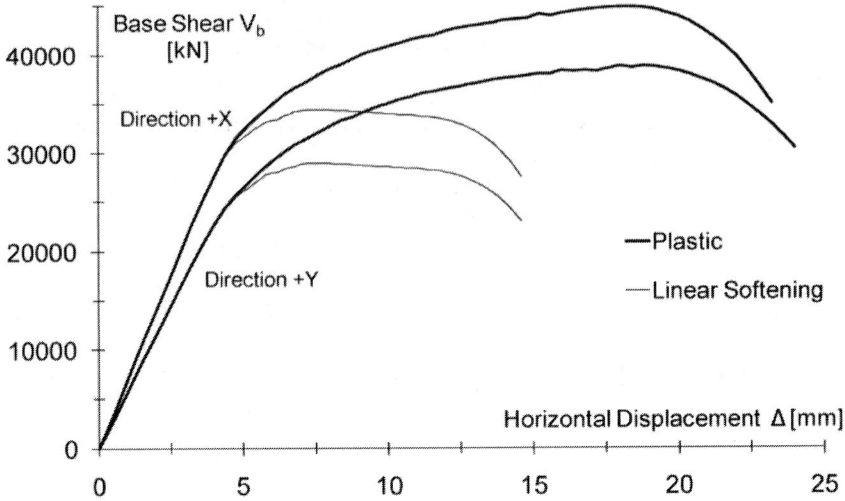

Figure 12. Capacity curves in direction +X and +Y for two different constitutive laws under tension.

In conclusion, the analyses evidences that the behavior of the whole structure is strongly influenced by the shear stiffness reduction, β, and by the softening branch of the tensile behavior. It is clear that a correct choice of β and of the post-peak behavior under tension significantly influences the safety verifications, with the displacement capacity being very different.

The exam of the strain distribution shows that, under the assumption of linear softening after the tensile strength, most of the panels are able to reach the assigned ultimate tensile strain (0.05%). Moreover, most of the compressed panels are in the elastic field because the strains are lower than 0.4%, while only in few cases the strains are proximal to the ultimate one (1.5%). These results confirm that the crisis of the structure is due to the tension behavior. By contrast, for the case of plastic behavior under tension, in most panels the tensile strains overcome the value of 0.15% that is three times larger than the maximum value (0.05%) assumed in the case of softening. This allows the entire structure to have larger displacements and to reach a slightly higher strain also under compression; nevertheless, most of the compressed panels remain in the elastic field with the strains being lower than 0.4%.

3.5.3. Assessment of q-Factor

On the basis of the numerical capacity curves it was possible to estimate the behavior q-factor [27,36] of the building according to the graphic procedures illustrated in Figure 13a,b for the direction +X and + Y, under the hypothesis of linear softening under tension and β = 0.01. In particular the q-factor, q_0 is defined as:

$$q_0 = q_\mu \rho_i \qquad \rho_i = F_u/F_1 \qquad q_\mu = F_e/F_u \tag{2}$$

where ρ_i is the ratio of the maximum load given by the capacity curve, F_u, to the limit elastic one, F_1, and q_μ is the ratio of the theoretical elastic load corresponding to the failure condition, F_e, to F_u. The values of q_μ, ρ_i, and q_0 calculated according to the graphical procedure of Figure 13 are 2.94, 1.31 and 3.84, respectively, for the direction X and 2.89, 1.40 and 4.06 for the direction Y.

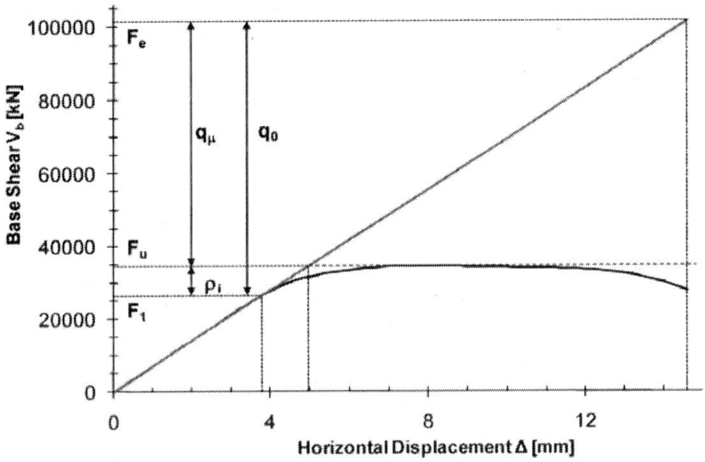

(a)

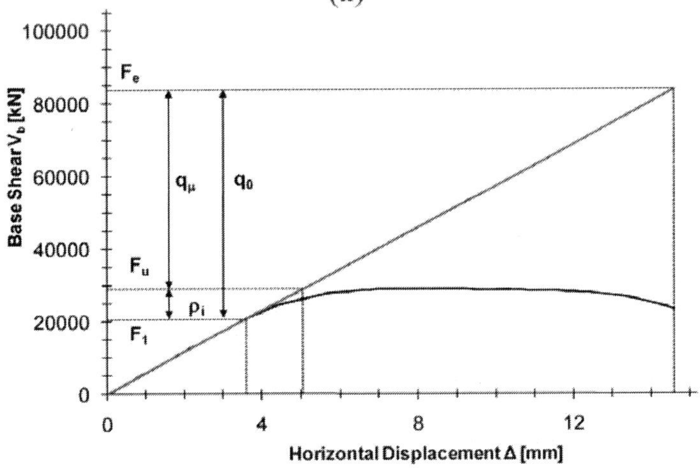

(b)

Figure 13. Graphic evaluation of q-factor in the case of linear softening after tensile strength: (**a**) direction +X; (**b**) direction +Y.

The European code [36] gives a value of q_0 ranging between 1.5 and 2.5.

The Italian code [27] gives a value of $q_0 = 2.0\ \alpha_u/\alpha_1$ for ordinary masonry constructions. Assuming $\alpha_u/\alpha_1 = 1.8$ for masonry buildings with two or more stories, as the examined one, the q-factor is $q_0 = 3.6$.

The values of q_0 calculated from the numerical curves ($q_{0x} = 3.84$ and $q_{0y} = 4.06$) are, thus, meanly 10–15% greater than the ones given by the Italian code, while they are significantly larger than the ones furnished by the European code [36]. Lower values of q_0 mean lower reduction of the seismic action in the case of linear analyses and, thus, are safer. Thus, in view of these results, the assumption of a linear softening under tension seems to be more consistent compared to the hypothesis of plastic behavior that leads to estimating a very large ductility of the structure and, thus, an even greater q-factor.

3.5.4. Safety Verification of the Building According to Code Indications

The capacity displacement curves reported in Figure 12 represent a characteristic of the structure and can be used to assimilate the behavior of a complex system characterized by Multi Degrees Of Freedom (MDOF system) to that of an equivalent non-linear Single Degree Of Freedom oscillator (SDOF system). Hence in the pushover procedure, after the first step leading to the definition of the capacity curve, the ultimate displacement available for the structure (MDOF system) has to be compared with a target design displacement, calculated according to national or European code indications [27,36].

The second step consists in defining the capacity displacement curve of the equivalent SDOF system that is usually assumed to have a bilinear shape. The equivalence is established by the principle that the area above the bilinear curve of the SDOF system is equal to the area above the true capacity curve of the structure. To this aim, firstly, the capacity curve of the MDOF structure has to be normalized by defining the force F^* and the displacement d^* as follows:

$$F^* = \frac{F_b}{\Gamma} \qquad d^* = \frac{d_n}{\Gamma} \tag{3}$$

where F_b and d_n are the base shear force and the control node displacement of the structure in its capacity curve and Γ is a transformation factor associated to the first modal shape. For the examined building the factor Γ is 1.29 and the 1st modal shape is the main one for both directions.

In Figure 14 for the direction ±X (note that the building shows a lower displacement capacity in direction ±X than in direction ±Y) the capacity displacement curves of the structure normalized to Γ and the bilinear ones of the equivalent SDOF systems are reported for both cases of linear softening and plastic behavior after the tensile strength. The minimum values of the maximum displacement d^* in direction ±X are equal to 11.20 mm and 17.74 mm under the assumption of linear softening or plastic behavior, respectively. This means that the maximum peak ground acceleration that the building can sustain is 0.74 g in the case of linear softening or 0.91 g in the case of plastic behavior under tension.

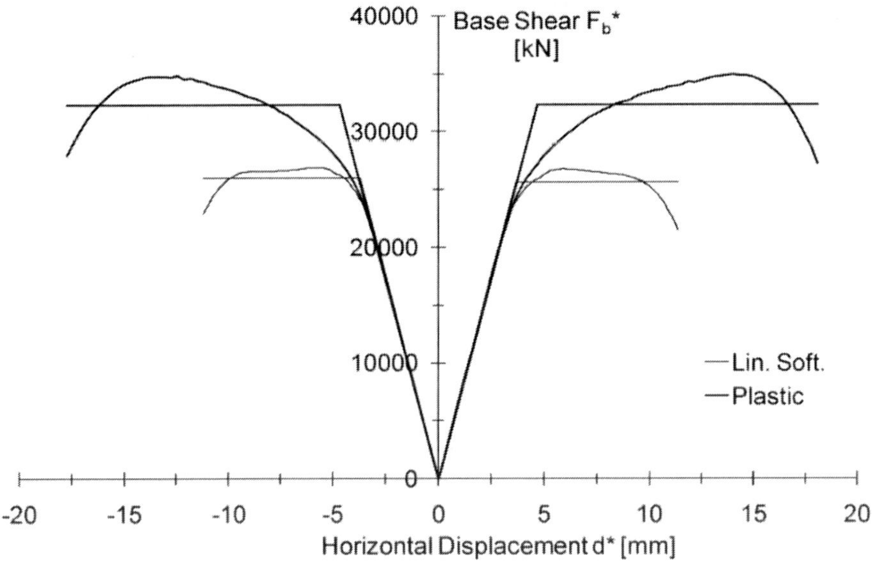

Figure 14. Capacity displacement curves of the structure normalized to Γ and of the SDOF equivalent system in direction ±X for the case of plastic behavior and linear softening under tension.

Considering that, according to the seismic risk map of Italy [43], the expected peak ground acceleration, a_g, for the site is $a_{g,475}$ = 0.262g assuming a return period of 475 years (probability of exceedance of 10% in 50 years) or $a_{g,975}$ = 0.325g assuming a return period of 975 years (probability of exceedance of 5% in 50 years), the building has a minimum safety factor of 2.8 and 2.3 for these two values of acceleration, respectively.

These values of peak ground acceleration can be modified according to an importance factor γ_I that in this case can be assumed to be equal to 1.0 considering the building in Class II (the classes are significant to the destination importance of the building and are numbered from I to IV); indeed the building has no importance for public safety in view of the consequences associated with a collapse, and its integrity is not of vital importance for civil protection during earthquakes [27,36].

4. Conclusions

The structural analysis of historical masonry buildings is particularly complex since each construction is a stand-alone system usually designed and erected for being a singular case. In this paper the procedure for assessing the seismic risk of a historical masonry building is applied to a case study that is of a very common typology of masonry building in Italy and, moreover, is located in an area with high seismicity. The main aim of the procedure is to identify the key parameters in the implementation of a FE numerical model aimed at allowing to carry out allowable analyses in the linear dynamic and non-linear static field.

The greatest difficulties in facing the study of the building were the definition of the mechanical properties of masonry, especially the strength and the elastic modulus. Furthermore, if the effect of the soil-structure interaction is accounted for, the type of modeling and the mechanical characterization of the soil are necessary. An important contribution to calibrating the FE linear model was given by the results of the dynamic *in situ* tests used as target of the numerical analysis in terms of fundamental frequencies. By means of the FE model, indeed, the influence of the uncertainties of masonry Young's modulus and soil deformability, which is simply taken into account by means of vertical linear elastic springs, on the dynamic behavior of the structure was studied; the following concluding remarks are highlighted:

- The influence of the deformability of the masonry and soil was found to be comparable, however, as it is well-known, the real uncertainty in evaluating soil deformability (*i.e.*, the coefficient of sub-grade reaction, k) is much higher than the one of the masonry;
- Concerting the properties of masonry, the non-linear static analysis demonstrated the major importance of the modeling of the constitutive relationship under tension, since the ductility of the entire building was directly related to the ductility developed by the masonry under tension;
- Regarding the behavior factor suggested by Italian and European codes, the numerical analysis provided higher values especially when the constitutive relationship of masonry under tension was assumed to be elastic-plastic, thus the assumption of a linear softening under tension seemed to be more adequate to obtain a safer prediction.

In conclusion, the present work enhanced the key parameters regulating the structural behavior of a historical masonry building under seismic actions (*i.e.*, the structural models) and the utility of the *in situ* dynamic tests.

Further studies are necessary to define the most suitable model to account for both the soil-structure interaction and the behavior of masonry under tension.

References and Notes

1. ICOMOS Charter. Principles for the Analysis, Conservation and Structural Restoration of Architectural Heritage; International Council on Monument and Sites (ICOMOS). In Proceedings of the 14th ICOMOS General Assembly, Victoria Falls, Zimbabwe; 2003.
2. *Linee Guida per la valutazione e riduzione del rischio sismico del patrimonio culturale allineate alle nuove Norme Tecniche per le costruzioni (D.M. 14 gennaio 2008); G.U.R.I.*; 2008; Volume 24, 1 29. (in Italian)
3. Binda, L.; Saisi, A.; Tiraboschi, C. Investigation procedures for the diagnosis of historic masonries. *Constr. Build. Mater.* **2000**, *14*, 199–233.
4. McCann, D.M.; Forde, M.C. Review of NDT methods in the assessment of concrete and masonry structures. *NDT E Int.* **2001**, *34*, 71–84.
5. Clark, M.R.; McCann, D.M.; Forde, M.C. Application of infrared thermography to the non-destructive testing of concrete and masonry bridges. *NDT E Int.* **2003**, *36*, 265–275.
6. Carpinteri, A.; Invernizzi, S.; Lacidogna, G. *In situ* damage assessment and non linear modelling of historical masonry towers. *Eng. Struct.* **2005**, *27*, 387–395.
7. Binda, L.; Zanzi, L.; Lualdi, M.; Condoleo, P. The use of georadar to assess damage to a masonry Bell Tower in Cremona, Italy. *NDT E Int.* **2005**, *38*, 171–179.
8. Carpinteri, A.; Invernizzi, S.; Lacidogna, G. Historical brick-masonry subjected to double flat-jack test: Acoustic emissions and scale effects on cracking density. *Constr. Build. Mater.* **2009**, *23*, 2813–2820.
9. De Sortis, A.; Antonacci, E.; Vestroni, F. Dynamic identification of a masonry building using forced vibration test.*Eng. Struct.* **2005**, *27*, 155–165.
10. Bennati, S.; Nardini, L.; Salvatore, W. Dynamical behaviour of a masonry medieval tower subjected to bell's action. Part I: Bell's action measurement and modelling. *J. Struct. Eng. ASCE* **2005**, *131*, 1647–1655.

11. Gentile, C. Operational Modal Analysis and Assessment of Historical Structures. In Proceedings of the 1st International Operational Modal Analysis Conference, Copenhagen, Denmark, 26–27 April 2005.
12. Gentile, C.; Saisi, A. Ambient vibration testing of historic masonry towers for structural identification and damage assessment. *Constr. Build. Mater.* **2007**, *21*, 1311–1321.
13. Ivorra, S.; Pallarés, F.J. A Masonry Bell-Tower Assessment by Modal Testing. In Proceedings of the 2nd International Operational Modal Analysis Conference (IOMAC), Copenhagen, Denmark, 30 April–2 May 2007.
14. Ceroni, F.; Pecce, M.; Voto, S.; Manfredi, G. Historical, architectural and structural assessment of the Bell Tower of Santa Maria del Carmine. *Int. J. Archit. Herit.* **2009**, *3*, 169–194.
15. Rainieri, C.; Fabbrocino, G. Operational modal analysis for the characterization of heritage structures. *Geofizika* **2011**,*28*, 109–126.
16. Tomazevic, M. The Computer Program POR; Report ZRMK, Ljubljana, Slovenian, 1978; (in Slovenian).
17. Tomazevic, M. Dynamic modelling of masonry buildings: Storey mechanism model as a simple alternative. *Earthq. Eng. Struct. Dyn.* **1987**, *15*, 731–749.
18. Magenes, G.; Della Fontana, A. Simplified non-linear seismic analysis of masonry buildings. *Proc. Br. Mason. Soc.* **1998**,*8*, 190–195.
19. Lourenço, P.B. Anisotropic softening model for masonry plates and shells. *J. Struct. Eng.* **2000**, *126*, 1008–1016.
20. Massart, T.J.; Peerlings, R.H.J.; Geers, M.G.D. Mesoscopicmodelling of failure and damage induced anisotropy in brick masonry. *Eur. J. Mech. Solids* **2004**, *23*, 719–735.
21. Calderini, C.; Lagomarsino, S. A continuum model for in-plane anisotropic inelastic behaviour of masonry. *J. Struct. Eng.* **2008**, *134*, 209–220.
22. Chen, S.-Y.; Moon, F.L.; Yi, T. A macroelement for the nonlinear analysis of in-plane unreinforced masonry piers. *Eng. Struct.* **2008**, *30*, 2242–2252.
23. Ceroni, F.; Pecce, M.; Manfredi, G. Modelling and seismic assessment of the bell Tower of Santa Maria del carmine: Problems and solutions. *J. Earthq. Eng.* **2010**, *14*, 30–56.
24. De Luca, A.; Giordano, A.; Mele, E. A simplified procedure for assessing the seismic capacity of masonry arches. *Eng. Struct.* **2004**, *26*, 1915–1929.

25. Lagomarsino, S.; Galasco, A.; Penna, A. Non Linear Macro-element Dynamic Analysis of Masonry Buildings. In Proceedings of the ECCOMAS Thematic Conference on Computational Methods in Structural Dynamics and Earthquake Engineering, Rethymno, Crete, Greece, 13–16 June 2007.
26. Aerial view of historical centre of Benevento, Italy form satellite. Available online: http://maps.google.it/maps(accessed on 15 January 2012).
27. NTC 2008. *Norme Tecniche per le Costruzioni*, Decreto Ministeriale del 14/01/2008. In *G.U.R.I.*; 2008 2 4; Volume 29. (in Italian).
28. OCPM 3431. Primi elementi in materia di criteri generali per la classificazione sismica del territorio nazionale e di normative tecniche per le costruzioni in zona sismica. *Ordinanza del Presidente del Consiglio dei Ministri del 03/05/2005. In G.U.R.I.*; 10 05 2005; 107. (in Italian).
29. Circolare 617. *Istruzioni per l'applicazione delle "Nuove norme tecniche per le costruzioni di cui al D.M. 14 gennaio 2008*. Ministero dei Lavori Pubblici, Roma, 02/02/2009 (in Italian).
30. Eurocode 8. *Design of Structures for Earthquake Resistance—Part 3: Assessment and Retrofitting of Buildings*; Committee for Standardization: Brussels, Belgium, 2005; EN 1998-1..
31. Marcari, G.; Fabbrocino, G.; Lourenço, P. Mechanical Properties of Tuff and Calcarenite Stone Masonry Panels under Compression. In Proceedings of the 8th International Masonry Conference, Dresden, Germany, 4–7 July 2010.
32. Binda, L.; Pina-Henriques, J.; Anzani, A.; Fontana, A.; Lourenço, P.B. A contribution for the understanding of load-transfer mechanisms in multi-leaf masonry walls: Testing and modelling. *Eng. Struct.* **2006**, *28*, 1132–1148.
33. Augenti, N.; Parisi, F. Constitutive models for tuff masonry under uniaxial compression. *J. Mater. Civ. Eng.* **2010**, *22*, 1102–1111.
34. Improta, L.; di Giulio, G.; Iannaccone, G. Variations of local seismic response inBenevento (Southern Italy) using earthquakes and ambient noise recordings. *J. Seismol.* **2005**, *9*, 191–210.
35. DIANA Finite Elements Analysis. In *User's Manual—Release 9.4—Element Library, Material Library*; TNO Building and Construction Research, Department of Computational Mechanics: A.A. Delft, Netherlands, 2009.

36. Eurocode 8. *Design of Structures for Earthquake Resistance—Part 1: General Rules, Seismic Actions and Rules for Buildings*; Committee for Standardization: Brussels, Belgium, 2005; EN 1998-1.
37. Terzaghi, K.; Peck, R.B. *Soil Mechanics in Engineering Practice*; Wiley: New York, NY, USA, 1948.
38. Ceroni, F.; Sica, S.; Pecce, M.; Garofano, A. Effect of Soil-Structure Interaction on the dynamic behavior of masonry and RC buildings. In Proceedings of the 15th World Conference on Earthquake Engineering (WCEE), Lisbon, Portugal, 24-28 September, 2012.
39. Eurocode 2. *Design of Concrete Structures—Part 1-1: General Rules and Rules for Buildings*; SPRING Singapore: Singapore, 2004; ENV 1992-1-1: 2004: E..
40. Selby, R.G.; Vecchio, F.J. Three-dimensional Constitutive Relations for Reinforced Concrete. In *Technical Report 93-02*; Department Civil Engineer, University of Toronto: Toronto, Canada, 1993.
41. Crisfield, M.A. *Non-Linear Finite Element Analysis of Solids and Structures*; John Wiley & Sons: Hoboken, NJ, USA, 1991; Volume 1.
42. Galasco, A.; Lagomarsino, S.; Penna, A. On the Use of Pushover Analysis for Existing Masonry Buildings. In Proceedings of the 1st European Conference on Earthquake Engineering and Seismology, Geneva, Switzerland, 3–8 September 2006.
43. Stime di pericolosità sismica per diverse probabilità di superamento in 50 anni: Valori di a_g. Available online: http://esse1.mi.ingv.it/d2.html (accessed on 15 January 2012, in Italian).

CHAPTER 7

Optimal Design and Earthquake-Resistant Design Evaluation of Low-Rise Framed RC Structure

Reyes Indira Herrera[1*], Juan Carlos Vielma[1], Ronald Ugel[1], Yolsanie Martínez[1], Alex Barbat[2]

[1]Structural Engineering Department, School of Civil Engineering, Lisandro Alvarado University, Barquisimeto, Venezuela;

[2]Technical University of Catalonia, Barcelona, Spain

Abstract

This study evaluates the seismic response of an existing two stories RC building using non-linear analysis. The original model was resized and two buildings were designed using two different methodologies to fulfill the Venezuelan codes requirements for a high seismic hazard. An elastic analysis was applied to the original building in order to verify interstory drifts; the resizing building was designed under requirements of strong column-wake beam condition. A third building was modeled according to the seismic displacement-based design procedure. A nonlinear static analysis and 2D and 3D dynamic analyses were performed, obtaining capacity curves, structural ductility, structural performance point, global and interstory drifts for each building. Torsional effects for the resizing building were also computed from 3D analysis. In the original building a weak seismic behavior was obtained, while resized buildings presented a good seismic performance under the Limits States evaluated in this study.

Keywords: Resized Building; Linear Analysis; Nonlinear Analysis; Seismic Response; Torsional Effects

1. Introduction

Along its history, Venezuela has been severely affected by destructive earthquakes [1]. Approximately 80% of the population lives in seismically active areas, where have occurred destructive earthquakes even in recent times [2]; The seismic hazard, inadequate design and construction of buildings as well as the damage occurred from previous earthquakes, demonstrate a high vulnerability in existing buildings. It is thus essential to continuously make progress and research in the field of earthquake engineering and earthquake resistant design code changes. Upgrades require the evaluation of predictions of the expected damage to structures at the time of an earthquake of a certain severity to occur. From this prediction it can be defined solutions for the reduction of structural vulnerability [3].

The damage occurred in buildings after an earthquake indicates the need for reliable methodologies for the evaluation of seismic behavior of the existing buildings. Ac-cording to current technical and scientific advances, seismic evaluation of reinforced concrete (RC) structures can be done by two different approaches: empirical methods and mechanical methods [4]. The current tendency of earthquake engineering in the evaluation of structural behavior is the application of simplified mechanical methods based on performance, involving the capacity spectrum [5], because there are developed refined models and detailed analysis.

This study used a mechanical method that involves non-linear analysis with deterministic and probabilistic approaches, as well as procedures of analysis based on Limits States defined by displacement [6], in order to evaluate the behavior of a low rise RC building with asymmetry in plant, designed according to Venezuelan codes [7-9] and subjected to seismic action effect. Through the use of mathematical models and computational tools, seismic behavior of the building is obtained in a suitable way.

2. Methodology

2.1. Case of Study

A two story RC framed building with residential use is analyzed, (**Figure 1**(a)), which contains internal staircase and 220 m^2 total plant area. It was designed a mathematical model, called original building (OB), with plan asymmetry (**Figure 1**(b)) and slab armed in direction X with a 25 cm depth one-way ribbed slab. It was designed a second model adjusted to seismic resistance requirements, called resizing building (RB), which presents equal geometrical and mechanical characteristics of OB model but considering the "strong column-weak beam" condition. It was also used the method based on displacement, Priestley et al. [10] for the design of a third model called displacement based design building (DBDB). These three models differ only in the dimensions of its structural elements (**Table 1**).

2.2. Nonlinear Analysis

The structures are modeled by incorporating the structural response when it incurs in the material inelastic range, produced by high deformations caused by accidental excitations (earthquakes) [11]. The analyses were performed using ZEUS-NL soft-ware [12], which allows to model complex structures with "n" number of finite elements, thus to know the elements in the building which are most vulnerable to dam-age. Each building is modeled in two dimensions, spitting each frame to get a more detailed response for the seismic behavior of each frame; a 3D dynamic analysis was applied to the ER model.

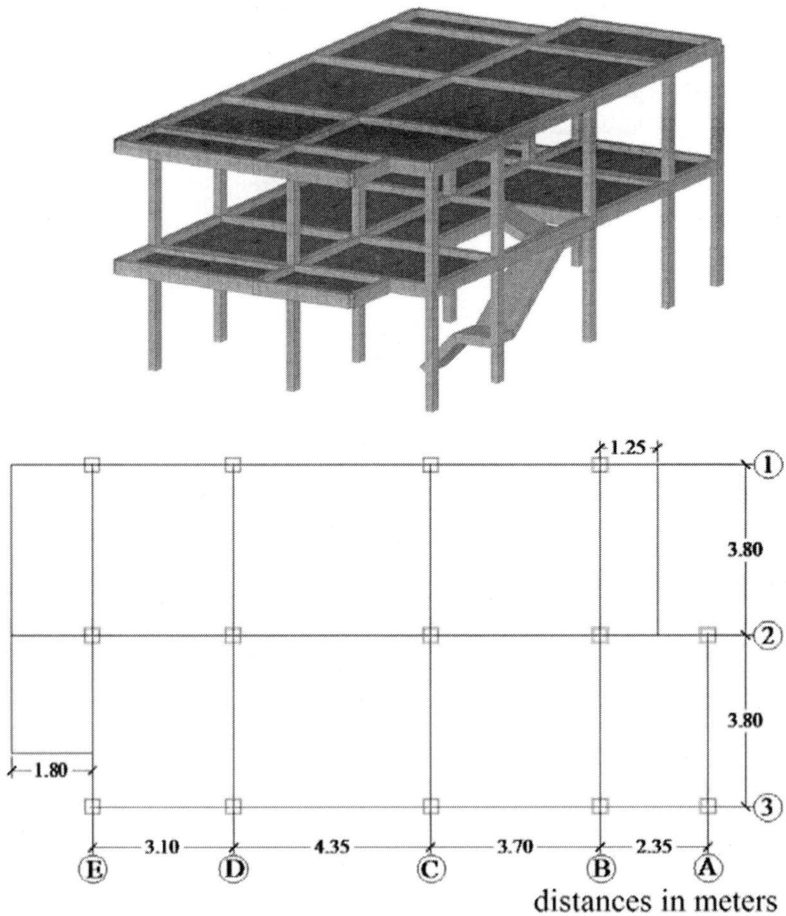

Figure 1. Low rise RC framed building. (a) 3D representation; (b) Asymmetric plant.

Table 1. Geometric characteristics of elements from each modeled building.

Building	Axis X beams (cm)	Axis Z beams (cm)	First level columns (cm)	Second level columns (cm)
EO	20 × 35	20 × 35	20 × 30	20 × 30
ER	20 × 45	20 × 35	30 × 30	30 × 30
DBDB	20 × 40	20 × 40	35 × 35	30 × 30

The static Pushover analysis is performed once the frames have been subjected to action of gravity loads, based on the pseudo-static application of lateral forces equivalent to displacements of seismic action [5]. The pattern of representatives lateral seismic loads is increasing loads with height (triangular distribution) applied in the monotonic form until the structure reaches its maximum capacity [13].

This procedure applies a solution of equilibrium equations in an incremental iterative form. In small increments of linear loads, equilibrium is expressed as:

$$K_t \Delta_x + R_t = \Delta F \tag{1}$$

where K_t = Tangent stiffness matrix, R_t = Restorative forces at the beginning of the increased load. Restorative forces are calculated from:

$$R_t = \sum K_t, K\Delta_u \tag{2}$$

while this procedure is applied, the resistance of the structure is evaluated from it is balance internal conditions, updating at each step the tangent stiffness matrix. Unbalanced loads are applied again until it can satisfy a criterion of convergence. Then, a new load increase is applied. The increases are applied until a predetermined displacement is reached or until the solution diverges.

From the capacity curve provided in this analysis, it is determined the structural ductility (μ) by the quotient between the ultimate displacement and cadence point displacement, as shown in the following expression:

$$\mu = \Delta_u / \Delta_y \tag{3}$$

where Δ_u = Ultimate displacement, Δ_y = Cedence point displacement.

Incremental dynamic analysis (IDA) is an analysis method that can be used to estimate structural capacity under seismic loads. It provides continuous response of the structural system from elastic range until it reaches collapse. In this method the structure is subjected to one or more seismic records scaled to intensity levels that increase progressively. The maximum values of response are plotted against the intensity of seismic signal [14,15]. The procedure to perform IDA from the seismic signal is:

- To define a seismic signal compatible with the project scenario;
- To define the scaled earthquake intensity in monotonic form;
- To define the extent of damage or damage Limit States;
- To study a seismic record (IDA) for the dynamic analysis of a structural model parameterized to measure earthquake intensity;

The IDA curve is a graphical representation of the extent of damage in a structure under one or more seismic intensities.

For the dynamic analysis the structures were subjected to seismic action (see **Table 2**) defined by accelerograms built on the basis of a likely value of maximum acceleration of the soil and the hazard level associated with the location of the structure and other seismic characteristic parameters of the design, [16]. These accelerograms called "synthetic accelerograms" are generated through the implementation of a set of earthquakes with wide frequency content, using the PACED program [17], based on the code's elastic design spectrum [7]. For the dynamic analysis of the three buildings (OB, RB, DBDB), it were used 3 synthetic accelerograms with duration of 60, 80 and 100 seconds.

These three earthquakes were applied to all frames from the three buildings evaluated, in order to obtain maximum displacement that can be reached by each one. In the software used [12], it was required the implementation of dynamic loads in direction X and the assignation of a control node located in the gravity center of the covert plan.

The 3D non-linear dynamic analysis is based on the procedure raised in [16]. The RB building is analyzed, defining its geometry, materials and sections, serviceability loads in Y direction in all beams-columns joints, and dynamic loads on built-in nodes with directions and combinations shown in **Table 3**. Ribbed slabs armed in one direction were modeled as rigid diaphragms in its plane (**Figure 2**).

Once built the model, it was applied the earthquake R1_3 with the combinations shown in **Table 3**, for the interstory drifts and maximum torsional moments on supports. These combinations are based on the Venezuelan seismic code [7] and following established by [18] about the seismic response of asymmetric structural systems in the inelastic range.

Table 2. Limit States and seismic hazard level.

Analysis Earthquake	Limit State	Return Period (years)	Occurrence Probability in 50 Years	Interstory Drift δ (%)
Frequent	Serviceability	95	50%	$\delta < 0.5$
Rare	Reparable Damage	475	10%	$\delta < 1.5$
Very Rare	Collapse Prevention	2475	2%	$\delta < 3.0$

Table 3. Applied seismic combinations.

Nº	Seismic combination
1	100% (X)
2	100% (Z)
3	100% (X) y 30% (Z)
4	100% (Z) y 30% (X)

3. Results

Verification of interstory drifts of the OB building, shows that they exceed the limit established in [7], while in the RB model it was obtained that it meets the code's parameters, which limits the inter storey drift to 0,018. On the other hand, in the DBDB building were not performed drifts verifications, since it was designed based on the method performed in [10], where the generated seismic forces are originally limited to not exceed the limit value of drift specified in the applied code.

3.1. Nonlinear Static Analysis

To determine the values of structural ductility it was necessary to plot the idealized curve in function of the capacity curve, in order to know the point at which the structure begins to yield.**Figure 3** shows an example of the normalized capacity curve with the idealized curve of Frame C of OB.

Structural ductility for each evaluated building values are presented in **Table 4**.

From the obtained capacity curves there were identified the Performance point (Pp) of every frame of each evaluated building. **Table 5** presents the values of Pp for frames of evaluated buildings.**Figure 4** shows the Pp of Frame C from OB building.

3.2. Nonlinear Dynamic Analysis

There were determined global and interstory drifts of each frame from all three models studied. Both types of drifts were calculated on the basis of the application of synthetic accelerograms with different intensities, representing the lateral forces applied to frames in order to to generate their respective maximum displacements. Figures 5-7 show the graphics with the evolution of the global (Δ/H) drifts ex-pressed as a percentage respect to time(s) of the frame C from OB, RB and DBDB models for a seismic threat of 0.3 g.

Figure 2. Rigid diaphragms in 3D RB framed building.

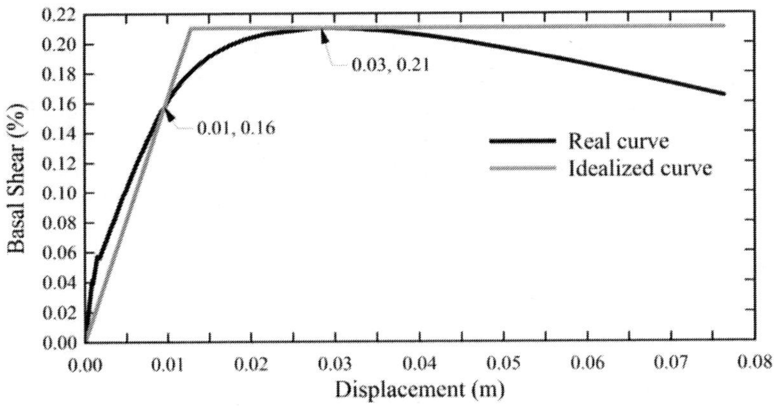

Figure 3. Normalized and idealized capacity curves. Frame C. OB building.

Table 4. Structural ductility results.

	FRAME	BUILDING		
		EO	ER	*DBDB*
STRUCTURAL DUCTILITY	**A**	5.56	5.52	4.77
	B	2.22	6.04	5.38
	C	2.17	4.69	5.25
	D	2.21	5.54	5.59
	E	2.23	7.07	6.06
	1	2.66	5.29	6.69
	2	2.20	4.17	5.92
	3	2.83	5.95	6.24

Table 5. Performance points (Pp) of studied buildings frames.

FRAME	Pp (cm)		
	EO	ER	*DBDB*
A	5.94	2.42	2.52
B	13.89	9.47	7.43
C	15.22	9.50	9.38
D	14.01	9.50	7.57
E	13.45	9.55	6.60
1	12.62	9.35	6.07
2	15.74	11.48	9.29
3	10.92	7.57	4.23

Figures 8-10 show the results for interstory drifts of frame C from OB, RB and DBDB buildings, taking into account the R1_3 earthquake with duration of 60 seconds. Similarly, interstory drifts for applied earthquakes, R1, R2 and R3 with its three intensities, were obtained. It was verified for each Limit State considered in this study. **Table 6** reflects the result of interstory drifts of buildings in study for earthquake R1, taking into account the three levels of hazard, 0.5%, 1.5% and 3%, for the Limits States considered.

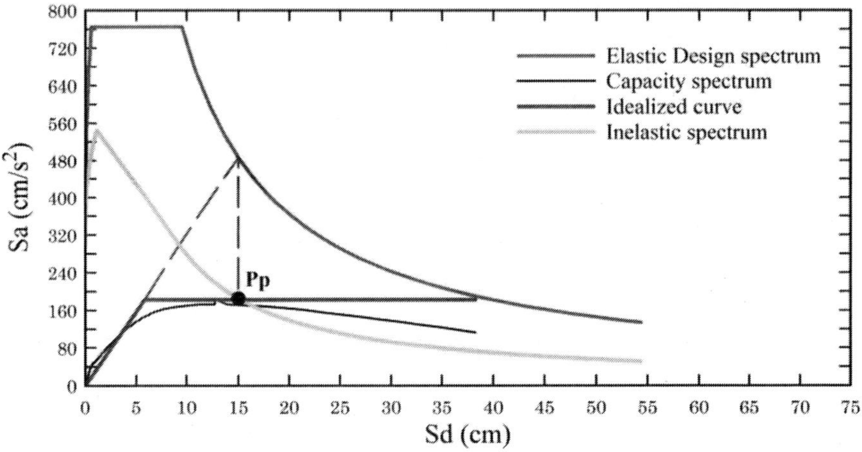

Figure 4. Performance point of Frame C. OB Building, determined by N2 procedure.

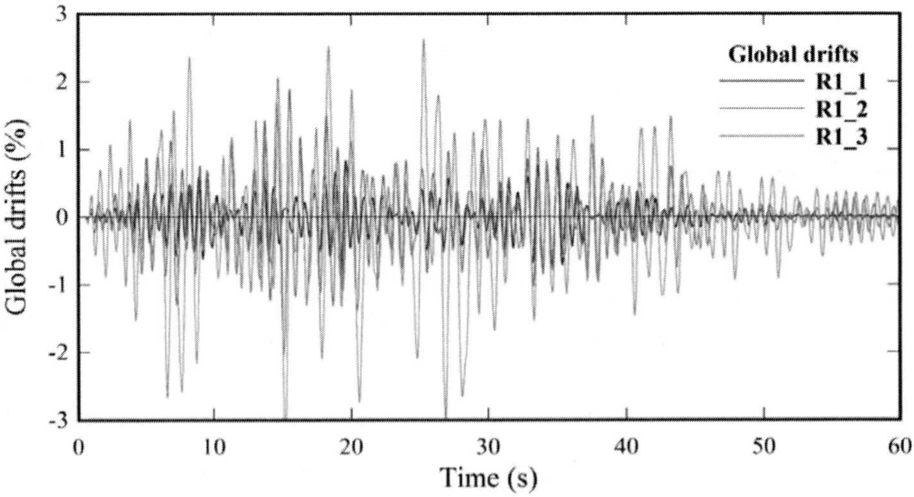

Figure 5. Global drifts. R1 earthquake. Frame C. OB building.

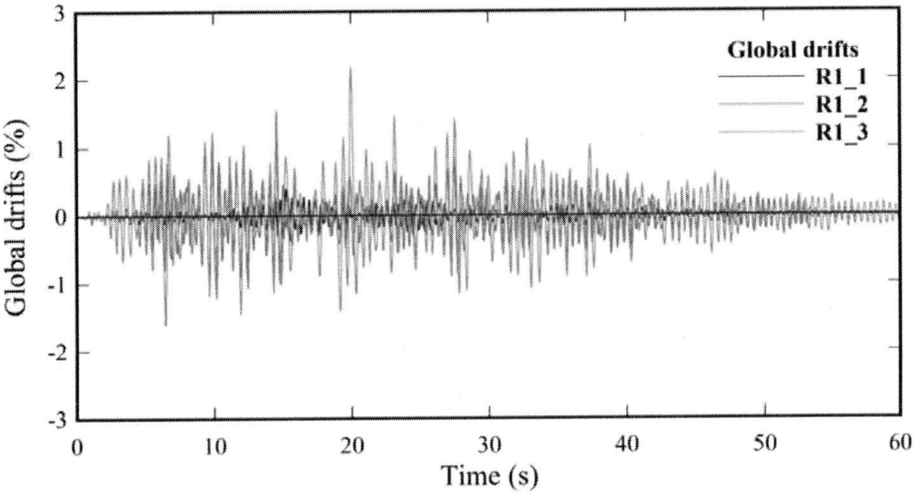

Figure 6. Global drifts. R1 earthquake. Frame C. RB building.

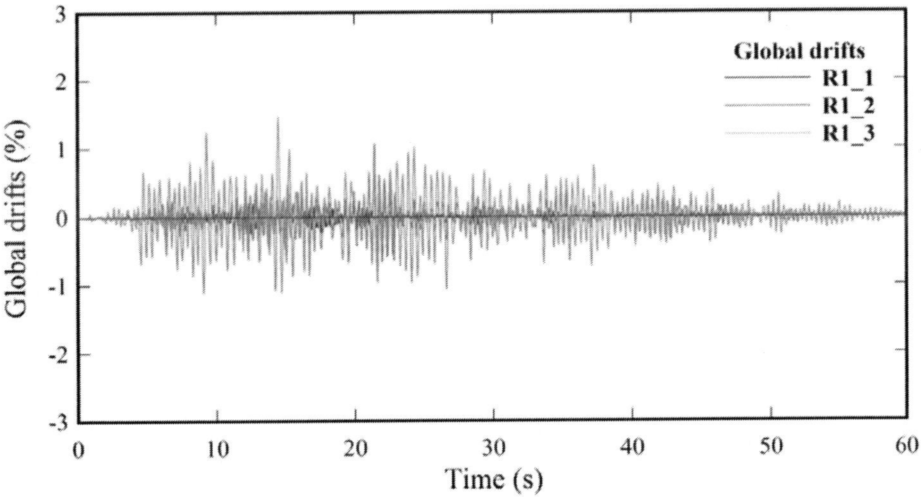

Figure 7. Global drifts. R1 earthquake. Frame C. DBDB building.

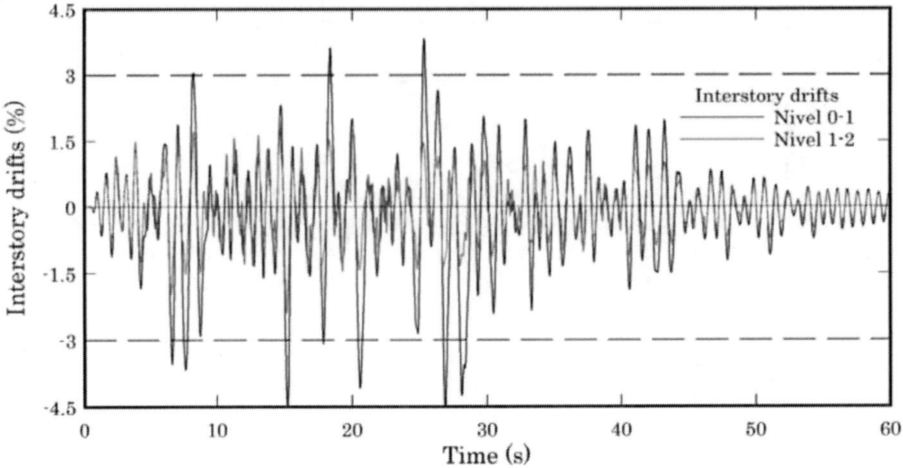

Figure 8. Interstory drifts. R1_3 earthquake. Frame C. OB building.

3.3. 3D Nonlinear Dynamic Analysis

Interstory drifts in frames of RB, were obtained by applying the R1_3 earthquake for the combinations 1 and 2

(**Table 7**). Figures 11 and 12 show the drifts of frame C. The remaining frames have shown a similar behavior.

According to results obtained, interstory drifts in 2D and 3D modeled buildings differ greatly from each other. In the 2D model greaters drifts were obtained. Figures 9 and 11 show interstory drifts of frame C of RB building for earthquake R1_3 and combination 1 (**Figure 9** is in 2D representation and **Figure 11** is in 3D representation). Also it was determined the maximum torsional moments in each column before the implementation of R1_3 earthquake in all supports for the four combinations described in **Table 7**. In **Figure 13** have been plotting torsional moments in function of time for the four combinations, where nodes appointed by n111 until the n513 are corresponding to supports, while **Figure 14**shows the maximum torsional moment for each column of the three-dimensional analysis. The torsional moments for the other seismic combinations used in this study were obtained using the same procedure.

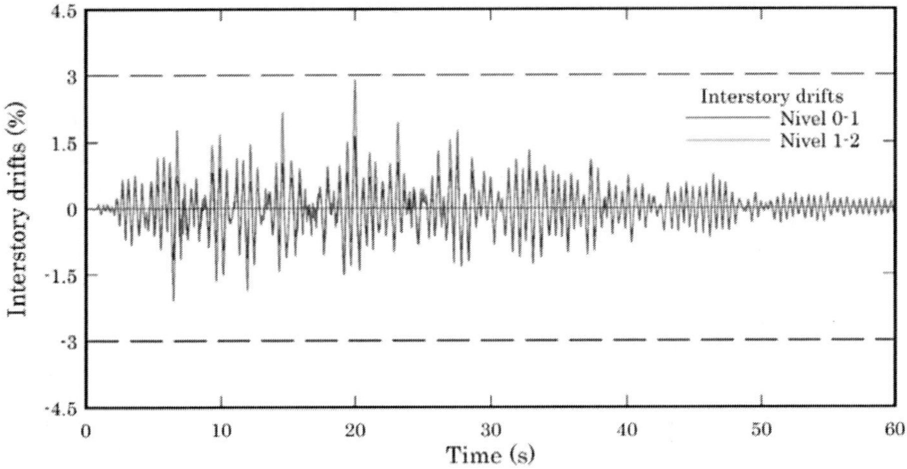

Figure 9. Interstory drifts. R1_3 earthquake. Frame C. RB building.

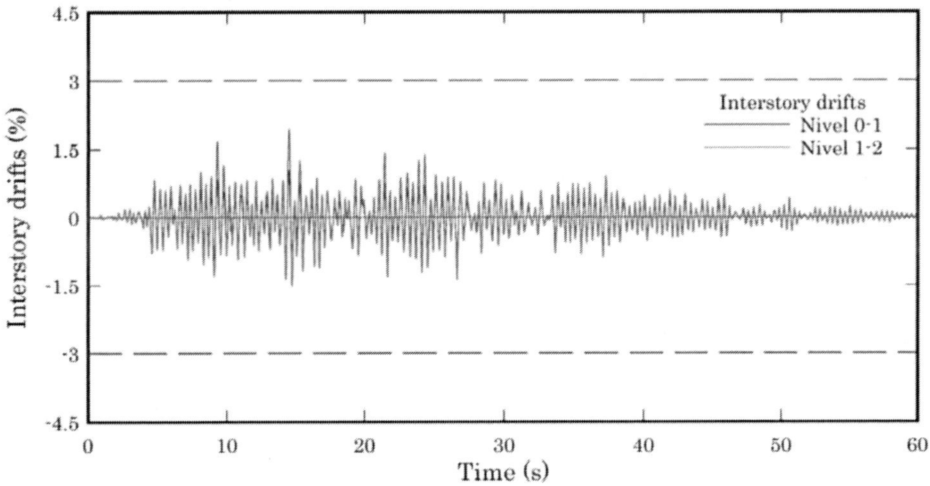

Figure 10. Interstory drifts. R1_3 earthquake. Frame C. DBDB building.

Table 6. Interstory drifts verification. R1 earthquake. OB, RB and DBDB building.

FRAME	LIMITS STATES								
	OB			*RB*			*DBDB*		
	SLS	RDLS	PCLS	SLS	RDLS	PCLS	SLS	RDLS	PCLS
A	✗	✗	✗	✗	✓	✓	✓	✓	✓
B	✗	✗	✗	✓	✓	✓	✓	✓	✓
C	✗	✗	✗	✓	✓	✓	✓	✓	✓
D	✗	✗	✗	✗	✓	✓	✓	✓	✓
E	✗	✗	✗	✗	✓	✓	✓	✓	✓
1	✗	✗	✗	✗	✓	✓	✓	✓	✓
2	✗	✗	✗	✗	✓	✓	✓	✓	✓
3	✗	✗	✗	✗	✓	✓	✓	✓	✓

SLS: Serviceability Limit State; **RDLS:** Reparable damage Limit State; **PCLS:** Prevention of Collapse Limit State; ✗ : No meet the norm, ✓ : Checks the norm.

Table 7. Maximum torsional moments for seismic combinations.

SEISMIC COMBINATION	NOD-COLUMN DESCRIPTION	MAX. TORSIONAL MOMENT (N.M)
1	Corner column. n513	64225
2	Corner column. n512	76000
3	Corner column. n513	41000
4	Corner column. n512	65000

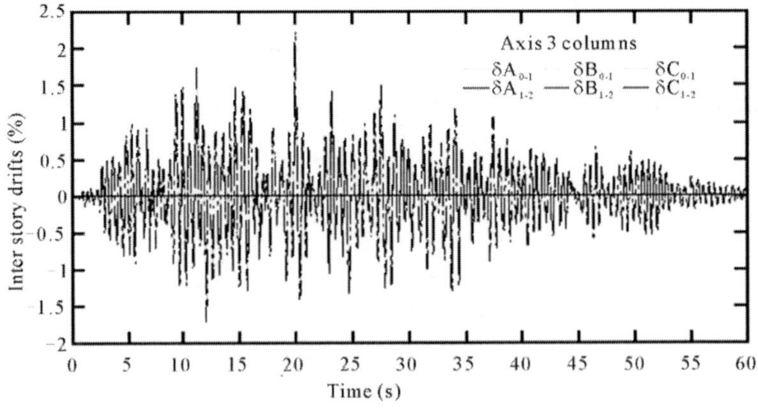

Figure 11. Interstory drifts. Frame C of RB building. Earthquake 100% (X).

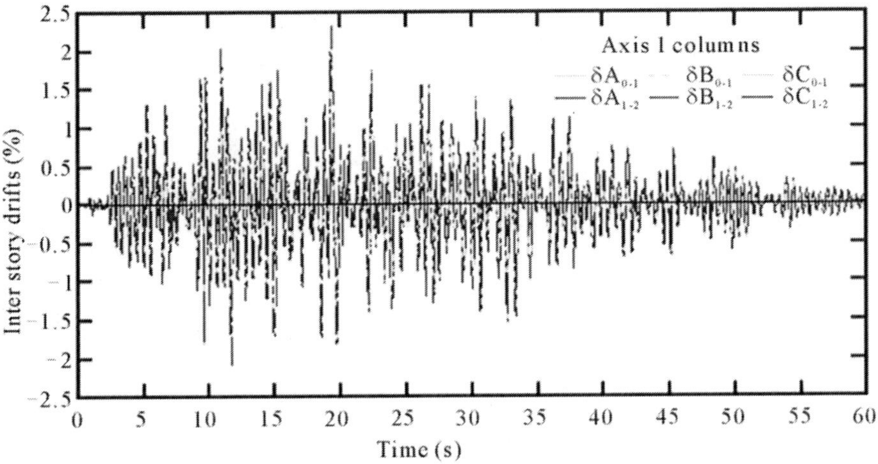

Figure 12. Interstory drifts. Frame C of RB building. Earthquake 100% (Z).

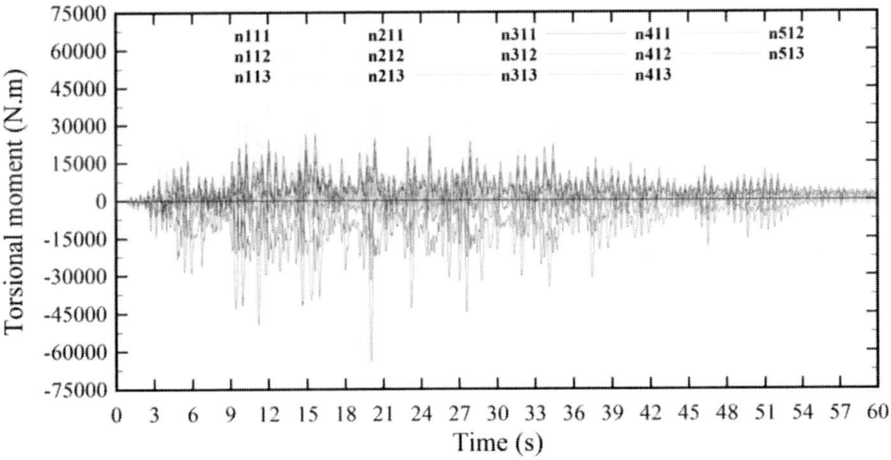

Figure 13. Torsional moments for earthquake 100% (X).

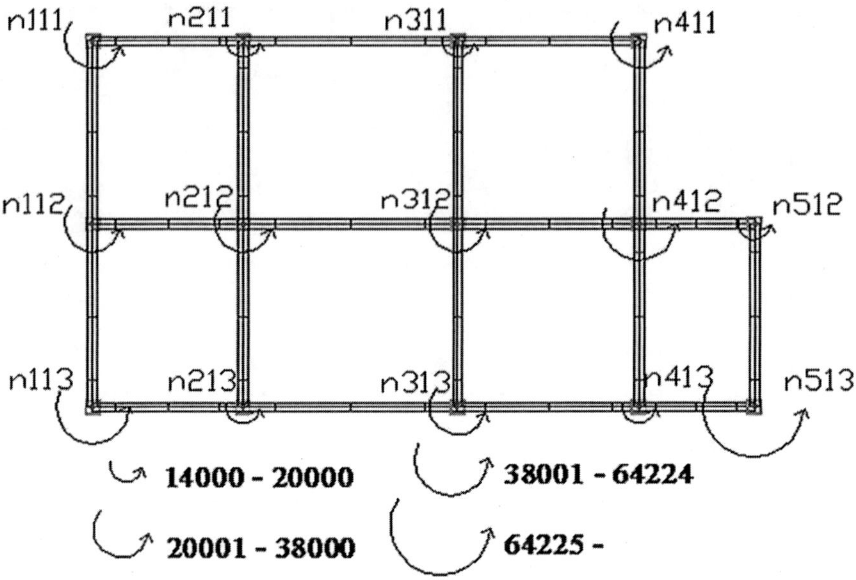

Figure 14. Torsional moments for earthquake 100% (X). Plant detail.

4. Conclusions

In order to determine or estimate the seismic response of the considered building, analytical methods were considered seismic hazard level and structural regularity basis. The elastic analysis applied to the OB building identified elastic displacements above the maximum value of inter story drift established in Venezuelan seismic code [7]. From the resizing model RB the inter story is seen to satisfy the maximum value established in the code. Thus, the cross-sections of the structural elements of OB are insufficient to properly control the damage caused by seismic forces.

From dynamic analysis obtained global and interstory drifts for all three evaluated models determining the dynamic response of these structures and control-ling the damage level reached in them. With the global drifts the threshold of the collapse Limit State, which corresponds to the maximum value of 2.5%. RB and DBDB buildings reached drifts values below this limit, proving good seismic performance on both buildings; OB presented drifts values which exceeded this limit. In the verification of interstory drifts it was generally noted that interstory drifts of OB building were longer than the considered by hazard levels, while the two resized buildings reached values within the thresholds established for each Limit State.

Three-dimensional dynamic analysis applied to RB building allowed determine that interstory drifts values were under the threshold of the

Limit States considered. On the other hand, in order to know the maximum torsional moments for each column in this model, there were applied four seismic combinations where it was noted that there was greater torsion in the case of the component of the earthquake in Z-direction. Based on these results it was demonstrated the structural asymmetry of the assessed building since the center of mass does not coincide with the center of rigidity, determining that the greatest torsional moments are on outer columns and inner corners.

Interstory drifts of RB building obtained from 2D and 3D nonlinear dynamic analysis, it was noted that 2D model provided greater drifts values than the 3D model drifts. This is a logical and expected result since the 3D dynamic analysis considers the rigid diaphragm, which introduces restrictions to the number of degrees of freedom in the structure.

Inelastic static analysis is more reliable than linear methods in the prediction of the parameters of response of buildings, although this method has no response on the effects of higher modes of vibration. A more reliable and sophisticated method is the 2D no linear dynamic analysis, where it can be better determined the likely behavior of the building in response to the earthquake. However, the uncertainties associated with the definition of accelerograms used in these analysis and properties of coplanar structural models can be reduced with the implementation of the dynamic 3D analysis because there are considered factors associated with structural redundancy and are used more actual values in terms of rigidity of resistant structural lines.

References

1. Grases, J., Altez, R. and Lugo, M. (1999) Destructives earthquakes catalogue. Natural Sciences, Physics and Mathematics Academy, Caracas.
2. Pérez, O. and Mendoza, J. (1998) Seismicity and tectonics in Venezuela and surroundings areas. Earth Physics, 10, 87-110.
3. Barbat, A., Mena, U. and Yépez, F. (1998) Probabilistic evaluation of seismic risk in urban zones. International Magazine for Numerical Methods for Calculus and Engineering Projects, 14, 247-268.
4. Calvi, G., Pinho, R., Magenes, G., Bommer, J., Restrepo, L. and Crowley, H. (2006) Development of seismic Vulnerability Assessment Methodologies over the past 30 years. ISET Journal of Earthquake Technology, 43, 75-104
5. Fajfar, P. (2000) Nonlinear analysis method for performance based seismic design. Earthquake Spectra, 16, 573- 591.

6. Vielma, J.C., Barbat, A.H. and Oller, S. (2011) Framed structures earthquake resistant project. International Center for Numerical Methods in Engineering Monograph, Barcelona.
7. COVENIN (2001) Earthquake resistant design code 1756:01. Ministerio de Fomento, Caracas.
8. COVENIN (2006) Design and construction of buildings with structural concrete 1753:06. Ministerio de Fomento, Caracas.
9. COVENIN (1988) Minimum actions and criteria for buildings projects code 2002:88. Ministerio de Fomento, Caracas.
10. Priestley, M., Calvi, G. and Kowalski, M. (2007) Displacement-based seismic design of structures. IUSS Press, Pavia.
11. Vielma, J.C., Barbat, A.H. and Oller, S. (2009) Nonlinear structural analysis: Application to evaluating the seismic safety. Nova Science Publishers, New York.
12. Elnashai, A., Papanilolau, V. and Lee, D. (2011) ZEUS-NL: A system for inelastic analysis of structures. User Manual. Mid-America Earthquake Center Report No. MAE, Champagne.
13. Papanikolaou, V. and Elnashai, A. (2005). Evaluation of conventional and adaptive pushover analysis I: Methodology. Journal of Earthquake Engineering, 9, 923-941.
14. Mwafy, A. and Elnashai, A. (2001). Static pushover versus dynamic collapse analysis of RC buildings. Engineering Structures, 23, 407-424.
15. Vamvatsikos, D. and Cornell, C.A. (2002) Incremental dynamic analysis. Earthquake Engineering and Structural Dynamics, 31, 491-514.
16. Kappos, A. and Stefanidou, S. (2010). A deformationbased seismic design method for 3D R/C irregular buildings using inelastic dynamic analysis. Bulletin of Earthquake Engineering, 8, 875-895.
17. UCLA-CIMNE (2009). Compatible accelerograms with elastic design spectrums generation programm (PACED). International Center for Numerical Methods in Engineering, Venezuela.
18. Fajfar, P., Marusic, D. and Perus, I. (2005) Torsional effects in the pushover-based seismic analysis of buildings. Journal of Earthquake Engineering, 9, 831-854.

CHAPTER 8

Seismic Performance of Historical and Monumental Structures

Halil Sezen[1] and Adem Dogangun[2]

[1] Department of Civil, Environmental, and Geodetic Engineering, The Ohio State University, Columbus, Ohio, USA

[2] Department of Civil Engineering, Uludag University, Bursa, Turkey

1. Introduction

Historical, religious and monumental structures and their susceptibility to damage in recent earthqaukes in Turkey are presented and discussed in this chapter. Turkey has a very large number of historical structures and is located in one of the most seismically active regions of the world. Some of these historical and monumental masonry and reinforced concrete structures suffered substantial damage or collapsed during two major earthquakes in 1999. The Kocaeli (M_w7.4) and Düzce (M_w7.2) earthquakes occurred on August 17 and November 12, 1999, and ruptured approximately 110 km and 40 km of the 1550-km-long North Anatolian fault, respectively. This chapter describes briefly the construction materials and techniques for historical religious and monumental structures and state-of-practice in Turkey, and presents dynamic analyses of a masonry minaret example. The seismic performance of the mosques and minarets (tall slender towers) during the 1999 earthquakes is presented.

2. Seismic Design and Construction Practice in Turkey

Prior to the 1999 earthquakes, two codes governed the design and construction of reinforced concrete and masonry buildings in Turkey: Turkish Earthquake Code (1998) and the Turkish Building Code, TS-500 (1985). The earthquake code included procedures for calculating earthquake loads on buildings. The ductility requirements and details described in the earthquake code were rarely observed in religious or monumental structures inspected by the authors after the 1999 earthquakes. Details of the seismic design and building construction practice prior to the 1999 earthquakes are provided in Sezen et al. (2003).

The descriptions of ground motion characteristics, structural damage, and performance of structures during these earthquakes are provided in Sezen et al. (2003). Response spectra for selected acceleration histories for 5% damping are presented in Figure 1. The ground motions included in Figure 1 were recorded at: SKR station in Adapazari (Peak Ground Acceleration, PGA 0.41g, stiff soil); YPT station in Yarimca (PGA 0.23g, soft soil); and DZC station in Düzce (PGAs: EW-Aug.17 0.36g, NS-Aug.17 0.31g, EW-Nov.12 0.54g, and NS-Nov.12 0.35g, soft soil). Figure 1 also provides a comparison between the linear elastic acceleration response spectra calculated for rock and that calculated for soft soil sites using the provisions of the Uniform Building Code (UBC, 1997) and those of the Turkish seismic code (1998) with 5 percent damping for the highest seismicity in the United States and in Turkey, respectively. From Figure 1, it can be concluded that, for a structure with given periods of vibration, the difference between the base shear calculated using the 1998 Turkish seismic code and the base shear demand obtained from the recorded acceleration response spectra do not differ significantly. If the static lateral load distribution over the height is suggested by the code is assumed to be credible, structures designed and detailed according to the Turkish code should not have collapsed or suffered severe damage during the 1999 earthquakes.

Specifically, according to Figure 1, a structure with a fundamental period of 0.5 seconds would be subjected to seismic forces larger than those specified in the code. The period of older masonry structures with thick and shorter walls tends to be relatively small, probably 0.5 seconds or less. On the other hand, the minarets or very slender towers tend to be very flexible with relatively large fundamental periods. Figure 1 shows that the recorded spectral accelerations are significantly large at large periods. The authors' field observations after the 1999 earthquakes, especially the November 12 Duzce earthquake (DZC), showed more damage in slender structures like minarets. In addition, among other factors leading to inadequate performance, the extent of damage observed in most reinforced concrete structures including minarets after the 1999 earthquakes is probably related to poor engineering and lack of conformance to the relatively new Turkish seismic code.

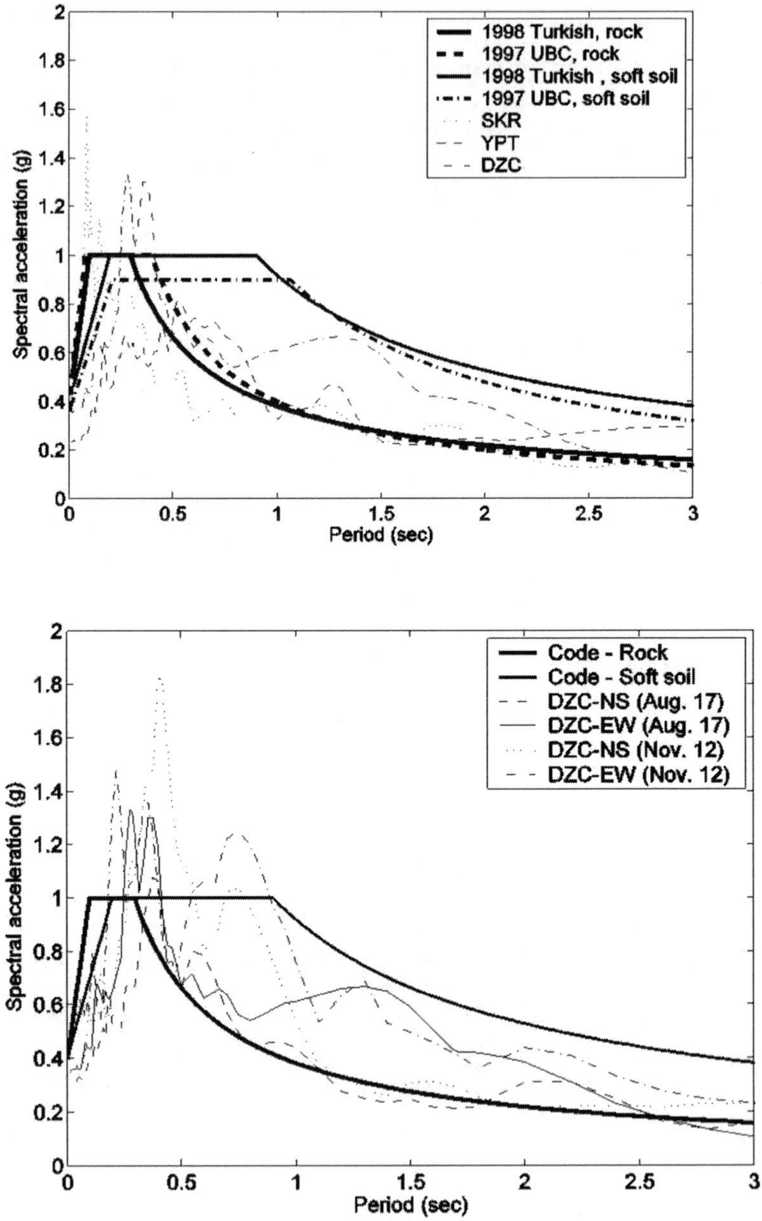

Figure 1. Absolute Acceleration Response Spectra

2.1. Construction Materials and Deterioration of Historical Structures

In the region affected by the earthquakes, in general, construction materials include stone or brick masonry and wood for older mosques and minarets, steel and reinforced concrete in more recently constructed structures. Typically locally available stone blocks were used. A special mortar called Horasan mortar was used in historical masonry structures. The Horasan mortar is made by mixing lime with baked clay powder that could be obtained by grinding clay tiles or bricks. The Horasan mortar was commonly used in the Middle East, where use of earthenware, pottery and brick construction were widespread (Çamlıbel 1988).

Hollow clay tiles or lightweight concrete blocks are used as infill wall material in recently built reinforced concrete (RC) moment frame mosques. Solid clay bricks were used in infill walls of RC mosques built before early 1970s. Almost all historical religious structures in Turkey were constructed using cut stone, masonry blocks or combination of these two materials. Timber construction is rare and reinforced concrete is relatively recent. The structural and geometrical properties of each masonry structure depend on many factors including the structural knowledge and applications at the time of construction, experience of the architect or engineer, seismicity of the region, and availability of construction materials in that area. Recent earthquakes in Turkey have shown that most masonry monumental structures as well as buildings (Ural et al. 2012) in high seismic regions are vulnerable to structural damage and collapse.

Historical structures deteriorate over time mainly due to environmental effects, and therefore may experience failure or collapse under gravity loads or seismic loads lower than those predicted by the design codes. As discussed in Dogangun and Sezen (2012) and Sezen and Dogangun (2009), in many cases, the following factors trigger or exacerbate the damage: (1) Surface or rain water runoff. If the structure and drainage system are not maintained properly, grass or fungus may grow and weaken the structural materials. Water accumulated on or penetrated into structural members may cause cracks due to freezing and thawing. (2) Soil settlement and relative movement of foundation. (3) Insufficient material strength. Layers of clay or other impure materials inside stone blocks may eventually lead to wearing, spalling or cracking. In stone masonry structures, the properties of the mortar significantly influence the strength of the entire structural component such as a load bearing wall. Deterioration of mortar binding the stone blocks, especially poor quality mortar including mud or low quality lime, can reduce the strength and stiffness of the wall considerably. (4) Other problems. Historical structures can be subjected to various environmental and loading conditions depending on their use and geographic location. For example, timber is more susceptible to humidity and temperature variations. Loading from continuous traffic and heavy trucks can lead to vibrations and excessive loads on foundations because the streets and other structures in historical cities are not designed for modern day traffic. Similarly, the use or occupancy of the structure may change and create larger unexpected loads. Parts of older structures are sometimes used as storage, in which the magnitudes of loads are usually much higher. Other local and environmental effects, such as acid rains, may adversely affect construction materials.

Most of the factors presented here result in gradual deterioration of materials or the load carrying structural system, which can be prevented as the damage progresses and becomes visible in many cases. On the other hand, structural damage, failure or total structural collapse occurs suddenly during moderate or strong earthquakes. Thus, it is essential to evaluate the capacity of existing historical structures and to retrofit them before an expected earthquake strikes.

2.2. Construction Practice and Techniques

Overall plan dimension of a typical mosque generally varies between 12 and 25 meters. Depending on the plan dimensions, height ranges from 7 m to 15 m not including the dome. Height of the dome, a half sphere, is typically half of the plan dimension (Figure 2). Lateral seismic loads are typically resisted by relatively thick unreinforced stone masonry walls in historical mosques. Addition to load bearing masonry walls, most mosques include a few columns typically carrying the gravity loads.

Minarets can be separate or contiguous and integral with the mosque structure, and are typically built using stone, brick, wood or reinfroced concrete. They typically include cylindrical or polygonal body/shafts, one or two balconies, and a conical roof or spire (Figure 2). In masonry minarets or slender tower structures, rather small tensile strength of mortar placed between the masonry blocks presents a major problem in regions of high seismicity. The brick or stone blocks have fairly large compressive strength, however unreinforced masonry lacks tensile strength required to resist bending moments imposed by the lateral earthquake loads. Older masonry minarets were typically constructed using stone blocks or solid clay bricks or a combination of two, whereas unreinforced lightweight stone blocks are preferred in new construction.

Figure 2. Typical mosque and its minarets in Turkey

After a major earthquake in 1509, Ottoman architects tackled the problem of constructing tall earthquake resistant minarets (Oğuzmert, 2002). They started to use a special technique for linking adjacent stone blocks with iron bars and clamps in the vertical and horizontal directions as shown inFigure 3 (Doğangün et al. 2007). Use of iron clamps in the two perpendicular directions (transverse and vertical) has improved the lateral load carrying capacity of slender masonry minarets significantly under earthquake loads. The clamps and vertical bars were placed inside anchorage holes in the stone blocks, and melted lead was poured inside the hole to provide bond between the stone and iron clamp or vertical bars. Depending on the properties and dimensions of the stone units, different clamps were developed. For example, as shown in Figure 4a, curved clamps were used within the circular stone wall on the minaret perimeter. Sharper and thinner clamps in thin stone blocks (Figure 4b) and shorter clamps were used if low tensile stresses are expected (Figure 4c). Approximately 2000 kilograms of this heavy metal, lead was used for the construction of a typical masonry minaret. Lead performs as intended for a very long time because it does not corrode or is hardly ever influenced by the adverse environmental conditions.

3. Post-Earthquake Surveys

The data presented here are based on observations from two surveys conducted after the August 17 and November 12, 1999 earthquakes. Damage to historical and recently constructed mosques and minarets was documented to investigate the seismic performance of these structures. The main objectives of these surveys were to provide detailed information about the characteristics of the observed damage, and to study the relative vulnerability of these historical and modern structures to strong ground motions. The parameters considered in the survey included: type of construction material, minaret height, location and description of damage, and location (coordinates, if available). Vast majority of the surveyed minarets and mosques were located in the cities of Düzce and Bolu. The peak ground accelerations recorded in Düzce during both earthquakes were larger than 0.30g (Figure 1), whereas that recorded in Bolu was reported as 0.82g after the November 12, 1999 earthquake.

Structural performance levels and corresponding representative sample damage descriptions for mosques and minarets are shown in Table 1. It should be noted that non-structural damage is irrelevant for minarets. In general, damage to the non-structural components of the mosques was insignificant as compared to the total structural damage.

A total of 59 sites were visited after the October 12, 1999 Duzce earthquake (second earthquake). The name, location, construction date, and observed damage levels are provided in Tables 2 and 3. The first 22 mosques listed in Table 2 were located in the city of Duzce and neighboring town of Kaynasli. Mosques numbered 23 through 44 were located in the city of Bolu.

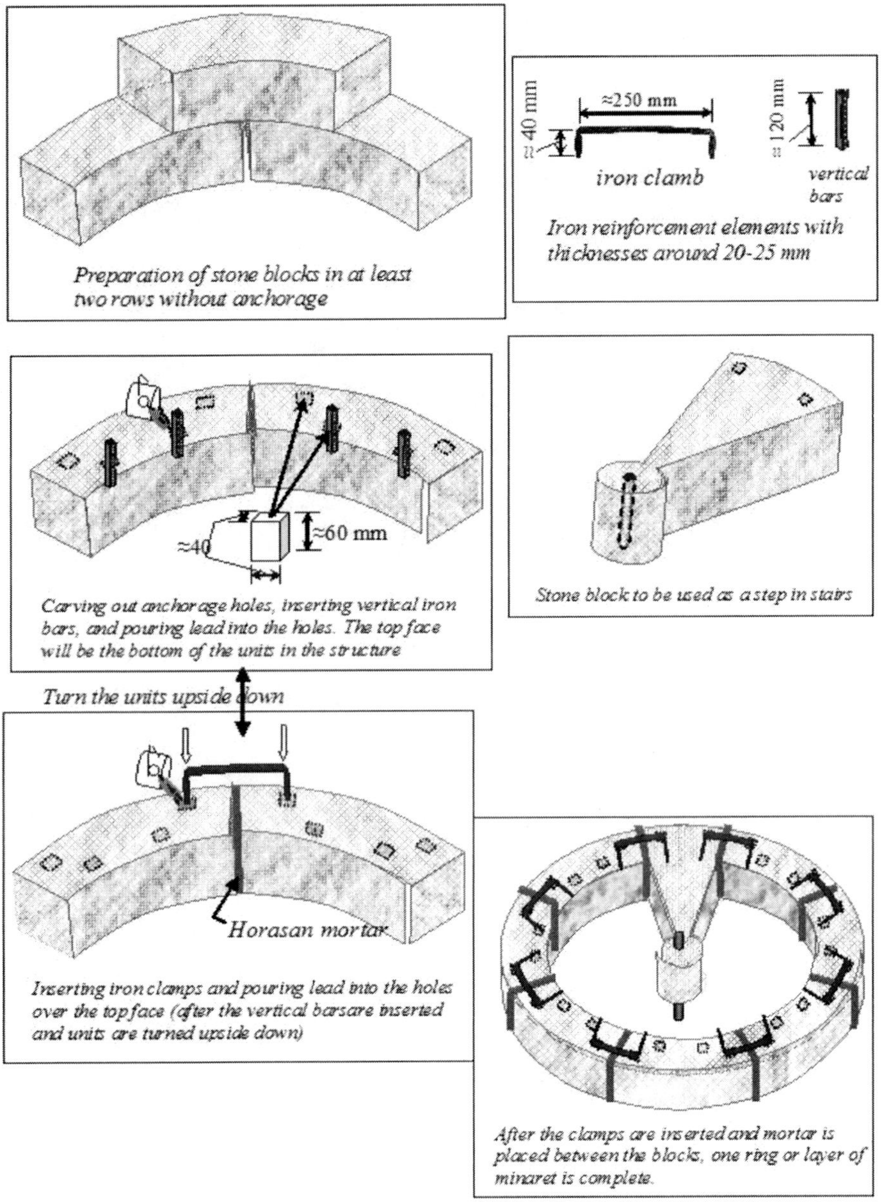

Figure 3. Construction of traditional Turkish minarets using stone blocks reinforced and anchored with iron bars and clamps (Dogangun et al. 2007)

(a)

(b)

(c)

Figure 4. Variations in iron clamps used in historical stone masonry walls

Table 1. Structural damage description and classification for mosques and minarets

Performance level	Damage classification	Sample damage description (mosque)	Sample damage description (minaret)
I	None	Negligible	Negligible
II	Light	Minor cracks in primary structural components	Minor cracks in masonry or RC minarets
III	Moderate	Significant cracks in RC members or masonry walls	Significant cracks especially around the minaret base
IV	Major	Hinge formation and wide cracks in primary RC members Infill wall collapse	Permanent visible drift Wide cracks and concrete spalling
V	Collapse	Partial or total collapse	Collapse

Table 2. Damage to the minarets and mosques surveyed in Duzce and Bolu

Name	Location (coordinates)		Constructi on date	Damage level	
	North	East		Mosque	Minaret
1.Sirali Koyu	40°49.246’	31°11.544’	1987	Light	Collapse
2.Kocyazi Koyu	40°50.569’	31°10.249’	1977-79	Light	Collapse
3.Karaca	40°50.733’	31°09.950’	1970s	Collapse	Collapse
4.Hamidiye	40°50.838’	31°09.518’	1980s	Light	Light
5.Rumelipalas	40°50.877’	31°08.444’	1972	None	None
6.Uzun Mustafa	40°50.755’	31°08.736’	1989	None	Collapse

7.Kultur Mahallesi	40°50.623’	31°09.257’	-	None	Moderate
8.Otopark	40°50.446’	31°09.119’	1971-73	None	Moderate
9.Aydinpinar	40°49.796’	31°09.190’	1994	Light	Collapse
10.Yesil	40°49.966’	31°09.480’	1990	None	Collapse
11.Asar	40°50.012’	31°09.650’	1977	None	Moderate
12.AzmimilliYeni	40°49.897’	31°10.087’	1988	None	Collapse
13.Mimar Sinan Nur	40°50.006’	31°10.236’	1988-91	None	Collapse
14.Maresal F.Cakmak	40°49.928’	31°10.525’	1990	Light	Collapse
15.Huzur	40°49.680’	31°11.324’	1986	None	None
16.Topalakli	40°49.606’	31°11.539’	1951	Light	None
17.Kirazli Koyu	40°48.918’	31°12.844’	1965	None	None
18.Doganli Koyu	40°48.182’	31°14.190’	1981-94	None	None
19.Uckopru Merkez	40°47.662’	31°15.018’	1985	None	Light
20.Yesiltepe	40°46.690’	31°17.467’	1986-90	Collapse	Collapse
21.Karacaali	40°46.496’	31°18.204’	1992-94	None	Moderate
22.Kaynasli merkez	40°46.406’	31°19.138’	-	Collapse	Collapse
23.Sanayi	40°44.280’	31°37.562’	1988	None	None
24.Kultur	40°44.507’	31°36.340’	1990	None	Light
25.Oksuztekke	40°44.488’	31°35.851’	1993	Major	Collapse

26.Ozayan	40°44.638'	31°35.405'	1996	None	Moderate
27.Beskonaklar Yeni	40°44.263'	31°35.599'	1997	Moderate	-
28.Pasakoy Berberler	40°43.942'	31°34.096'	1987	None	None
29.Pasakoy Eniste	40°43.803'	31°34.689'	1997	None	Light
30.Sumer	40°43.575'	31°35.587'	1992-95	None	Light
31.Sumer Mah. Yeni	40°43.468'	31°35.297'	1983	None	None
32.Karacayir Siteler	40°43.577'	31°36.033'	1998-99	None	-
33.Semsi AhmetPasa	40°43.852'	31°36.635'	14th cent	Major	Collapse
34.Sarachane	40°43.935'	31°36.513'	17th cent	Light	None
35.Yildirim Bayezid	40°44.040'	31°36.576'	1804	Major	None
36.Kadi	40°43.901'	31°36.459'	1499	Major	Collapse
37.Aslahaddin	40°43.981'	31°36.732'	1978	None	None
38.Balci	40°43.885'	31°37.111'	1990	None	None
39.Aktas	40°43.783'	31°36.650'	1900s	Major	None
40.Karacayir	40°43.825'	31°36.515'	1946	None	None
41.Kabaklar	40°44.859'	31°36.111'	1981	None	None
42.Camli	40°44.243'	31°35.841'	1980s	Major	None
43.Sultanzade	40°44.249'	31°35.817'	1930s	Major	Major
44.Yesil	40°44.170'	31°36.170'	1966	Major	None

TABLE 3. Damage to the other mosques and minarets visited (coordinates not available)

No.	Name	Location	Mosque	Minaret
45	Cumhuriyet Mah.	Duzce, downtown	None	Collapse
46	Merkez	Duzce, downtown	Major	Collapse
47	Cedidiye Merkez	Duzce, downtown	Light	Collapse
48	Yuvacik	Yuvacik village, near Golcuk	Major	-
49	Asagi Yuvacik	Yuvacik village, near Golcuk	None	None
50	Yeni	Adapazari	None	Collapse
51	Yalova	Yalova, downtown	Light	None
52	Izmit (1)	Izmit, next to highway E5	Light	Collapse
53	Izmit (2)	Izmit, next to highway TEM	None	Light
54	Izmit (3)	Izmit, downtown	Major	Light
55	Golyaka	Golyaka, downtown	Major	Collapse
56	Suleymanbey	5 km east of Golyaka	Collapse	-
57	Golcuk (1)	4 km west of downtown	Major	-
58	Golcuk (2)	Near Ford plant, west of city	Collapse	Moderate
59	Dariyeri Hasanbeyi	East of Duzce (a village)	Major	Light

Before mid-1960s, major construction materials were wood and stone or brick masonry. Most of the recently constructed mosques and minarets were reinforced concrete. All of the reinforced concrete mosques were built after 1965. In older mosques, solid bricks were used in infill walls. Hollow clay tiles were used as infill material in the mosques built after late 1970s. As shown in Figure 5, 84 percent of the minarets surveyed in Duzce and Kaynasli were reinforced concrete, whereas only 46 percent of the minarets were reinforced concrete in Bolu. Unreinforced stone masonry was commonly used in old minarets as well as in the minarets constructed in recent years.

4. Observed Earthquake Damage

The Figure 6 shows the damage distribution for the mosques and minarets surveyed. Damage distribution for RC mosques and minarets are presented in Figures 6c and 6d separately.

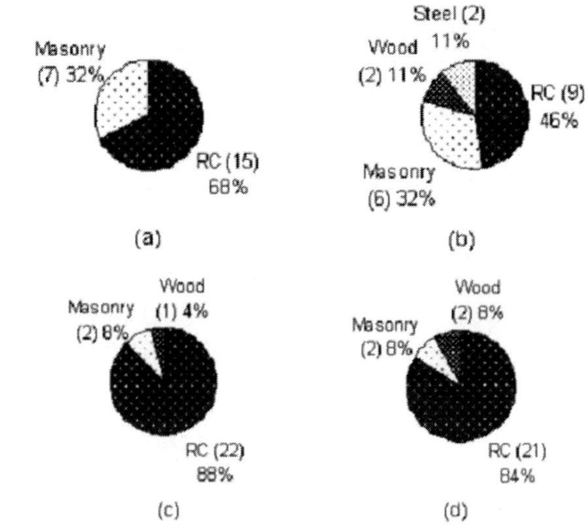

Figure 5. Type of construction for (a) mosques, (b) minarets in Bolu; and (c) mosques, and (d) minarets in Duzce and Kaynasli (number of mosques/minarets is given in parenthesis)

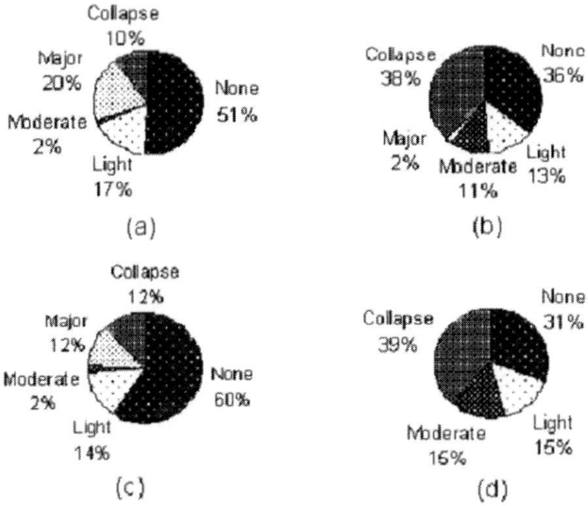

Figure 6. Damage distribution for (a) all mosques and (b) all minarets, and damage distribution for (c) RC mosques and (d) RC minarets

Comparison of Figures 6a and 6c indicates that the damage in RC mosques was less as compared to other structural systems. More than ten percent of the mosques surveyed collapsed. Three of the 26 mosques surveyed in Duzce and Kaynasli region collapsed. There was no collapsed mosque in Bolu, but five of the 21 mosques surveyed had to be closed after the October 12, 1999 earthquake. Closed Mosques in Bolu were Semsi Ahmet Pasa (Imaret), Yildirim Bayezid, Kadi, Camli and Yesil. Note that the first three of these mosques are at least 200 years old.

As illustrated in Figures 6b and 6d, percentage of all damaged minarets and only RC minarets are similar. Almost forty percent of the minarets collapsed, and approximately one third of the minarets were undamaged. Failure plane for almost all collapsed RC minarets was within 1.5 meter long region above the minaret base or pyramid-shaped transition segment (Figure 7a), where the longitudinal reinforcing bars were usually spliced. Horizontal circumferential cracks and spalling of concrete were commonly observed at the bottom of the cylindrical body of RC minarets (Figure 7a). Frequently, flexural cracks, concrete crushing or spalling was observed in this region of the damaged RC minarets. As shown in Figure 7, less frequently collapse or damage occurred within the transition segment and middle of the cylindrical body or near the top of the minaret.

(a) (b) (c)

Figure 7. Collapse of a minaret near the bottom of cylinder (left), minaret damage within transition segment (middle), and collapsed minaret at mid-height of cylinder (right)

The ratio of collapsed or damaged unreinforced solid brick or stone masonry minarets was much larger than that of RC minarets. As listed in Table 4, majority of the visited masonry minarets collapsed (Sezen et al. 2003, and Firat 1999). Note that most of these minarets were either very new or few hundred years old (Tables 2 and 3). A minaret may either have an independent foundation (referred to as Type I minaret hereafter) or the base of the minaret may be attached to the roof of the mosque (referred to as Type II minaret). The minarets in Figures 2 and 7 are Type I and II minarets, respectively. In Type II minarets, the minaret and mosque structure respond independently. Large deformatiosn or failure in one structure does not affect the other one.

Table 4. Masonry minarets surveyed after the 1999 earthquakes

Name	City	Location or coordinates	Type	Observed damage
Aziziye Merkez	Düzce	40.50N-31.08E	II	Failed at the bottom of cylinders and collapsed
Cedidiye Merkez	Düzce	40.50N-31.09E	I	Failed at the bottom of cylinders and collapsed
Oksuztekke	Bolu	40.44.488N-31.35.851E	I	Minaret segment above the 2nd balcony level collapsed
Semsi Ahmet Pasa (Imaret)	Bolu	40.43.852N-31.36.635E	I	Failed near the bottom of the cylinder and collapsed
Sarachane	Bolu	40.43.935N-31.36.513E	I	Cracks in stone blocks in the mosque – no observed minaret damage
Yildirim Bayezid	Bolu	40.44.040N-31.36.576E	II	Dislocation of stone blocks in the mosque – no obserevd damage in minarets
Kadi	Bolu	40.43.901N-31.36.495E	I	Severe damage to mosque, minaret collapsed

Lateral laod resisting mechanism of minarets are quite different from that of other structures. The height of center of minaret's mass can be very high above ground, resulting in large bending moments and shear forces. Masonry minarets without reinforcement or clamps (Figures 3 and 4) and with weak mortat are the most vulnerable against earthquakes. Masonry minarets typically either collapsed or suffered minor or no damage. This suggests that these minarets had no ductility or very limited deformation capacity. If the demand due to lateral earthquake forces is less than minaret's ultimate strength, minaret behaves elastically and suffers no damage. However, as soon as lateral demand exceeds the elastic capacity, the minaret collapses. Unlike RC minarets, masonry minarets failed at different locations along the height of the minaret inconsistently (Figure 8). The Imaret and Kadi mosques and possibly their minarets were 600 and

500 year old, respectively. As shown in Figures 8aand 8b, their minarets collapsed at the bottom of the cylindrical body. On the other hand, the Oksuztekke minaret collapsed at its mid-height (Figure 8c). Oksuztekke mosque was constructed only six years before the 1999 earthquakes (Table 2).

4.1. Earthquake Damage in Historical Mosques

Among the historical mosques surveyed, Düzce Merkez, 600-year-old Imaret, 500-year-old Kadi, 300-year-old Sarachane, and 200-year-old Yildirim Bayezid mosques observed to suffer significant structural damage after the 1999 earthquakes (Dognagun and Sezen, 2012). The Yıldırım Bayezid mosque was originally built in 1382 and was burned down in the 19th century. A new structure was constructed after the fire, and it was severely damaged during a 7.3 magnitude earthquake in 1944. The mosque was damaged during the 1999 earthquakes and was closed for a period of time. On the south side of the mosque, portion of the walls above and below the windows were subjected to larger shear stresses (compared to solid wall sections) during the strong ground shaking. Higher shear demand in those parts of the relatively thick walls created serious cracks and openings between the stone blocks (Figure 9). As shown in Figure 9 (middle photo), portion of the walls above and below the windows act, in a sense, like short columns during the earthquake. Larger shear demand in those parts of the relatively thick walls created wide cracks between the stones.

(a) (b) (c)

Figure 8. Failures near the bottom and mid-height of the masonry cylindrical minaret body: (a) Imaret, (b) Kadi, and c) Oksuztekke mosques in Bolu

CITRON

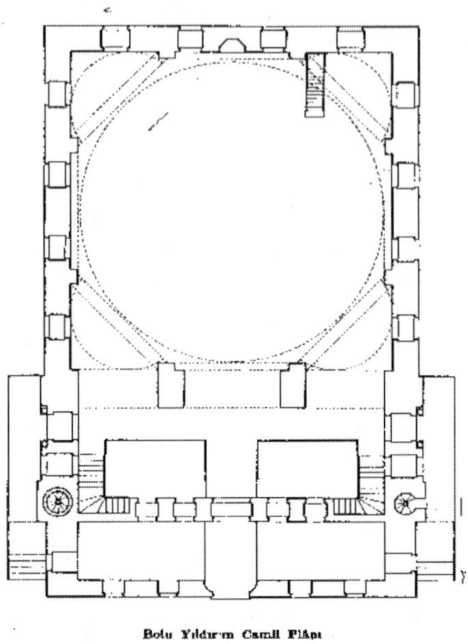

Figure 9. Wall damage and plan view of historical Yildirim Bayezid mosque in Bolu

Another historical mosque, Kadi (Figures 8b and 10a) sustained substantial damage including large cracks around historical entrance door, severe cracks and stone dislocations in the 1.5 m wide unreinforced stone masonry perimeter walls. Damage was mostly concentrated below or above the windows. The missing two key keystones and dislocated stones on top of the upper window are visible in Figure 10a. The reduced wall area along the vertical section through the windows was stressed more, causing considerable damage in those wall areas. The stone masonry minaret collapsed right above its base because the minaret base was integral with minaret walls (Figure 8b) and was quite stiff compared to the cylindrical minaret body.

The Imaret mosque, one of the oldest structures in the region, and its minaret were built using stones and small solid bricks bounded by a thick layer of mortar. The mortar between the bricks is typically as thick as the bricks. Old brick masonry minaret collapsed and Imaret mosque was closed after the 1999 earthquakes due to cracks in the walls (Sezen et al. 2003). The 1999 earthquakes caused very limited damage to the 300-year-old Sarachane mosque and there was no visible damage to its minaret. Some large cracks were observed in the walls at locations similar to those observed in other mosques discussed in this paper. Figure 10b and 10c show examples of couple such cracks; one immediately above a window and another in a corner near the roof.

(a) (b) (c)

Figure 10. a) cracks developed in the walls and above windows of Kadi; and b, c) damage in in the walls of Sarachane mosque

5. Dynamic Analysis of Representative Masonry Minarets

Although numerous collapses and structural damage to minarets and mosques were documented after strong earthquakes, only few researchers investigated the seismic behavior and performance of minarets (Sezen et al. 2008, Portioli et al. 2011, Turk and Cosgun 2012, Altunisik 2012, Oliveira et al. 2012). It is essential to understand the dynamic behavior of these structures to improve the life safety and to preserve and strengthen the historical monumental structures. In this research, as an example, dynamic modeling and seismic analysis of unreinforced masonry minarets is presented.

Recently, a study was conducted by the authors to investigate the seismic response of reinforced concrete (RC) and masonry minarets (Dogangun et al. 2008, and Sezen et al. 2008). Modal analysis and dynamic time history analyses were conducted. Strength and deformation capacities of the RC and unreinforced masonry models are influenced by unique differences in material properties, including weight, modulus elasticity, and Poisson's ratio. For example, the flexural strength of an unreinforced masonry model is significantly lower as the reinforcing steel in a similar RC model provides tensile resistance. While Sezen et al. 2008 investigates distinct parameters affecting the behavior of RC minarets using a single model, the example presented below examines the effect of masonry material properties and minaret height on the minaret response using three different models. Detailed modeling properties and analysis results are presented in Dogangun et al. (2008).

Although the height of a minaret varies greatly depending on many factors such as the location or construction materials used, the height of a typical minaret with a single balcony over its height is about 20 to 25 m. In this research, two generic minarets with a height of 25 and 30 m and double balconies (Minaret I and Minaret II) and another 20 m tall minaret with a single balcony (Minaret III) were modeled and analyzed. As shown in Figures 11 and 12, only the height of the cylindrical minaret body was varied while the geometrical properties of the base or boot, transition segment, balconies and conical cap were kept the same in all three models. It should be noted that not all conical caps at the top of the masonry minarets are constructed using masonry materials. For uniformity, all minarets including the conical caps are modeled with the same masonry material in this study.

The finite element models of the three minarets are shown in Figure 12, and significant mode shapes of a typical minaret are shown in Figure 13 (Sezen et al. 2008). The computer program, ANSYS was used to model and analyze the minarets (see Dogangun et al. 2008 for details). In the models, the modulus of elasticity of uncracked section, Poisson's ratio and unit weight of masonry material (ordinary limestone) are taken as 3000 MPa, 0.2, and 20 kN/m^3, respectively. In the models, linear elastic material models and five percent damping ratio is used in all dynamic time history analyses. It is assumed that the minaret is located in a high seismic region Zone 1 in Turkish Earthquake Code (TEC, 2007). Considering the TEC, a structural behavior factor, R of 3 and an importance factor, I of 1.2 can be used for such structures.

Dynamic time history analyses are carried out for each minaret model using two ground motions recorded during the 12 November, 1999 Duzce and 17 August, 1999 Kocaeli earthquakes (seeDogangun et al. 2008 for details). The input ground motion was applied only in one horizontal direction. The modal periods of the minaret models (calculated from the modal analysis) and their contribution to the total dynamic response are calculated. The modal periods are greatly affected by the height of the minaret which affects the total mass and flexural stiffness. The calculated first or fundamental periods were 1.41, 0.90, and 0.51 seconds for the minaret models I, II, and III, respectively. The calculated modal response quantities indicate that the contribution of higher mode effects to total dynamic response is significant. The first mode contribution to the total response is about 47 or 49 percent. The torsional or the fifth mode has virtually no effect on the total response of the almost symmetrical minaret structures.

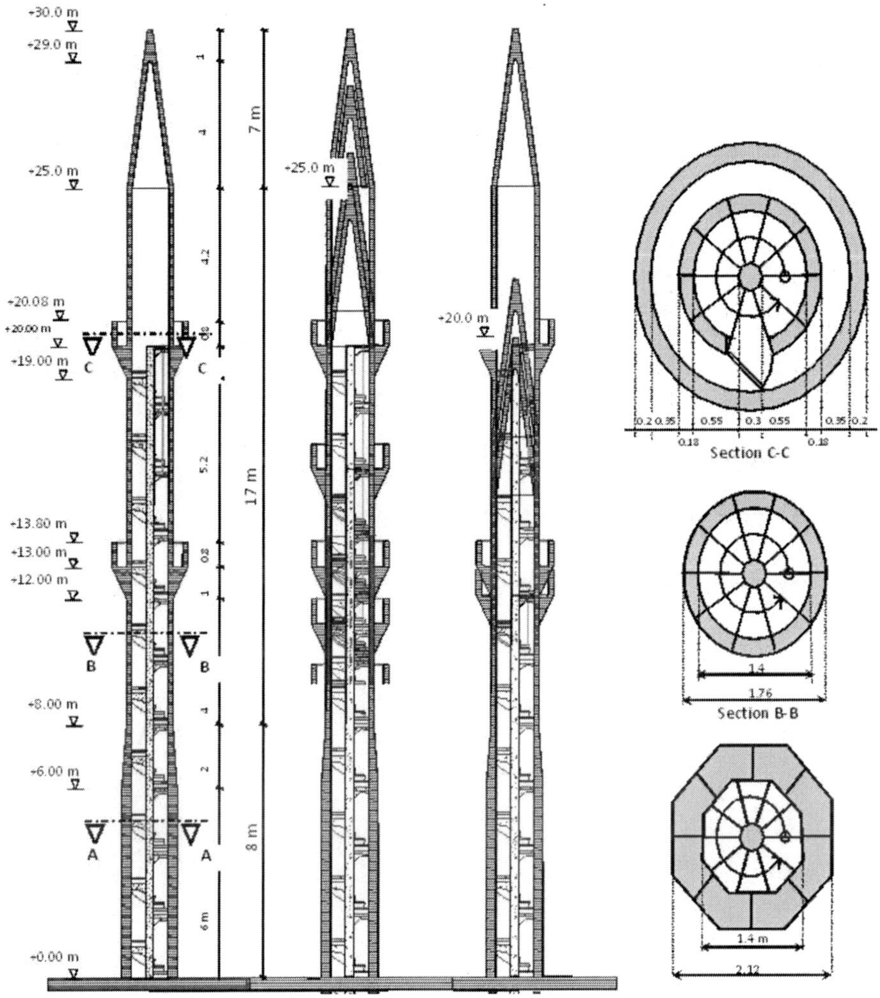

Figure 11. Three minaret models and their geometrical and cross-sectional properties

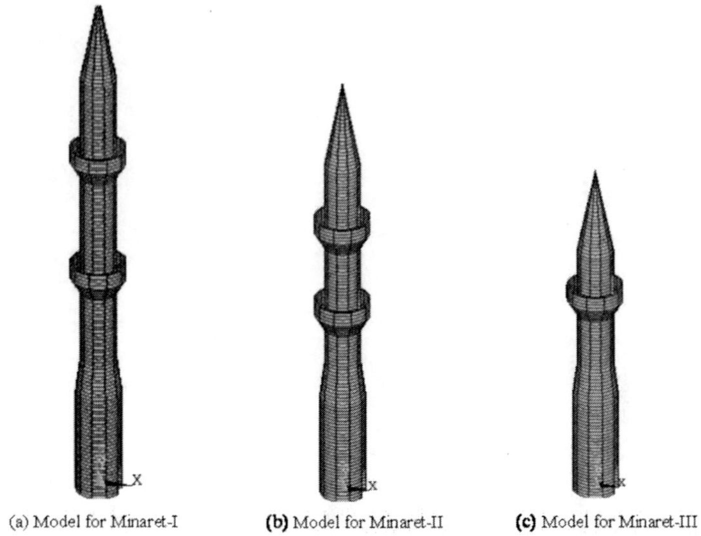

Figure 12. Finite element models of the three minarets

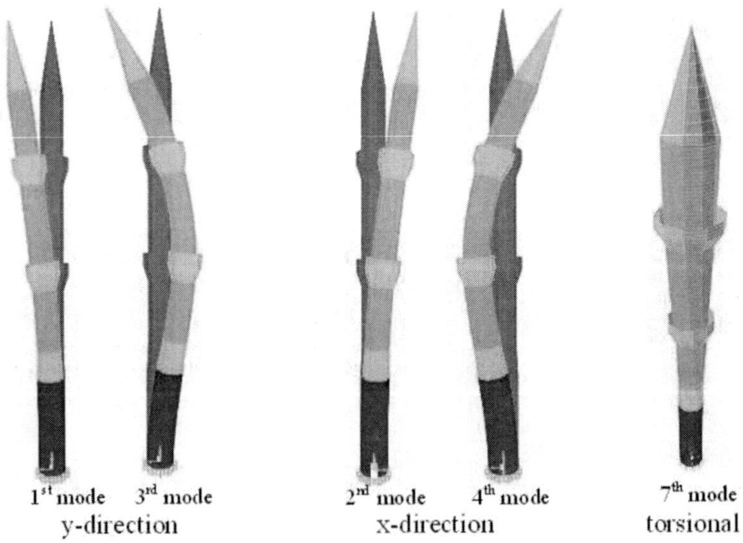

Figure 13. First two translational and the seventh mode shapes for a typical model

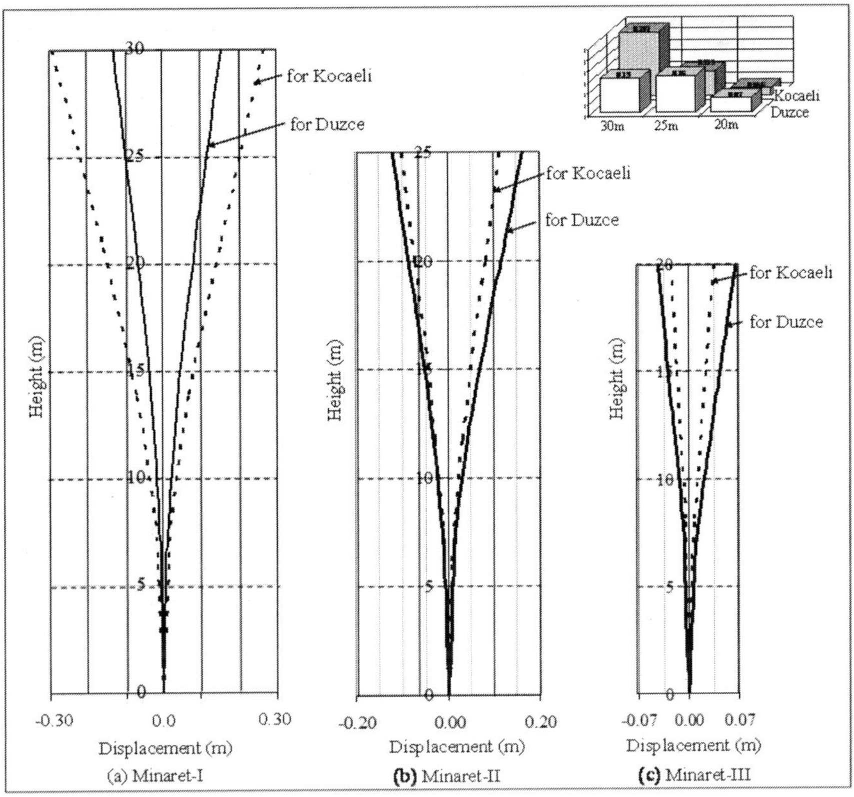

Figure 14. Lateral displacement distribution over the height of minarets for Duzce and Kocaeli ground motions

Figure 14 shows the calculated maximum elastic lateral displacement distribution along the length of three minaret models subjected to 1999 Duzce and Kocaeli earthquake ground motions. As shown inFigure 14 the deflected shapes of the minarets appear to show flexure dominated response with the largest displacements calculated at the top. Although the minarets act as cantilevers, the deformations are much smaller over the height of relatively stiff 6 m high base or boot. The displacements start to increase above the transition segment at about 8 m from ground level. The maximum dynamic displacement of Minaret I is 88% larger for the Kocaeli input motion (0.291 m) than that for the Duzce ground motion (0.154 m), although the maximum acceleration of the Duzce motion was larger. For the other two shorter minarets Minaret I and II, the maximum calculated displacements are larger for the Duzce ground motion. The maximum dynamic lateral displacement of 0.29 m calculated for the 30 m tall minaret under Kocaeli earthquake exceeds the TEC (2007) code limit of 0.20 m for the Minaret I model (Dogangun et al. 2008).

The most critical axial stress contours calculated during the dynamic analysis of minaret models are shown in Figure 15. The maximum tensile and compressive stresses occur at a height of 8 m, immediately above the transition zone (Figure 12). This is consistent with the most common minaret failure mode observed in the field after the earthquakes (Figures 7a, 8a and 8b). The minaret models used here were stand-alone fixed-ended cantilevers, which did have any lateral support over their height. Nevertheless, in some cases one side of the minaret boot or base (bottom 6 m of the minaret models) are supported by or attached to the mosque structure (Type II in Table 4, e.g., Figures 8a and8b). Such a partial lateral support, even on one side, provides added stiffness near the bottom of the minaret. This will further increase failure potential at the top of the transition zone as shown in Figures 7a, 8a, 8b and 14.

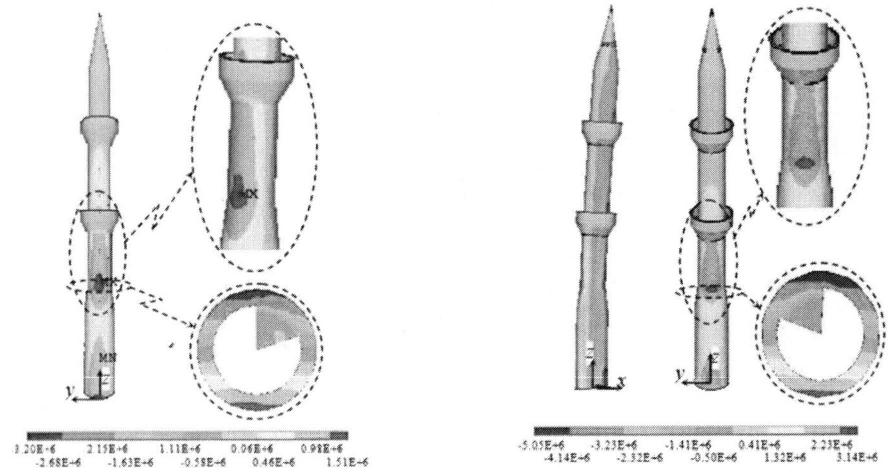

(a) (b)

Figure 15. Maximum stress contours for Minaret-I under: a) Duzce, and b) Kocaeli records

6. Summary and Conclusions

Structural damage and failures observed in 59 historical and monumental structures (mosques and/or minarets) were documented after the August 17 and November 12, 1999 earthquakes. The reported damage distribution in minarets in the cities of Düzce and Bolu gives an indication of the extent of damage. Ten percent of all visited mosques collapsed while 20% suffered severe damage. Of all visited minarets, 38 percent collapsed (Figure 6). Five out of seven masonry minarets (approximately 70%) in the cities of Düzce and Bolu reporeted in this chapter collapsed (Table 4).

Historical structures with heavy and stiff walls are subjected to larger lateral earthquake forces due to their small periods of vibration. The missing keystones in the upper-floor windows and wide cracks between the stones in the walls above and below window openings were the most common damage types.

The location of the failure in the minarets that collapsed during the 1999 earthquakes was generally at the region near the bottom of the cylinder where a transition was made from a square to a circular section. At the bottom of the cylinder body the lateral stiffness and strength are smaller compared with those of the transition region or the minaret base. Consistent with the observed damage, dynamic analysis results from three generic finite element minaret models showed the largest bending stresses in the same region of the minaret models. For minarets with the base or boot attached to the mosque structure (Type II), additional rigidity and stiffness provided by the mosque prevents deformation and damage within the base. In such cases the failure mostly occurs just above the transition region or base of the cylindrical body.

References

1. Turkish Earthquake Code.1998and 2007).Specification for structures to be built in disaster areas; Part III-earthquake disaster prevention. Ministry of Public Works and Settlement, Government of Republic of Turkey.
2. Turkish Building Code. TS-5001985Building Code Requirements for Reinforced Concrete. Turkish Standards Institute, Ankara, Turkey (in Turkish).
3. H. Sezen, A. S. Whittaker, K. J. Elwood, K. M. Mosalam, 2003Performance of Reinforced Concrete and Wall Buildings during the August 17, 1999 Kocaeli, Turkey Earthquake, and Seismic Design and Construction Practice in Turkey. Engineering Structures, 25103114
4. Uniform Building Code.1997Structural Engineering Design Provisions, 2International Conference of Building Officials, ICBO. Whittier, California.
5. N. Çamlıbel, 1998Sinan Mimarlığında Yapı Strüktürünün Analitik İncelenmesi, Doçentlik Tezi, İTÜ, Fen Bilimleri Enstitüsü, İstanbul (in Turkish).
6. A. Ural, A. Dogangun, H. Sezen, S. Angin, 2012Seismic Performance of Masonry Buildings during the 2007 Bala, Turkey Earthquakes. Natural Hazards, 60310131026
7. H. Sezen, A. Doğangün, 2009Vulnerability and conservation of historical masonry structures. Prohitech 2009, Protection of Historical Buildings, Rome, Italy, June 2124

8. M. Oğuzmert, 2002Dynamic behaviour of masonry minarets, M.Sc. Thesis, Istanbul Technical University, Turkey.
9. A. Dogangun, H. Sezen, O. I. Tuluk, R. Livaoglu, R. Acar, 2007Traditional Turkish Monumental Structures and their Earthquake Response. International Journal of Architectural Heritage, 13251271
10. H. Sezen, G. Y. Firat, M. A. Sozen, 2003Investigation of the Performance of Monumental Structures during the 1999 Kocaeli and Duzce Earthquakes. Fifth National Conference on Earthquake Engineering, Istanbul, Turkey, 2630May
11. Y. G. Firat, 2001A Study of the Structural Response of Minarets in the 1999 Anatolian Earthquakes. M.S. Thesis. Purdue University. West Lafayette, Indiana.
12. A. Dogangun, H. Sezen, 2012Seismic Vulnerability and Preservation of Historical Masonry Monumental Structures. Earthquakes and Structures, 318395
13. H. Sezen, R. Acar, A. Dogangun, R. Livaoglu, 2008Dynamic Analysis and Seismic Performance of Reinforced Concrete Minarets." Engineering Structures, 30822532264
14. F. Portioli, O. Mammana, R. Landolfo, F. M. Mazzolani, L. Krstevska, L. Tashkov, K. Gramatikov, 2011Seismic Retrofitting of Mustafa Pasha Mosque in Skopje: Finite Element Analysis. Journal of Earthquake Engineering, 154620639
15. A. M. Turk, C. Cosgun, 2012Seismic Behaviour and Retrofit of Historic Masonry Minaret. Gradevinar, 641
16. A. C. Altunisik, 2012Comparison of earthquake behavior of reinforced concrete minarets using fiber-reinforced polymer composite. The Structural Design of Tall and Special Buildings, DOI:tal.731
17. C. S. Oliveira, E. Çaktı, D. Stengel, M. Branco, 2012Minaret behavior under earthquake loading: The case of historical Istanbul. Earthquake Engineering and Structural Dynamics, 411939
18. A. Dogangun, R. Acar, H. Sezen, R. Livaoglu, 2008Investigation of Dynamic Response of Masonry Minaret Structures. Bulletin of Earthquake Engineering, 63505517

CHAPTER 9

On the Anomalies in ULF Magnetic Field Variations Prior to the 2008 Sichuan Earthquake

Qi Li[1], Alexander Schekotov[2], Tomokazu Asano[3], Masashi Hayakawa[3,4,5]

[1]Institute of Geophysics, China Earthquake Administration, Beijing, China

[2]Institute of Physics of the Earth, Russian Academy of Sciences, Moscow, Russia

[3]Earthquake Analysis Laboratory, Information Systems Inc., Minato-ku, Japan

[4]Hayakawa Institute of Seismo Electromagnetics, Co. Ltd., University of Electro-Communications (UEC), Chofu, Japan

[5]UEC, Advanced Wireless Communications Research Center, Chofu, Japan

Abstract

There have been reported several papers on the ionospheric F region perturbations prior to the 2008 Sichuan earthquake (EQ) (magnitude 8.0), but it seems that very few reports have been published on the characteristics of ground-based ULF (ultra low frequency) magnetic field variations for this EQ. This paper deals with two different aspects of ground-based ULF magnetic field variations: 1) ULF radiation from the lithosphere, and 2) depression of ULF horizontal magnetic field as a signature of lower ionospheric perturbations. ULF data from two Chinese stations [Chengdu (epicentral distance, 80 km) and Xichang (about 300 km away from the EQ epicenter)] are analyzed, with paying attention to the local nighttime period (LT = 22 h to 02 h, UTC = 14 h to 18 h) in order to avoid man-made noise. We have analyzed powers of the horizontal component (H^2), vertical component (Z^2), polarization as their ratio (Z^2/H^2), depression of the horizontal component (as an inverse of horizontal magnetic field component power, $1/H^2$) and δDep as a variation of depression at a particular frequency of 0.01 - 0.02 Hz (10 - 20 mHz). It is then found that there seems no clear signature of lithospheric ULF radiation. Whilst, the most evident fact is the finding of depression of ULF horizontal magnetic field at Chengdu a few days before the Sichuan EQ, which suggests that the lower ionosphere was perturbed before the EQ. The characteristics of the lower ionospheric perturbations are compared with those of upper ionospheric perturbations reported before.

Keywords: Earthquake, Earthquake Prediction, ULF Magnetic Variation, Lithospheric ULF Radiation, ULF Depression

1. Introduction

There has been accumulated a huge amount of evidence on the presence of electromagnetic precursors to the earthquakes (EQs) during the last few decades [1] -[6] . Some of those precursors are considered to be very promising for the short-term EQ prediction, especially perturbations in the lower ionosphere [7] and in the upper F region [8] , because there have been established close statistical correlations between those ionospheric perturbations and EQs based on the long-term data. Furthermore, much progress has been achieved on the mechanisms of those different seismogenic effects, though not completely understood [5] [6] .

In addition to the statistical studies, case studies for some EQs are still very important to understand the whole view of different electromagnetic signatures and then the lithosphere-atmosphere-ionosphere (LAI) coupling mechanism. We take the 2008 Sichuan EQ in China here as one of the most destructive EQs during the last decade. Several papers have already been published on this Sichuan EQ [9] -[13] . These are all based on the observational results on the perturbations in the upper ionosphere (F region) from the ground-based ionospheric sounders and GPS TEC measurements. Further studies on the perturbations on electron density, O^+ densities, ULF and ELF wave activity in the ionosphere have been performed by using the in-situ observations by different authors including [14] -[18] . As compared with various electromagnetic and plasma anomalies in association with the 2008 Sichuan EQ, there have been very few reports on the ground-based wave measurements. Especially, the analysis of geomagnetic ULF data is very lacking [19] , so that this paper presents our results of extensive analyses on the variation of ground-based ULF magnetic field changes.

As is known, lithospheric ULF radiation has been observed prior to large EQs, such as Spitak EQ [20] [21] , Loma Prieta EQ [22] , and Guam EQ [23] . Recent studies on lithospheric ULF radiation have been summarized in [24] [25] . Another new nonconventional effect in ULF is discovered by Schekotov et al. [26] [27] , which is vice versa of the above lithospheric radiation. That is the depression of ULF horizontal magnetic field component before an EQ, which is explained in terms of the enhanced absorption of magnetospheric ULF waves through the perturbed lower ionosphere [28] . This effect of seismo-ionospheric signatures has been recently confirmed even for the 2011 Tohoku EQ [28] [29] . These two aspects in ULF are treated together, by using the ULF data.

2. Description of the Sichuan EQ and of the ULF data

The 2008 Sichuan EQ or the Great Sichuan EQ happened at 02:28:01 pm Chinese Standard Time (or 06:28:01 UTC) on 12 May 2008 in the Sichuan province. Its magnitude was Ms = 8.0 (or Mw = 7.9) and its depth was 19 km. This EQ had a lot of casualties; about 70,000 people were killed, and about 18,000 were left missing.

The epicenter of this EQ is located at the geographic coordinates (31°01'05''N, 103°36'05''E), which is shown in Figure 1 as the center of the largest circle. This EQ is also known as Wenchuan EQ after its epicentral location in the Wenchuan County, Sichuan, but we call it "Sichuan EQ" in this paper.

The EQ epicenter was 80 km west-northwest of Chengdu, the provincial capital. The EQ was felt also in nearby countries and as far away from both Beijing and Shanghai (1500 and 1700 km respectively away from the epicenter), where office buildings swayed with tremor. Strong aftershocks, some exceeding the magnitude 6, continued to hit the area even months after the main shock as shown in Figure 1 by several circles, causing new casualties and damage.

Figure 1 illustrates the locations of two stations of observing ULF magnetic field variations: Chengdu (abbreviated as CDP) and Xichang (XIC) as black diamonds. The magnetometer at those stations are of the conventional fluxgate type, and its sampling frequency is 1 Hz. The resolution in ULF measurement is 0.1 nT.

3. Data Quality Analysis

The subjects of seismogenic effects using ULF data are, 1) seismogenic ULF radiation from the lithosphere, and 2) depression of ULF magnetic fields (horizontal component) as a signature of lower ionospheric perturba- tions.

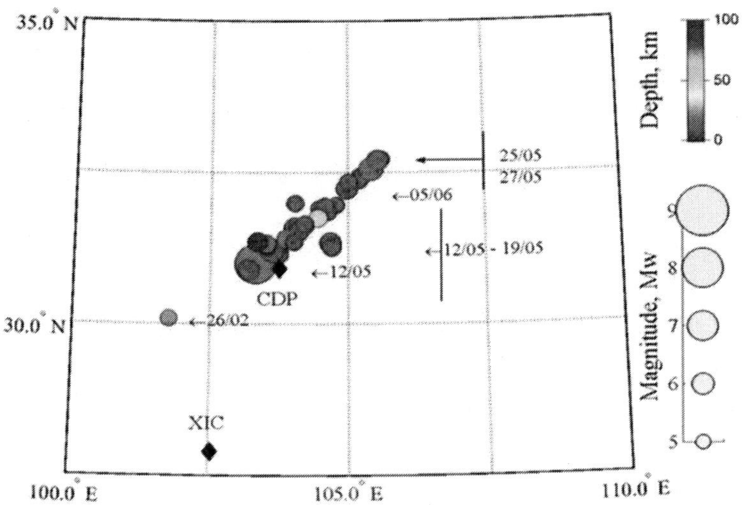

Figure 1. Map of the region with EQs and ULF observation stations. The center of the largest circle is the epicenter of the 2008 Sichuan EQ, and other several circles, its aftershocks. Two ULF stations are indicated by black diamonds: Chengdu (abbreviated as CDP) and Xichang (XIC)

The success of signal detection is known to depend not only on the nature and amplitude of observed signals, but also on the level of interferences at a station. The latter phenomenon seems to be very important especially in the case of studying field depression, when its value depends on minimal values of the magnetic field.

Figure 2(a) and Figure 2(b) illustrate examples of ULF signal in time and frequency domains on a particular day of 15 April 2008 at the two stations: CDP (a) and XIC (b). Top panels illustrate one-day evolutions of ULF magnetic field (horizontal component H) in a particular frequency range of 10 - 50 mHz at those two stations. The second panels correspond to the vertical Z component filtered in the same frequency range. The third panels illustrate the dynamic (frequency-time) spectra of H component at the two stations in the frequency range of f = 3 - 100 mHz. The bottom two refer to the corresponding spectra of Z component. As you can see from these dynamic spectra, they suggest high interferences, in general, in the form of strong pulses which occur very regularly every hour and caused by ionospheric sounding transmitters. It seems that the level of interferences is higher at XIC than at CDP. The background level at these stations indicates not only strong pulses, but also some noise-like signals even between those pulses. Position of the local midnight LMN = 24-Lon/15 is shown by the vertical dashed lines; here Lon is the longitude of the site. It is the time when depression may reach maximal values [26] .

Figure 3 illustrates a comparison of spectra of the horizontal magnetic field (H) averaged over nighttime on the same day at those stations. The nighttime interval is defined by LT = 22 h to 02 h (UT = 14 h - 18 h), and the spectra averaged over night in Figure 3 is estimated on the basis of the information only between the pulsed interferences. Also the data from a Japanese standard magnetic observatory, Kakioka (KAK) (geographic coordinates, 36°13'56"N, 140°11'11"E) on the same day are plotted in Figure 3 for the sake of comparison. The interference is found to be much higher at Chinese stations than at Kakioka, especially in the frequency range above 0.01 Hz. Because this frequency range above 0.01 Hz is known to be optimal in the study of ULF depression [3] [26] [27] , this will cause some uncertainties on the possible results on the ULF depression.

4. Analysis Results

Figure 4 and Figure 5 illustrate the temporal evolutions of magnetic field characteristics at both stations (CDP and XIC) for an interval from January to June, 2008. We tried to analyze as long period as possible, but the observation itself at CDP started in the beginning of January 2008. Moreover some of intervals of data were lost due to extreme interferences. They are marked by horizontal dashed red lines. In each figure, the top panel shows the variations of seismic activity (Kls) (as a function of EQ magnitude and epicentral distance [3]) and geomagnetic activity (Dst) (in nT). The magnetic field intensity (or power) (either horizontal (H^2) or vertical

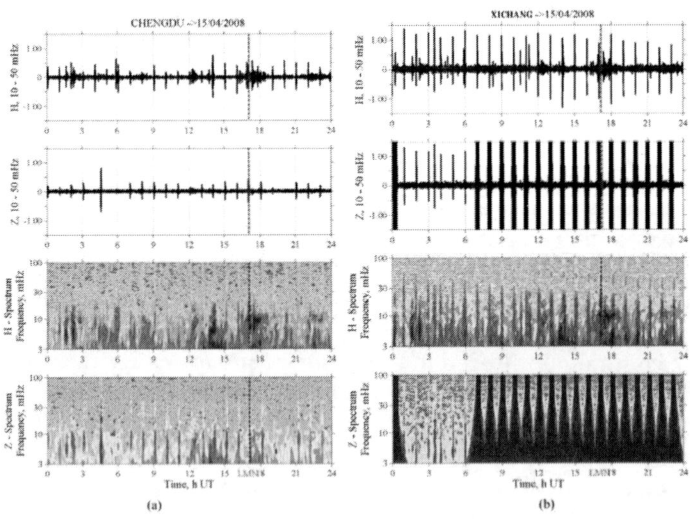

Figure 2. Examples of daily ULF background in CDP (a) and XIC (b) on a particular day of 15 April 2008. Top panels refer to the H component (at 10 - 50 mHz), the second two, the Z component (at the same frequency), the third two represent the dynamic spectra, and the bottom two refer to the Z component.

(Z^2)) is the one averaged over the above nighttime interval between the pulsed interferences, so that we have one datum per day. The second and third panels refer to the vertical (Z^2) and horizontal (H^2) magnetic field components, respectively. The fourth refers to the polarization as defined by [23] as the ratio of vertical to horizontal components (Z^2/H^2). The fifth and sixth panels illustrate magnetic field depression equal to the inverse of horizontal magnetic field ($1/H^2$) and the variation of ULF depression (δDep) respectively. The definition of δDep is given in [27] [30] , but it is the ratio of daily variation of depression relative to its mean value for previous N days to the mean value. In other words it is a normalized variation of depression. In our case N is taken equal to 30 days. The frequency band is 0.01 - 0.02 Hz (10 - 20 mHz) after the extensive comparison of results in different frequency ranges. We have applied a special algorithm of data processing only to the intervals between the pulsed interferences. Moments of the possible foreshock which occurred on 26 February and the main shock on 12 May, 2008 are shown by vertical red dashed lines.

We summarize the observational facts as follows from Figures 3-5.

1) The obtained flat spectra at the Chinese stations at the frequencies larger than 0.01 Hz (10 mHz) make it probably more difficult for us to detect the ULF depression. Those interferences may be caused by not only local sources including PCs, pulsed power supplies etc., but also by inappropriate installation of sensors.

2) Pulsed interferences are likely to be caused by a local ionospheric sounding transmitter.

3) High level of interferences does not give a possibility to estimate a comparative value of depression. We conclude that the level of depression at XIC is always smaller due to the higher level of background at this station.

4) Variation of vertical component (Z) is not found to exhibit any clear peculiarities.

5) Variation of horizontal component (H) shows peaks 10 - 12 days before the February 25 EQ (M = 5.1) and before the relevant main shock on 12 May, respectively, especially notably at CDP.

6) Polarization does not exhibit any reliable peculiarities, but shows a peak just before the EQ.

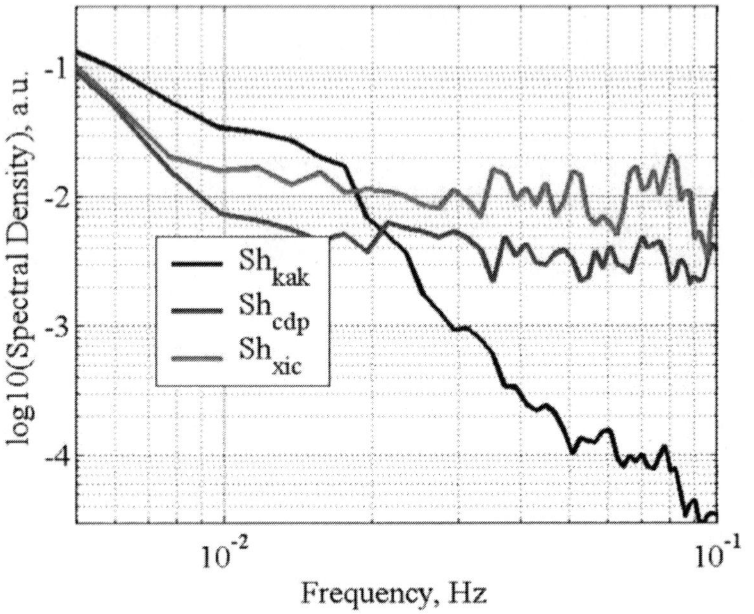

Figure 3. Nighttime averaged spectra of horizontal magnetic field at two Chinese stations (CDP and XIC) together with that at a Japanese station of Kakioka (KAK). Estimations of Chinese spectra are made only for the interval between the pulses.

7) Depression and δDep are found to show a significant increase in the vicinity of (just before, a few days before) the EQs occurred on 26 February and the main shock on 12 May.

It is very unfortunate that the noise level or interference at our two Chinese stations is considerably higher as compared with that in Japan (KAK), especially at the higher frequency range above 0.01 Hz (10 mHz). The most optimal frequency range to detect the ULF depression is found to be 0.03 - 0.05 Hz (30 - 50 mHz) [3] [26] , so that it is unlikely that high noise levels at our two Chinese stations are favorable for the study of ULF depression, so that we are obliged to choose the appropriate frequency to study both effects of lithospheric radiation and ULF depression.

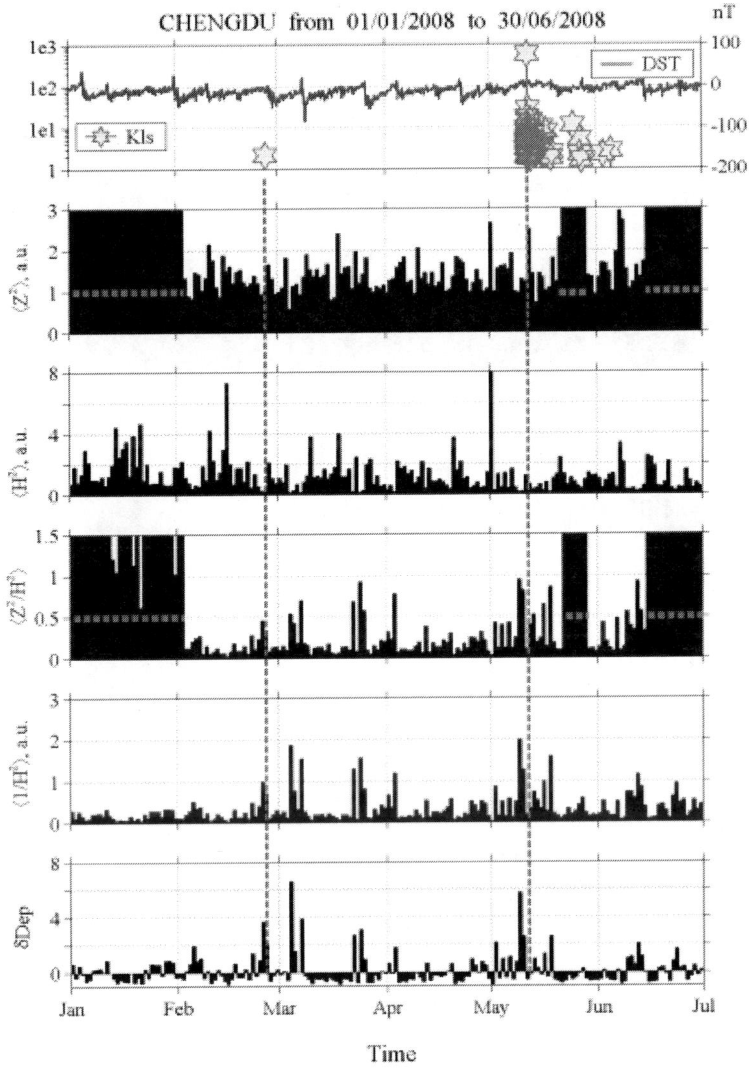

Figure 4. The result at CDP. Top panel illustrates the temporal evolutions of seismicity (Kls) and geomagnetic activity Dst (in nT). The second and third panels are the temporal evolutions of the powers of vertical (Z^2) and horizontal (H^2) magnetic field components. The fourth panel corresponds to the polarization as the ratio of vertical to horizontal components (Z^2/H^2). The fifth and the last panels refer to the temporal evolutions of Dh as the inverse of H^2 and the corresponding δDep, respectively. The time intervals of unreliable results due to the incorrect data by high interferences, are indicated with red dashed horizontal lines. Then, the moments of the (possible) foreshock occurred on 26 February and the main shock on 12 May are shown by vertical red dashed lines

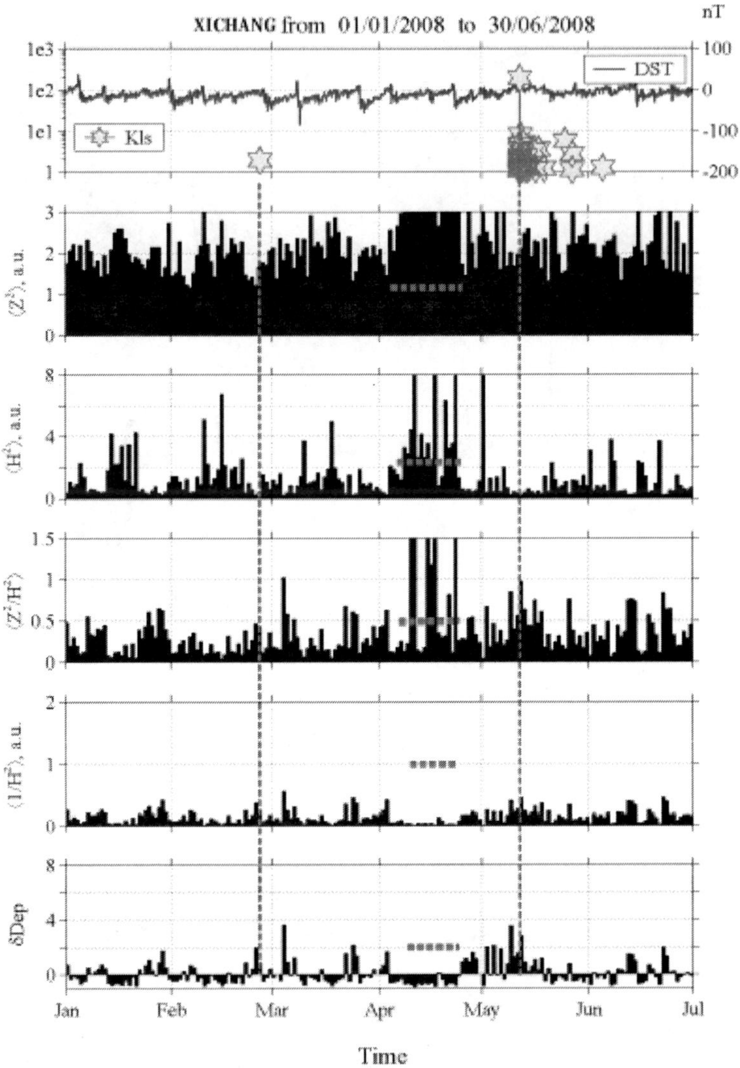

Figure 5. The same as Figure 4, but for XIC.

5. Conclusions and Discussion

Using the ULF data during several months around the 2008 Sichuan EQ, we performed detailed studies of temporal evolutions of magnetic field variations; horizontal (H^2), vertical (Z^2), inverse horizontal ($1/H^2$), and ULF depression (δDep) in different frequency bands. After having analyzed different frequency ranges, we have shown only the results at a particular frequency of 0.01 - 0.02 Hz (10 - 20 mHz) because we could identify something in this frequency range and also this range seems to be the upper limit with smaller interferences as shown in Figure 3. We know that there are two possible effects in ULF range: 1) lithospheric radiation from the EQ hypocenter, and 2) depression of horizontal component as an ionospheric signature. Main conclusions from the present analyses can be summarized as follows for each item.

1) There is no clear (or definite) evidence on the presence of lithospheric ULF emissions.

2) A significant effect seems to be detected in the ULF depression and δDep a few days before the main shock on 12 May.

First, we discuss the first topic of lithospheric ULF radiation. When studying the radiation, we try to find anomalous enhancement in the magnetic field components. We have found that the horizontal magnetic field component (H) exhibited, at both stations, peaks prior to the 26 February EQ (M = 5.1) (probably a foreshock) and also before the main shock on 12 March. These peaks are found to take place only on a particular day as summarized as Point (5) and to be simultaneous at the two stations of CDP and XIC as seen from Figure 4 and Figure 5. Also, the vertical component (Z) showed the corresponding peak on the same day. All of these suggest that these peaks are not local, but rather global in nature over the spatial scale of at least 300 km. The most probable origin of such a global disturbance of this spatial scale may be the effect of geomagnetic storms, but the comparison with Dst in Figure 4 shows no connection between the two. So, though the origin of these peaks is not well understood, it is not probable that they are related to seismic activity. Then, it seems that the vertical (Z) component at CDP does not show any significant changes, and no clear signature is seen at XIC. This inference might be furthermore supported by the previous observational fact [24] that the range of detection of seismogenic ULF radiation is on the order of 100 km for an EQ with M = 7, though there is no information on this value for M = 8. Of course, there is no reason to deny that those peaks are of seismic origin. Whereas, the polarization as suggested by [23] is found to be very effective in distinguishing seismogenic from other effects including geomagnetic activity [31] , and it shows a significant peak a few days before the EQ. This enhancement seems to be significant, but not so conspicuous.

The most evident observational result from this paper is the ULF depression (the second topic) as shown as δDep in both Figure 4 and Figure 5. We consider its evolution in a separate Figure 6. Herein top panel shows the evolution of seismicity and geomagnetic activity and next two panels are the copies of bottom panels of Figure 4 and Figure 5. They reflect the evolutions of δDep at the stations CDP and XIC accordingly. Especially, at the station (CDP) closest to the EQ epicenter, we have observed clear depression a few days before the main shock on 12 May, and its value amounts to about 6. Furthermore, this ULF depression seems to be seen on the same days even at the station of XIC which is about 300 km from the EQ epicenter. Then, the similar ULF depression is seen at the station of CDP for the February 26 EQ, just before the EQ, and this kind of behavior is also seen at the station of XIC. The way of appearance of these depression effects seems to be a natural phenomenon, and these results would suggest that those are a seismogenic effect. This February EQ is likely to be a precursor to the following seismic events, and the growth of depression at CDP is characterized as an onset of seismic activity in this region. Finally, a few days before the main shock we observed a precursor to the main shock a few days before the EQ as a depression in ULF horizontal magnetic field as a signature of lower ionospheric perturbations.

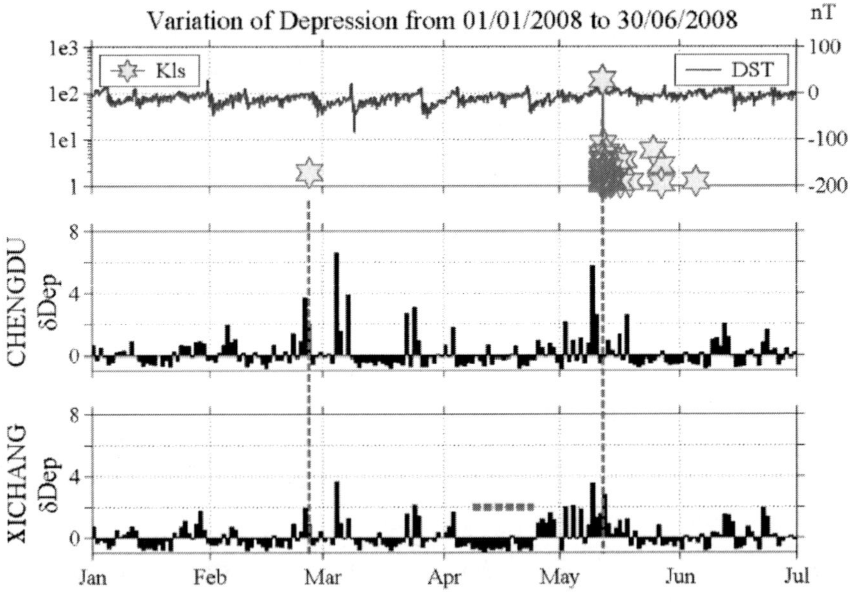

Figure 6. Evolutions of δDep at two stations (CDP and XIC) during a whole period of January-June, 2008. The top panel illustrates the temporal evolution of seismic (Kls) and geomagnetic (Dst) activities. Again vertical red dashed lines indicate the moments of the foreshock on 26 February and the main shock on 12 May.

There has been published one paper [19] on the ULF variations for this Sichuan EQ. The ULF data at several stations including the same station of CDP as in this paper, were analyzed, and his conclusion was as follows. He applied the previous analysis methods such as polarization [23] and wavelet method, but no clear anomalous magnetic variations were observed for this EQ. The only one thing seeming to be associated with the EQ might be some anomalous variations in the power of second eigenvalue in the principal component analysis in January-March, 2008. As the most important point when analyzing the ULF variations in a noisy station, we have to pay the greatest attention to the study of noise spectra. Our study in this paper has yielded nearly the same result as [19] : no clear and definite signature of lithospheric ULF radiation for this 2008 Sichuan EQ, with the conventional statistical analysis as in this paper. We need more sophisticated analysis methods based on critical concept, such as fractal analysis [32] and natural time analysis [33] -[35] . However, we have succeeded in identifying the depression of ULF horizontal magnetic field as the signature of lower ionospheric perturbation, which occurs a few days before the EQ.

Though there have been published several papers on the ionospheric plasma and wave characteristics before the Sichuan EQ on the basis of DEMETER satellite data [15] -[17] [36] , the ground-based observation is definite to provide us with the detailed temporal evolution of different seismogenic phenomena when the number of stations is sufficient to form a network. We will compare our results on the lower ionosphere with those in the upper (F region) ionosphere based on the ground-based ionospheric measurements. Based on the network of ionosonde sounding, Liu et al. [10] have found that anomalies in foF2 occurred on 9 May, 3 days before the EQ. Then, Liu et al. [11] have indicated that the anomaly in TEC observation showed an anomaly on 9 and 10 May, a few days before the EQ. The same conclusion was obtained by [37] . The most evident result from this paper is the depression of ULF horizontal magnetic field on 9, 10 May, a few days before the EQ. This result is known to be the consequence of the perturbation in the lower ionosphere [9] . With taking into account these results altogether, we may come to the conclusion that the ionosphere both in the lower part and in the upper F region, is perturbed nearly on the same days before the Sichuan EQ.

To our knowledge, this paper is likely to be the first attempt to compare the perturbations both in the lower and upper ionosphere before an EQ. The normal lead time of seismic-related F layer perturbation is 3 - 5 days [8] , whereas that for the lower ionospheric perturbation is about one week [7] . So, it seems that there may exist a significant difference of a few days between the lower and upper ionospheric perturbations. However, we have found close synchronism in temporal evolutions of ionospheric perturbations both in the lower and upper ionosphere, at last, for this Sichuan EQ, suggesting a common agent for both lower and upper ionospheric perturbations. This would provide us with a hint on the generation mechanism of seismo-ionospheric perturbations as discussed in [38] .

Acknowledgements

The authors are grateful to the referees for their useful suggestions on our paper.

References

1. Hayakawa, M. and Molchanov, O.A. (2002) Seismo Electromagnetics: Lithosphere-Atmosphere-Ionosphere Coupling. TERRAPUB, Tokyo, 477 p.
2. Pulinets, S. and Boyarchuk, K. (2004) Ionospheric Precursors of Earthquakes. Springer, Berlin, 315 p.
3. Molchanov, O.A. and Hayakawa, M. (2008) Seismo-Electromagnetics and Related Phenomena: History and Latest Results. TERRAPUB, Tokyo, 189 p.
4. Hayakawa, M. (2009) Electromagnetic Phenomena Associated with Earthquakes. Transworld Research Network, Trivandrum, 279 p.
5. Hayakawa, M. (2012) The Frontier of Earthquake Prediction Studies. Nihon-Senmontosho-Shuppan, Tokyo, 794 p.
6. Hayakawa, M. (2013) Earthquake Prediction Studies: Seismo Electromagnetics. TERRAPUB, Tokyo, 168 p.
7. Hayakawa, M., Kasahara, Y., Nakamura, T., Muto, F., Horie, T., Maekawa, S., Hobara, Y., Rozhnoi, A., Solovieva, M. and Molchanov, O.A. (2010) A Statistical Study on the Correlation between Lower Ionospheric Perturbations as Seen by Subionospheric VLF/LF Propagation and Earthquakes. Journal of Geophysical Research, 115, A09305.
8. Liu, J.Y. (2009) Earthquake Precursors Observed in the Ionospheric F-Region. In: Hayakawa, M., Ed., Electromagnetic Phenomena Associated with Earthquakes, Transworld Research Network, Trivandrum, 187-204.
9. Zhao, B., Wang, M., Yu, T., Wan, W., Lei, J., Liu, L. and Ning, B. (2008) Is an Unusual Large Enhancement of Ionospheric Electron Density Linked with the 2008 Great Wenchuan Earthquake? Journal of Geophysical Research, 113, A11304.
10. Liu, J.Y., Chen, Y.I., Chen, C.H., Liu, C.Y., Chen, C.Y., Li, J.Z., Xia, T.Q., Oyama, K.I., Hattori, K. and Lin, C.H. (2009) Seismoionospheric GPS Total Electron Content Anomalies Observed before the 12 May 2008

Mw 7.9 Wenchuan Earthquake. Journal of Geophysical Research, 114, A04320.

11. Liu, C.Y., Liu, J.Y., Chen, W.S., Li, J.Z., Xia, Y.Q. and Cui, X.Y. (2010) An Integrated Study of Anomalies Observed before Four Major Earthquakes: 2004 Sumatra M9.3, 2006 Pingtung M7.0, 2007 Chuetsu Oki M6.8, and 2008 Wenchuan M8.0. Journal of Asian Earth Sciences, 41, 401-409.
12. Pulinets, S.A., Boudur, V.G., Tsidilina, M.N. and Gaponova, M.V. (2010) Verification of the Concept of Seismoionospheric Coupling under Quiet Heliogeomagnetic Conditions, Using the Wenchuan (China) Earthquake of May 12, 2008, as an Example. Geomagnetism and Aeronomy, 50, 231-242.
13. Klimenko, M.V., Klimenko, V.V., Aakharenkova, I.E., Pulinets, S.A., Zhao, B. and Tsidilina, M.T. (2011) Formation Mechanism of Great Positive TEC Disturbances Prior to Wenchuan Earthquake on May 12, 2008. Advances in Space Research, 48, 488-499.
14. Zhang, X., Shen, X., Liu, J., Ouyang, X., Qian, J. and Zhao, S. (2009) Analysis of Ionospheric Plasma Perturbations before Wenchuan Earthquake. Natural Hazards and Earth System Science, 9, 1259-1266.
15. Sarkar, S. and Gwal, A.K. (2010) Satellite Monitoring of Anomalous Effects in the Ionosphere Related to the Great Wenchuan Earthquake of May 12, 2008. Natural Hazards, 55, 321-332.
16. Blecki, J., Parrot, M. and Wronowski, R. (2010) Studies of the Electromagnetic Field Variations in ELF Frequency Range Registered by DEMETER over the Sichuan Region Prior to the 12 May 2008 Earthquake. International Journal of Remote Sensing, 31, 3615-3629.
17. Walker, S.N., Kadirkamanathan, V. and Pokhotelov, O.A. (2013) Changes in the Ultra-Low Frequency Wave Field during the Precursor Phase to the Sichuan Earthquake: DEMETER Observations. Annales Geophysicae, 31, 1597-1603.
18. Ryu, K., Parrot, M., Kim, S.G., Jeong, K.S., Chae, J.S., Pulinets, S. and Oyama, K.I. (2014) Suspected Seismo-Ionos- pheric Coupling Observed by Satellite Measurements and GPS TEC Related to the M7.9 Wenchuan Earthquake of 12 May 2008. Journal of Geophysical Research, 119, 10305-10323.
19. Huang, Q.H. (2011) Retrospective Investigation of Geophysical Data Possibly Associated with the M_S 8.0 Wenchuan Earthquake in Sichuan, China. Journal of Asian Earth Sciences, 41, 421-427.
20. Kopytenko, Y.A., Matiashvily, T.G., Voronov, P.M., Kopytenko, E.A. and Molchanov, O.A. (1990) Discovering of Ultra-Low-Frequency Emissions

Connected with Spitak Earthquake and Its Aftershock Activity on Data of Geomagnetic Pulsations Observations at Dusheti and Vardzija. IZMIRAN Preprint N3, 888, 27 p.

21. Molchanov, O.A., Kopytenko, Y.A., Voronov, P.M., Kopytenko, E.A., Matiashvili, T.G., Fraser-Smith, A.C. and Bernardy, A. (1992) Results of ULF Magnetic Field Measurements near the Epicenters of the Spitak (M_s = 6.9) and the Loma-Prieta (M_s = 7.1) Earthquakes: Comparative Analysis. Geophysical Research Letters, 19, 1495-1498.
22. Fraser-Smith, A.C., Bernardy, A., McGill, P.R., Ladd, M.E., Helliwell, R.A. and Villard Jr., O.G. (1990) Low-Frequency Magnetic Field Measurements near the Epicenter of the M_s7.1 Loma-Prieta Earthquake. Geophysical Research Letters, 17, 1465-1468.
23. Hayakawa, M., Kawate, R., Molchanov, O.A. and Yumoto, K. (1996) Results of Ultra-Low-Frequency Magnetic Field Measurements during the Guam Earthquake of 8 August 1993. Geophysical Research Letters, 23, 241-244.
24. Hayakawa, M., Hobara, Y., Ohta, K. and Hattori, K. (2011) The Ultra-Low-Frequency Magnetic Disturbances Associated with Earthquakes. Earthquake Science, 24, 523-534.
25. Hattori, K. (2013) ULF Geomagnetic Changes Associated with Major Earthquakes. In: Hayakawa, M., Ed., Earthquake Prediction Studies: Seismo Electromagnetics, TERRAPUB, Tokyo, 129-152.
26. Schekotov, A., Molchanov, O.A., Hattori, K., Fedorov, E., Gladyshev, V.A., Belyaev, G.G., Chebrov, V., Sinitsin, V., Gordeev, E. and Hayakawa, M. (2006) Seismo-Ionospheric Depression of the ULF Geomagnetic Fluctuations at Kamchatka and Japan. Physics and Chemistry of the Earth, 31, 313-318.
27. Schekotov, A., Fedorov, E., Molchanov, O.A. and Hayakawa, M. (2013) Low Frequency Electromagnetic Precursors as a Prospect for Earthquake Prediction. In: Hayakawa, M., Ed., Earthquake Prediction Studies: Seismo Electromagnetics, TERRAPUB, Tokyo, 81-99.
28. Hayakawa, M., Rozhnoi, A., Solovieva, M., Hobara, Y., Ohta, K., Schekotov, A. and Fedorov, E. (2013) The Lower Ionospheric Perturbation as a Precursor to the 11 March 2011 Japan Earthquake. Geomatics, Natural Hazards and Risk, 4, 275-287.
29. Schekotov, A., Fedorov, E., Hobara, Y. and Hayakawa, M. (2013) ULF Magnetic Field Depression as a Possible Precursor to the 2011/3.11 Japan Earthquake. Journal of Atmospheric Electricity, 33, 41-51.

30. Schekotov, A., Izutsu, I. and Hayakawa, M. (2015) On Precursory ULF/ELF Electromagnetic Signatures for the Kobe Earthquake on April 12, 2013. Journal of Asian Earth Sciences, in Press.
31. Currie, J.L. and Waters, C.L. (2014) On the Use of Geomagnetic Indices and ULF Waves for Earthquake Precursor Signatures. Journal of Geophysical Research: Space Physics, 119, 992-1003.
32. Hayakawa, M. and Ida, Y. (2008) Fractal (Mono- and Multi-) Analysis for the ULF Data during the 1993 Guam Earthquake for the Study of Prefracture Criticality. Current Development in Theory and Applications of Wavelets, 2, 159-174.
33. Varotsos, P.A. (2005) The Physics of Seismic Electric Signals. TERRAPUB, Tokyo, 338 p.
34. Eftaxias, K., Potirakis, S.M. and Chelidze, T. (2013) On the Puzzling Feature of the Silence of Precursory Electromagnetic Emissions. Natural Hazards and Earth System Science, 13, 2381-2397.
35. Hayakawa, M., Schekotov, A., Potirakis, S. and Eftaxias, K. (2015) Criticality Features in ULF Magnetic Fields Prior to the 2011 Tohoku Earthquake. Proceedings of the Japan Academy: Series B, 91, 25-30.
36. Onishi, T., Berthelier, J.J. and Kamogawa, M. (2011) Critical Analysis of the Electrostatic Turbulence Enhancements Observed by DEMETER over the Sichuan Region during the Earthquake Preparation. Natural Hazards and Earth System Science, 11, 561-570.
37. Saradjian, M.R. and Akhoonzadeh, M. (2011) Prediction of the Date, Magnitude and Affected Area of Impending Strong Earthquakes Using Integration of Multi Precursors Earthquake Parameters. Natural Hazards and Earth System Science, 11, 1109-1119.
38. Kamiyama, M., Sugito, M., Kuse, M., Schekotov, A. and Hayakawa, M. (2014) On the Precursors to the 2011 Tohoku Earthquake: Crustal Movements and Electromagnetic Signatures. Geomatics, Natural Hazards and Risk.

CHAPTER 10

Innovative Procedures to Assess Seismic Behaviour of Existing Structures by Means of Non Linear Static Analysis: Polar Spectrum and Capacity Domains

Luigi Petti * and Ivana Marino

Department of Civil Engineering, University of Salerno, Via Ponte Don Melillo, Fisciano, SA 84084, Italy

Abstract

In the last few years, the need to evaluate the seismic performances of buildings on sustaining strong motion has encouraged the development of simplified non-linear static analyses. Several procedures are available today to assess the behavior of plane-frame systems or plan-regular framed buildings suitable for engineering purposes. Less accurate procedures are instead available for irregular structures. This study introduces new tools to assess the seismic performance of irregular structures by using capacity domains and polar spectra. In particular, the capacity domains, plotted in terms of base shear and node control displacements and obtained by means of static non-linear analyses, lead to the evaluation of the direction of least seismic capacity of the investigated structure. The polar spectrum, instead, leads to taking into account the directivity and site effects of seismic events. In particular, the polar spectrum represents the spectral seismic response evaluated for different in-plan directions.

Keywords: non linear static analysis; capacity domains; polar spectrum

1. Introduction

As known, the standard analysis for traditional seismic design is Linear Dynamic Procedure (LDP) with assigned response spectrum (Response Spectrum Analysis). More complex procedures, such as non-linear time history analysis (NDP), are seldom used because of the difficulties in the definition of an accurate hysteretic model to describe the behaviour of the materials under cyclic actions and the choice of a set of accelerograms that describes the real site conditions. Moreover, as it is well-known, NDP analyses present convergence difficulties and require major computational effort.

Non-linear static analysis is instead becoming the key method to evaluate the seismic response of existing structures. This method, which is able to evaluate the collapse mechanism of structures, is suitable for the analysis of existing structures, including the ones not explicitly designed for seismic actions.

However, non-linear static procedures can lead to unsuitable results when applied to irregular structures because of the difficulties in taking into account dynamic lateral-torsional effects and higher mode effects. As it is known, an asymmetric distribution of mass and stiffness, or of strength in plan, leads to high ductility demand for the elements near the soft or weak edge [1,2,3]. In recent years, several proposals have been put forward to extend traditional pushover analysis, calibrated on plan systems, to the assessment of three-dimensional models behavior [4,5,6,7,8,9,10,11,12,13,14,15]. At present no simplified procedure which can adequately account for the torsional behaviour of asymmetric structures has been implemented. The main questions concern the combination of the forces in the two directions, and the best way to consider extra ductility demand for the elements near the edges [16,17,18,19,20,21,22,23,24].

This report presents and discusses the main results of the new methodology to assess the seismic response of irregular structures by means of capacity domains which allow us to evaluate the direction of least seismic capacity of the investigated buildings and the polar spectrum which represents the seismic demand in the plan. In particular, the capacity domains, represented in terms of strength and displacement capacities, are evaluated using traditional pushover analyses by applying load distributions, triangular and uniform, at different angles in the plan to search for the least seismic-resistant direction. The polar spectrum, instead, is a new spectral representation which is able to investigate the seismic demand of each in-plane direction. In particular, this tool is based on the spatial spectral surface obtained by the spectral seismic response evaluated for different in-plan directions and the in-plan projection is defined "Polar Spectrum".

Both polar spectrum and capacity domains allow researchers to investigate the non-linear seismic capacities of irregular structures. As is well-known, indeed, the collapse conditions for a three-dimensional system are also governed by the spatial features of the seismic event. Therefore, the goal of the presented results is to investigate the seismic behavior of irregular structures by considering the least seismic-resistant directions of the investigated structure and the spatial features of the seismic event.

2. Description of the Benchmark Structures

For the scope of this report the benchmark structure proposed in the ReLUIS project (Earthquake Engineering Test Labs Network) is considered [25].

The structure, designed without seismic actions, is a five-storey L-shaped building with RC moment frames in two orthogonal directions, as represented in Figure 1. Details of the benchmark structure are contained in the ReLUIS project.

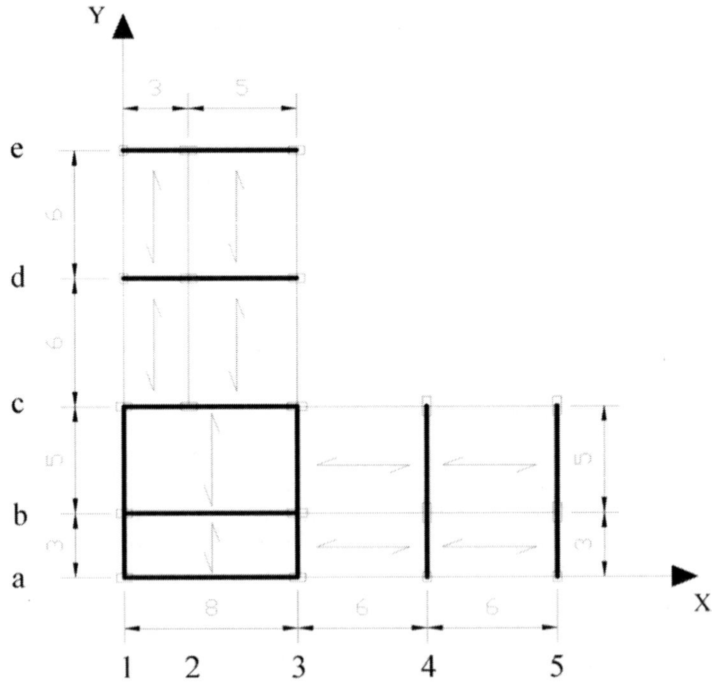

Figure 1. Benchmark structure plan.

A finite-element model has been implemented in Opensees software [26]. The floors are modeled by means of 50 mm elastic shells and the elastic modulus is set to the value of 30,000 MPa. Beams and columns are modeled by spread plasticity non-linear elements with fiber sections applied.

The collapse condition for the generic section is defined in terms of section curvature, evaluated by imposing limit strain values for the concrete fibers ($\varepsilon_c = 0.006$ in compression) and for the reinforced steel fibers ($\varepsilon_s = 0.03$ in tension).

The main dynamical properties of the model are reported in Table 1. In particular, each mode shape is described by the vibration period (T), the mass participation ratios (M%) and the sum of the participation ratios up to the considered modal shape (SUM).

Table 1. Modal properties for the Benchmark Structure.

Mode	T(s)	M%x	M%y	SumX	SumY
1	1.35	10.9%	52.1%	10.9%	52.1%
2	1.22	56.9%	19.6%	67.8%	71.8%
3	0.99	10.8%	9.2%	78.6%	80.9%
4	0.42	0.9%	8.5%	79.4%	89.4%
5	0.38	9.4%	1.9%	88.8%	91.2%
6	0.31	1.5%	1.1%	90.3%	92.4%
7	0.23	0.1%	4.1%	90.4%	96.5%
8	0.20	4.2%	0.5%	94.6%	97.0%
9	0.17	0.9%	0.5%	95.5%	97.5%

The modal analysis shows a flexible behavior and, moreover, the plan irregularity leads to latero-torsional behaviour for modes 1 and 2.

3. The Concept of Capacity Domains

In order to assess the non-linear static response of irregular structures it is possible to obtain capacity domains, in terms of control node displacement or base shear. In particular, such domains are obtained from the results of non-linear static procedures (pushover) carried out by varying the in-plan direction of the applied load distribution. The domains, represented in the following, depict the displacement of the control node or the base shear up to collapse.

The analyses are carried out by applying two load distributions along the height of the structure. The first one is a triangular distribution, taking into consideration the inertial masses and the height above the ground. The second one considers only the floor masses. The forces are applied at beam-column joints, proportionally to the joints mass.

For each applied load distribution, 24 capacity curves have been obtained by rotating the direction of action in the x-y plane, 15° at each step. The control node is assumed to be the centre of mass on the top floor in all the pushover analyses carried out.

Using the described limit values for section curvature, the global collapse is assumed when the first primary element exceeds its limit. Each pushover curve is bi-linearized according to the FEMA 356 procedure [27].

The capacity domains are compared to the non-linear dynamic response to recorded earthquakes, selected among those proposed by ReLUIS [28]. The records, considering both NS and EW components, are suitably scaled to study the structural response by varying ground acceleration as in the case of non linear incremental dynamical analysis.

The main results, for the considered accelerogram, are represented in the following. In particular, Figure 2 refers to the collapse of the analyzed benchmark structure by varying the direction of action in the plan in the case of triangular (TR) and uniform (UN) load distributions and represents the capacity domains. These domains are plotted in terms of displacement of the control node (U) and base shear (Vb). It is possible to observe for both domains a non-regular behaviour of the structure, above all in terms of limit displacements. In particular, the analyses show that the benchmark structure presents lower displacement capacities in the 150°–330° direction. Otherwise, structural behaviour seems to be more regular in terms of strength.

In order to evaluate the applicability of the capacity domains to accurately describe the collapse state of a structure, incremental dynamic analyses have been performed. For both the considered in-plan dispositions of the structures, 0° and 90°, the analyses have been carried out considering the Iceland 2000 earthquake (code 06334, maximum PGA equal to 7.07 m/s2). The Figure 3 and Figure 4 show the dynamic response (NDP) for the analyzed cases.

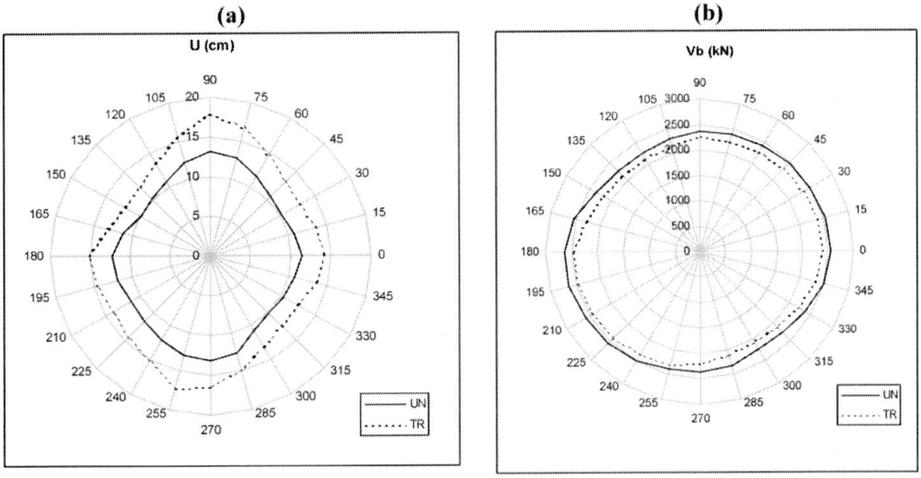

Figure 2. (**a**) Capacity Domains in terms of node control displacements at collapse; **b**Capacity Domains in terms of base shear.

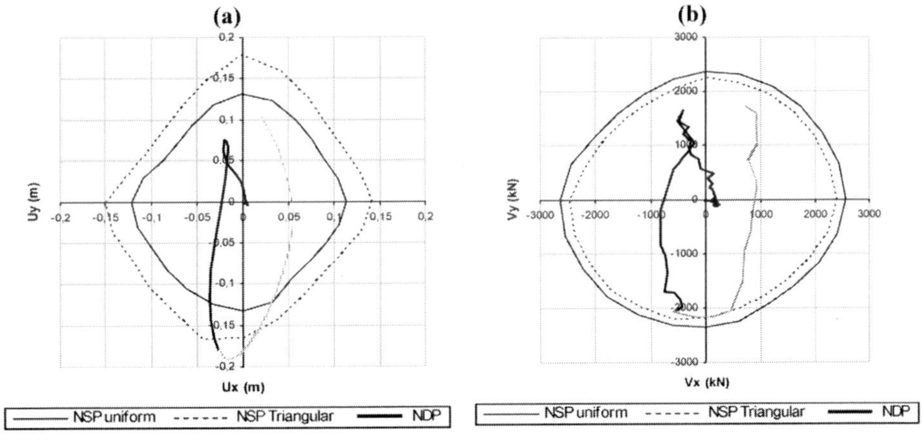

Figure 3. (**a**) Comparison between capacity domains and NDP-Collapse amplification factor 0.40-In plan orientation of benchmark structure 0°; (**b**) Comparison between capacity domains and NDP-Collapse amplification factor 0.40-In plan orientation of benchmark structure 0°.

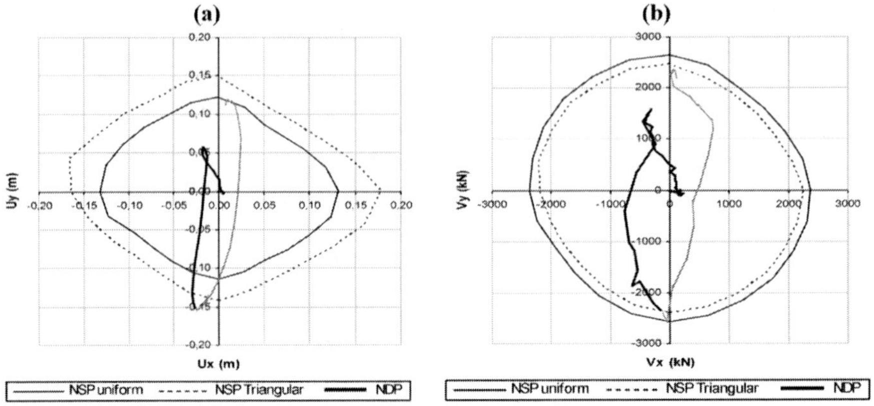

Figure 4. (**a**) Comparison between capacity domains and NDP-Collapse amplification factor 0.35-In plan orientation of benchmark structure 90°; (**b**) Comparison between capacity domains and NDP-Collapse amplification factor 0.35-In plan orientation of benchmark structure 90°.

As it is apparent, the collapse is attained in the direction of about 255° independently of plan orientation of benchmark structure. Moreover, the collapse occurs in correspondence with the capacity domain for triangular load distribution, and the collapsed element, column b-1 (Figure 1), the second column located along the longer side in direction y, at ground level, is the same for both dynamic and static analyses. For both analyses the collapse amplification factor in the case of NDP is represented. In the figures, the capacity domains are indicated with NSP (Non-linear Static Procedure) for the uniform and triangular load distributions.

As a further comparison between static and dynamic responses, Figure 5 shows storey displacements at collapse for both analyses in the considered plan directions of benchmark structure (0° and 90°).

Results show that the collapse mechanism is a global one, being the displacement path carried out by NDP well described by the one obtained by NSP with triangular load distribution. Moreover, as is apparent, the collapse as well as when read with the NDP procedure always intercepts the capacity domains corresponding to the triangular load distribution (Figure 3and Figure 4).

Figure 6 shows the comparison between the behaviour evaluated by means of pushover analyses and that obtained by incremental dynamic analysis.

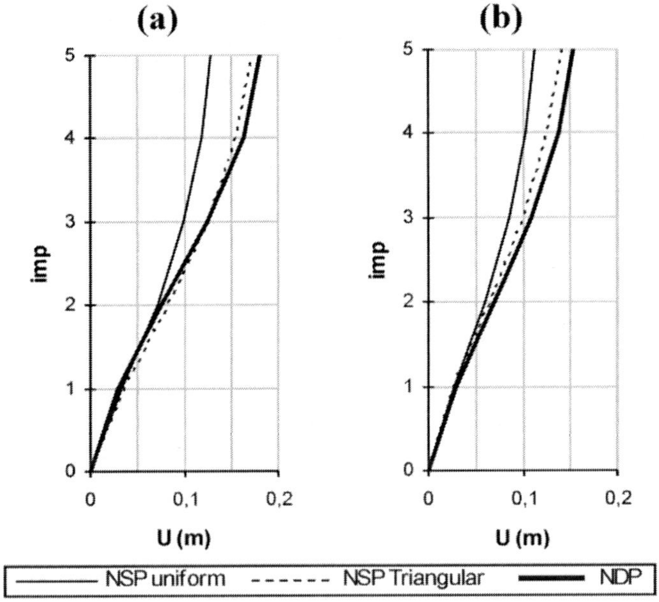

Figure 5. (**a**) Storey Displacement Comparison at collapse for in plan directions 0°; (**b**) at 90°.

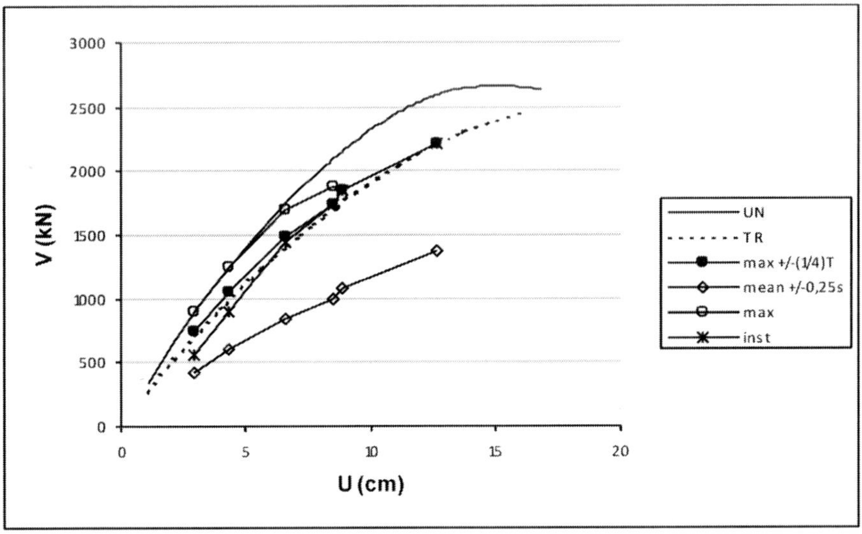

Figure 6. Comparison between Incremental Dynamical Analysis and static non linear analyses-In plan orientation of benchmark structure 0°.

In particular, the figures show the comparison between pushover curves obtained by static non linear analyses in the least resistance direction considering the uniform and triangular load distributions and the pushover curves obtained by incremental dynamic analyses considering several criteria for the graphical representation. Regarding the representation of the time history data, several proposals can be found in the literature [29]. Among these, in the present study, several criteria have been considered: maximum displacement *vs.* maximum base shear (max), maximum displacement *vs.* corresponding base shear (inst) and maximum displacement *vs.* mean base shear within the interval +/−0.25 s (mean +/−0.25 s). Moreover, a new criterion has been proposed herein: the maximum displacement *vs.* maximum base shear within the interval +/−(1/4) T (max +/−(1/4)T) with T the dominant natural period of vibration in the direction under examination. The proposed criterion leads to catching the maximum response in a semi-period of oscillation.

The results (Figure 6 and Figure 7) show that the static non-linear analyses carried out in the least seismic resistant direction will effectively describe the dynamic non-linear response for the investigated structure and considered seismic event. Moreover, the proposed criteria for comparing incremental dynamic and static non linear analysis seems to supply us with the best results for all investigated displacements.

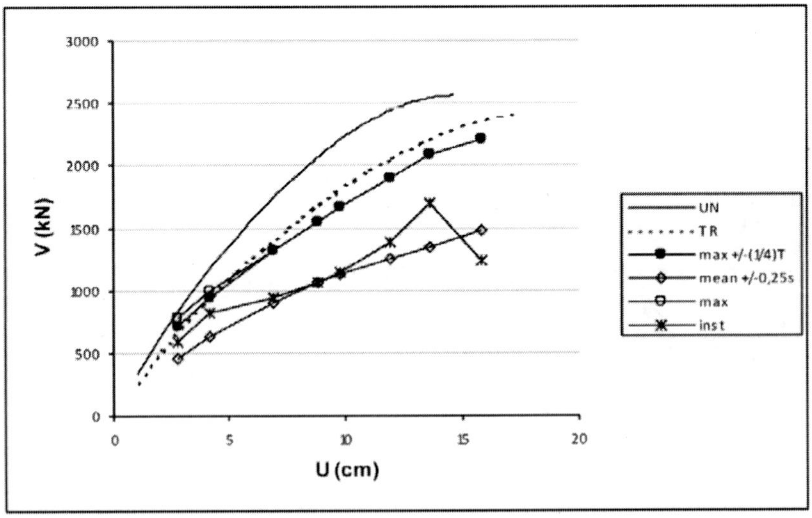

Figure 7. Comparison between Incremental Dynamical Analysis and static non linear analyses-In plan orientation of benchmark structure 90°.

4. The Concept of Polar Spectrum

As is known, seismic events generally present strong directivity effects due to site geo-morphological conditions, the source mechanism and the path from the source to the site of the seismic waves.

To take into account the directivity and site effects it is possible to consider the polar spectrum, a new tool, proposed by the authors, which is able to investigate each in-plan direction. In particular, this tool is based on the spatial spectral surface obtained by the spectral seismic response evaluated for different in-plan directions. The in-plan projection of this surface is defined "Polar Spectrum".

Figure 8 shows the polar spectrum in terms of pseudo-acceleration for the seismic event South Iceland aftershock (code 006334) with a damping factor $\xi = 0.05$ in the range of periods 0–2 seconds. The polar spectrum represents the projection on the horizontal plan of the spectral surface obtained by evaluating the spectra in each horizontal direction by considering the ground acceleration time histories only in these directions. For the scope, of each considered direction the accelerograms were evaluated by considering the NS and EW recorded components of the seismic event. Each radius of the polar spectrum thus represents, the response evaluated in that direction and each circumference, instead, the spectral demand for a fixed period in each direction. The origin corresponds to a 0 sec period. As it is easy to observe, the polar spectrum is a polar-symmetric figure.

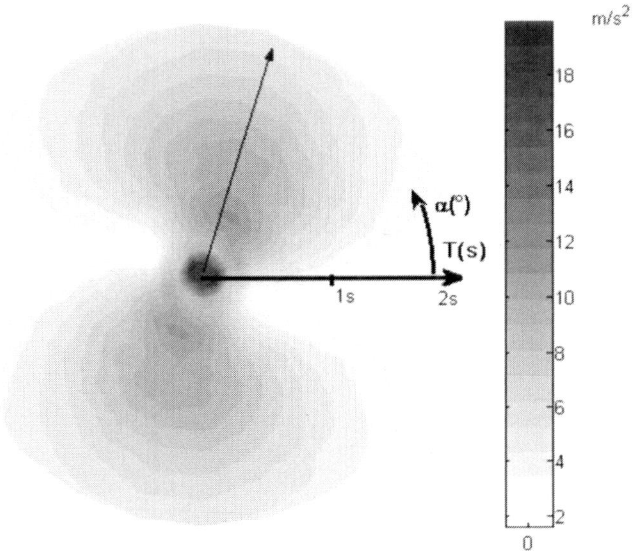

Figure 8. Polar Spectrum in terms of pseudo-acceleration for the South Iceland aftershock (code 006334).

The analysis of the polar spectrum shows that the considered seismic event presents a peak demand along the direction (75°–255°) regardless of the vibration period value.

The comparison between the polar spectrum of the considered seismic event and the results of Non-Linear Dynamic Procedure (NDP) is shown clearly, as the collapse of the benchmark structure is controlled in this case by the directivity effects more than by the seismic capacity of the analyzed structure.

5. Critical Alignment

The possibility of evaluating the worst structural response moving from observation of the polar spectrum suggests performing dynamic analyses with the structure oriented so that its least capacity direction (assessed by means of the capacity domains) is aligned with the maximum seismic demand (evaluated by means of the polar spectrum).

Figure 9 depicts the results in this case. In particular, the structure has been rotated 105° counter-clockwise in order to match the least resistant direction (α = 150°–330°) with the maximum spectral seismic demand direction (75°–255°).

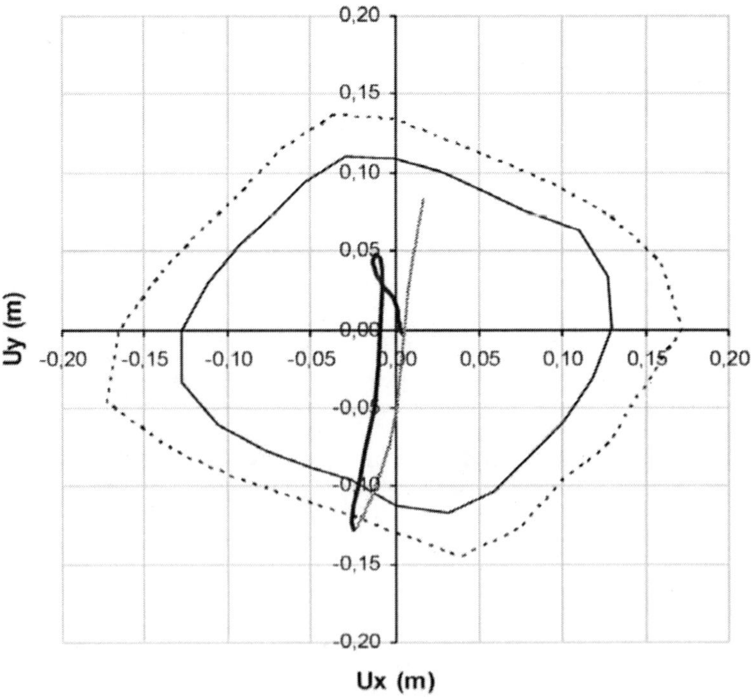

Figure 9. Dynamic analyses results, in plan orientation of the benchmark structure 105°, NDP collapse amplification factor 0.30.

The obtained results show that when the least seismic resistance direction is aligned with the maximum seismic demand, the minimum collapse amplification factor is reached.

6. Conclusions

In the field of seismic engineering, an existing open issue is the need to define a feasible procedure to evaluate the seismic performance of irregular structures. In this article, new tools, devised in order to assess the non-linear behaviour of an irregular building and the spectral seismic demand in the plan, have been presented. In particular, capacity domains are evaluated by means of pushover analyses by varying the load distribution in the plan and lead to the search for less seismic resistant directions. The polar spectra lead to full assessment of the spatial features of the seismic demand and represent an opportunity to investigate thoroughly the seismic demand and the event characteristics, being able to clearly evaluate directivity and site effects.

The results, for the investigated structure and the considered earthquake, show that the collapse condition, assessed by means of non linear dynamic procedures, turns out to be coherent with the capacity domains based on the results of traditional non linear procedures carried out by varying the angle of the application of the load distribution.

The non-linear dynamic analyses have shown that the directional components of the seismic event govern the collapse direction. Therefore, the worst condition for an investigated structure is obtained by aligning its least resistance capacity direction, evaluated on the basis of the capacity domains, with the maximum response demand direction identified by the polar spectrum.

Moreover, the study proposes a new criterion to compare incremental dynamic and static non linear analysis which seems to supply the best results for all investigated cases.

References

1. Fajfar, P.; Marusic, D.; Perus, I. Torsional effects in the pushover-based seismic analysis of buildings. *J. Earthq. Eng.* **2005**, *9*, 831–854.
2. Kan, C.L.; Chopra, A.K. *Linear and Nonlinear Earthquake Responses of Simple Torsionally Coupled Systems*; University of California: Berkeley, CA, USA, 1979.
3. Hejal, R.; Chopra, A. *Earthquake Response of Torsionally-coupled Buildings*; University of California: Berkeley, CA, USA, 1987.
4. Ayala, A.G.; Tavera, E.A. A new approach for the evaluation of the seismic performance of asymmetric buildings. In Proceedings of the

7th National Conference on Earthquake Engineering, Boston, MA, USA, 21–25 July 2002.
5. Bhatt, C.; Bento, R. Extension of the CSM-FEMA440 to plan-asymmetric real building structures. *J. Earthq. Eng. Struct. Dyn.* **2011**, *40*, 1263–1282.
6. Chopra, A.K.; Goel, R.K. A modal pushover analysis procedure to estimate seismic demands for unsymmetric-plan buildings. *Earthq. Eng. Struct. Dyn.* **2004**, *33*, 903–927.
7. Fajfar, P.; Kilar, V. Simple push-over analysis of asymmetric buildings. *Earthq. Eng. Struct. Dyn.* **1997**, *26*, 233–249.
8. Fajfar, P.; Kilar, V.; Marusic, D.; Perus, I.; Magliulo, G. The extension of the N2 method to asymmetric buildings. In Proceedings of the Fourth Forum on Implications of Recent Earthquakes on Seismic Risk, Tokyo, Japan, 27–29 May 2002; Technical Report TIT/EERG,02/1.
9. Fajfar, P.; Marusic, D.; Perus, I.; Kreslin, M. The N2 method for asymmetric buildings. In Proceedings of Nonlinear Static Methods for Design/Assessment of 3D Structure, Lisbon, Portugal, 5–6 May 2008.
10. Fujii, K.; Nakano, Y.; Sanada, Y. Simplified nonlinear analysis procedure for asymmetric buildings. In Proceedings of the 13th World Conference on Earthquake Engineering, Vancouver, BC, Canada, 1–6 August 2004.
11. Goel, R.K. Generalized pushover curves for nonlinear static analysis of three dimensional structures. In Proceedings of Nonlinear Static Methods for Design/Assessment of 3D Structures, Lisbon, Portugal, 5–6 May 2008.
12. Kilar, V.; Fajfar, P. On the applicability of pushover analysis to the seismic performance evaluation of asymmetric buildings. *Eur. Earthq. Eng.* **2001**, *15*, 20–31.
13. Moghadam, A.S.; Tso, K. Pushover analysis for asymmetric and set-back multistory buildings. In Proceedings of the 12th World Conference on Earthquake Engineering, Auckland, New Zealand, 30 January–4 February 2000; p. 1093.
14. Tso, W.K.; Moghadam, A.S. Seismic response of asymmetrical buildings using pushover analysis. In Proceedings of the International Workshop on Seismic Design Methodologies for the Next Generation of Codes, Bled, Slovenia, 24–27 June 1997.
15. Yu, Q.S.K.; Pugliesi, R.; Allen, M.; Bischoff, C. Assessment of modal pushover analysis procedure and its application to seismic evaluation of existing building. In Proceedings of the 13th World

Conference on Earthquake Engineering, Vancouver, BC, Canada, 1–6 August 2004.
16. Pettinga, J.D.; Priestley, M.J.N.; Pampanin, S.; Christopulos, C. The role of inelastic torsion in the determination of residual deformations. *J. Earthq. Eng.* **2007**, *11*, 133–157.
17. Zarate, A.; Ayala, A.G. Validation of single storey models for the evaluation of seismic performance of multi-storey asymmetric buildings. In Proceedings of the 13th World Conference on Earthquake Engineering, Vancouver, BC, Canada, 1–6 August 2004.
18. Faella, G.; Kilar, V. Asymmetric multistory R/C frame structures: Push-over versus nonlinear dynamic analysis. In Proceedings of the Eleventh European Conference on Earthquake Engineering, Paris, France, 6–11 September 1998.
19. Rutenberg, A.; Levy, R.; Magliulo, G. Seismic response of asymmetric perimeter frame steel buildings. In Proceedings of the 12th European Conference on Earthquake Engineering, London, UK, 9–13 September 2002.
20. Marino, I.; Petti, L. Seismic assessment of asymmetric structures behaviour by using static non-linear analysis. *J. Civ. Eng. Archit.* **2011**, *5*, 401–411.
21. Tso, W.K.; Myslimaj, B. Effect of strength distribution on the inelastic torsional response of asymmetric structural systems. In Proceedings of the 12th European Conference on Earthquake Engineering, London, UK, 9–13 September 2002.
22. Reyes, J.C.; Chopra, A.K. Evaluation of three-dimensional modal pushover analysis for unsymmetric-plan buildings subjected to two components of ground motion. *Earthq. Eng. Struct. Dyn.* **2011**, *40*, 1475–1494.
23. Rutenberg, A. EAEE Task Group (TG) 8: Behaviour of irregular and complex structures asymmetric structures-progress since 1998. In Proceedings of the 12th European Conference on Earthquake Engineering, London, UK, 9–13 September 2002.
24. Petti, L.; Marino, I.; Cuoco, L.; Palazzo, B. Assessment of seismic response of plan-asymmetric structures through simplified static nonlinear analyses. In Proceedings of the 14th World Conference on Earthquake Engineeering, Beijing, China, 12–17 October 2008.
25. ReLUIS project-Earthquake Engineering Test Labs Network Homepage. Available online: http://www.reluis.unina.it(accessed on 25 June 2012).

26. Open System for Earthquake Engineering Simulation Homepage. (OpenSEES). Available online: http://opensees.berkeley.edu/ (accessed on 25 June 2012).
27. FEMA (Federal Earthquake Management Agency) 356, Prestandard and Commentary for the Seismic Rehabilitation of Buildings, [DVD]. FEMA 356, Washington, DC, USA, 2000.
28. European Strong Motion Database Homepage. Available online: http://www.isesd.cv.ic.ac.uk (accessed on 25 June 2012).
29. Chopra, A.K. Application of modal pushover analysis to RC frames and incremental dynamic analysis. In Proceedings of Nonlinear Static Methods for Design/Assessment of 3D Structures, Lisbon, Portugal, 5–6 May 2008.

CHAPTER 11

Feasibility Study of Community Earthquake Warning System Proposed for Mbeya City and Surrounding Regions

Asinta Manyele, Alfred Mwambela

Department of Electronics and Telecommunications, University of Dar es Salaam, Dar es Salaam, Tanzania

Abstract

The Mbeya city and its surrounding regions (Rukwa, Katavi and Njombe) in south western Tanzania, fall among the fast growing economical regions in Tanzania. It is also unfortunately under potential for seismic risk due to its proximity to the western and eastern junction of the East African Rift (EAR) Valley. Construction of Community Earthquake Early Warning (CEEW) system model based on community owned MEMS accelerometer sensors in being proposed for the region. To optimize the warning time function and as the process of planning and designing the community-hosted seismic network, the paper presents the simulation of warning times that can be realized in this region based on the distribution of sensor stations in relation to the historical strong earthquakes and target sites. The distribution of sensor stations determines the detection and reporting time of the event, while location of earthquakes and position of target site determine the available warning time for the target to be protected. Testing the various sensor station configurations (regional, On-site and Hybrid) models of CEEWS by simulation of scenario earthquakes, the hybrid configuration that distributes sensor stations closer to the source and on target sites, was able to provide at least 5 seconds of warning times to various targets. This time has been demonstrated to be enough for shutting down hazardous industrial processes and for people to take cover at safer locations to reduce injuries.

Keywords: Community-Hosted Sensors, Sensor Station Configuration, Warning Times, Target Sites

1. Introduction

In the past few years, research and development of earthquake early warning (EEW) have shown significant potential for the mitigation of earthquake disasters [1] -[4] . Countries/areas such as US, Japan, Mexico, and Taiwan have established practical EEW function that protect sensitive structures and provide warning of few seconds to tens of seconds when strong earthquakes are detected [1] -[4] .

Two types of EEW systems (regional and on-site) are mostly in operation around the world and in both systems, warning time is a critical factor to consider as all the public would like to be informed ahead of strong earthquake to save lives. The distribution of sensor network when combined with the reliable communication infrastructure and data processing speed of EEW systems sets the limit to the warning times that can be available to monitor targets before the strong ground motion arrives.

The CEEW system in this study is based on the community sensor deployment and is limited to cities or urban areas where power and internet connectivity is currently possible in the south-western Tanzania (SWTZ) zone. To guarantee some earthquake warning times for the target sites, distribution of sensor stations relative to the locations of historical strong earthquakes in the region, is being evaluated through simulation. That is, varying the positions of sensor stations across SWTZ region, the resulting warning time at each target site as well as the expected level of ground shaking is explored for establishment of CEEWS in SWTZ region.

2. General Seismicity of the SWTZ Region and Location of Target Sites

2.1. Seismicity of SWTZ Region

The SWTZ region is seismically active due to its proximity to the western and eastern junction of the East African Rift (EAR) Valley [5] -[8] . This fact is demonstrated by the occurrences of several earthquakes with varying magnitudes as detailed in Figure 1 that maps the earthquake recorded in the region between 1973 and May 2012.

From Figure 1, the events that have been recorded in Tanzania for the period of 1973 to May 2012, using the United States Geological Survey (USGS) catalogue, are concentrated along the Lake Tanganyika, Lake Nyasa and in the North-Eastern zones, which is the location for the East African Rift (EAR) Valley system across Tanzania [1] .

2.2. Selected Target Sites

The South-Western Tanzania (SWTZ) extends from Njombe, including Mbeya city, Tunduma, Sumbawanga, Namanyere, Mpanda, and Kigoma townships. This region is part of seismically active region, the Western Rift Valley of Africa (WRA), a branch of East African Rift Valley system (EARS) where most recent strong earthquakes with some observable impacts to the communities have been observed [6] -[11] . The proposed CEEWS in this region aims at providing seismic protection to several urban settlements or population centers from earthquakes that may originate from the Lake Tanganyika and Lake Nyasa seismic sources. The Sumbawanga Township and Mbeya city are selected as major target sites in this study, and the other target sites considered for warning issuance whenever possible are Kyela, Njombe, Tunduma, Namanyere, Mpanda and Kigoma Townships.

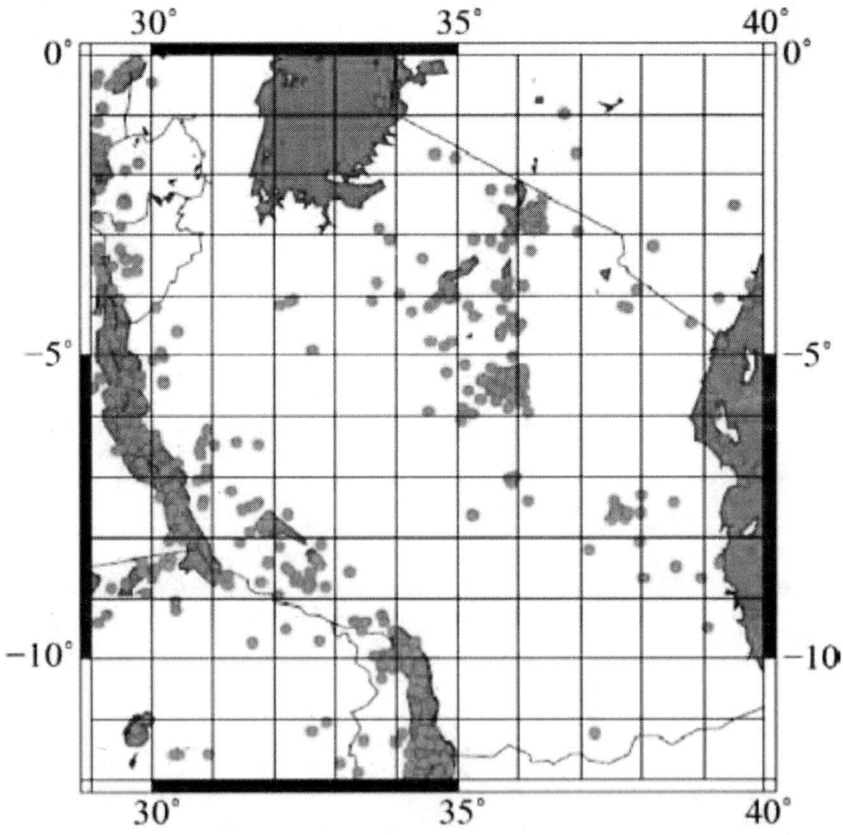

Figure 1. Event recorded between 1973 and May 2012 in Tanzania.

Table 1. Target sites and their epicentral distances to the scenario earthquakes.

Target Site	Magnitude of Scenario Earthquake	Distance, R
Kyela	6.8	606.2
	6.5	397.56
	6.0	65.18
	5.9	42
Njombe	6.8	651.28
	6.5	467.42
	6.0	176.55
	5.9	155.63
Mbeya city	6.8	493.19
	6.5	306.06
	6.0	184.03
	5.9	164.23
Sumbawanga	6.8	296.05
	6.5	100.67
	6.0	338.15
	5.9	327.97
Tunduma	6.8	487.86
	6.5	269.12
	6.0	146.34
	5.9	138.99
Mpanda	6.8	162.97
	6.5	187.95
	6.0	516.57
	5.9	502.79
Kigoma	6.8	147.71
	6.5	364.01
	6.0	740.39
	5.9	727.44
Namanyere	6.8	216.5
	6.5	67.55
	6.0	417.56
	5.9	407

2.3. Scenario Earthquakes Selected for Simulation

The scenario earthquakes considered in this study for warning time simulation includes: magnitude 6.5 earthquake of October 2000; magnitude 6.8 earthquake of December 5, 2005, both from Lake Tanganyika, magnitude 6.0 of February 19, 2009, and magnitude 5.9 earthquake of December 20, 2009 along Lake Nyasa seismic source [1] [6] [9] -[11] . The Sumbawanga Township is at a distance of about 100 km from the nearest Lake Tanganyika 6.5 earthquake epicenter, while Mbeya city is situated at a distance of about 164 km from the Lake Nyasa earthquake epicenters. Other target points considered for warning issuance whenever possible are Kyela, Njombe, Tunduma, Namanyere, Mpanda and Kigoma townships. With respect to the magnitude 5.9 earthquake from Lake Nyasa, distances from the earthquake to Kyela, Njombe and Tunduma are 42 km, 155 km, and 138 km, respectively. Considering the Lake Tanganyika earthquake epicenter (magnitude 6.5), Namanyere, Mpanda and Kigoma towns are at the distances of 67 km, 187 km and 364 km, respectively. Table 1 shows the distances of all the target sites to the epicenters of the four scenario earthquakes used in the simulation.

3. Sensor Station Configurations for Simulation

The community-hosted seismic sensor stations are positioned into regional, onsite, and hybrid configurations (see Figure 2) each at a time during simulation process. The regional earthquake warning method, positions at least five seismic sensors next to the seismic source, to rapidly detect an earthquake onset and contribute station observation data to estimate the earthquake source parameters. The onsite method, embed at least five sensor stations directly on the major target sites (Sumbawanga town and Mbeya city), so that during the earthquake occurrences the beginning of P-wave ground motion observed at these target sites can be used to predict the ensuring ground motion of S-waves at the same sites. For hybrid method, about 49 seismic sensor stations are proposed, where some sensors are embedded on the target sites and some sensors are installed near the seismic sources. Like the other two configurations, five sensor station data would be required to estimate earthquake source parameters.

In Figure 2, yellow triangles are target sites and green triangles are sensor stations/arrays for recording earthquake ground motion across the region.

4. Methodology

To accomplish the objective of the research, examining the warning time that can be realized at target sites for various scenario earthquakes, while adopting different sensor station configurations (see Figure 2), an algorithm for simulating the earthquake travel times across the region was developed for the study. This algorithm utilizes the event location, station location, target site position and the seismic waves travel velocities across the region to estimate the warning times at the target sites. The algorithm also, uses empirical peak ground acceleration (PGA) attenuation relationship across the region to predict earthquake ground shaking levels to be expected at each target sites. The earthquake simulations were performed using the four sensor positioning (Figure 2) and the four selected scenario earthquakes to estimate the achievable warning time at each target site. The optimal sensor positioning was to be the one that provides at least five seconds to major target sites for all scenario earthquakes.

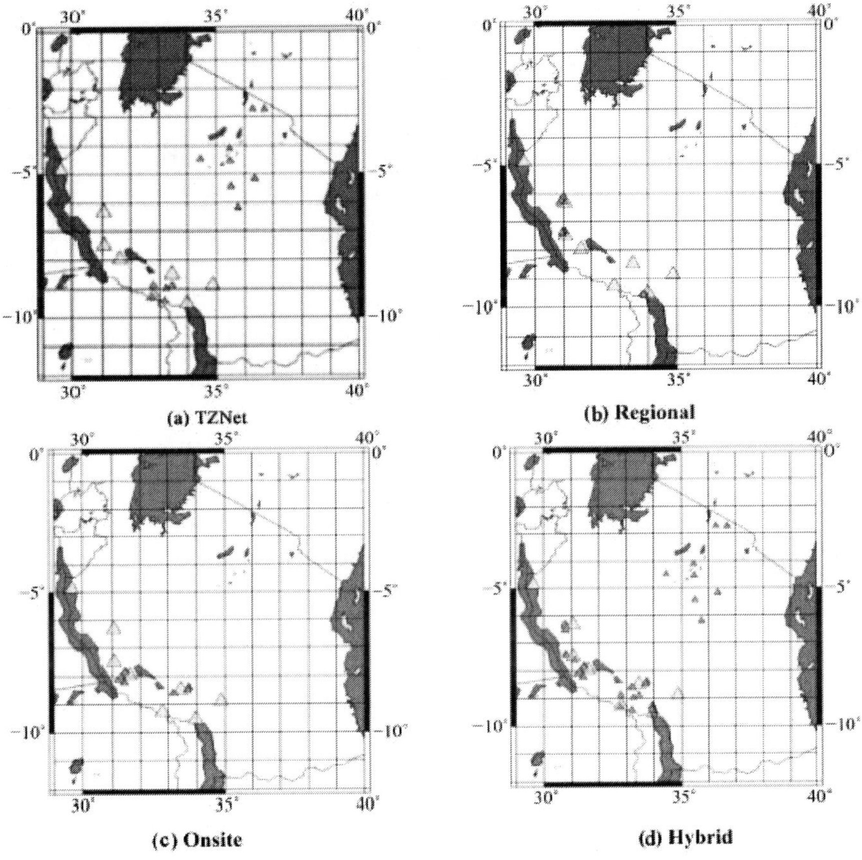

Figure 2. Sensor configurations considered in simulation.

4.1. Warning Time Estimation

To estimate warning times, Pand S-wave arrive times at different stations and target sites were measured for each simulation case. Assuming that the time needed for P-wave to reach sensor station and target site is T_p, the time needed to process data and estimate earthquake source parameter T_{pe}, the time for each sensor station to record sufficient P-wave data for earthquake parameter estimation process to be T_d, and that five stations must trigger to reduce false alarm with time for waiting for five triggers to be (T_{Tr}), Equations (1) is used to estimate time for reporting earthquake event.

$$T_r = T_p + T_{pr} + T_d + T_{Tr} \tag{1}$$

That is, to calculate the warning time, times for P-wave triggers from the first five closest stations to the eventas well as the waiting time for triggers acquisition were added as data processing time. In this study the time for processing data T_{pr} is assumes to be 10 seconds, while the time T_d to record sufficient P-wave data at each site is assumed to be 4 seconds. The time T_p depends on the distance from the event to recording point and the time T_{Tr} depends on the sensor station separation distances.

Then, warning time for each target site is estimated using Equation (2)

$$T_W = T_S - T_r \tag{2}$$

By simulating the four scenario earthquakes used in this study while alternating among the three proposed sensor network configurations, the measurements of seismic wave's arrival time and warning times at various target sites were measured for calculations of warning times.

4.2. PGA Predictions

Warnings are to be issued when the predicted PGA value at a station or target point exceeds the threshold value that is set at 0.001 g in this study. Thus, for each simulation case, the PGA values at each station and each target site were also predicted empirically using Equation (3).

$$\text{PGA}=1.42\exp(1.43\text{M})\text{R}^{-1.2}\left(0.719\ \ln(\tau)\right) \tag{3}$$

where PGA is peak ground acceleration, M is earthquake estimated magnitude, R is the epicentral distance and t is the P-wave observing time.

5. Results and Discussion

Each simulation process simulated an individual scenario event recorded in the SWTZ zone at a time through three steps that starts with significant detection of P-waves PGA at one station, estimation of distances between station and targets/other stations, predicting the expected S-waves PGA and arrival times at other stations in the network, and finally the checking of the alarm decision criteria (trigger for five stations and PGA levels) for CEEWS.

5.1. Simulated Warning Times Results

When all the scenario earthquakes were simulated with the proposed sensor positioning, available target warning times variations across the region were plotted as shown in Figure 3.

According to Figure 3, warning times for each sensor configuration increases with epicenter distances, but at varying rates. The Hybrid configuration show higher warning times at comparable epicentral distances, and the TZNet configuration shows the lowest warning times. The summary of the simulated warning times trends for the four sensor positioning and four scenario earthquakes are shown in Table 2 for the major target sites.

Using Table 2, only the hybrid sensor configuration guarantees the five warning times to the target sites for all four scenario earthquakes. That is, the hybrid configuration satisfies the criteria for the design of CEEWS in this region and is therefore considered as an optimal configuration of sensor stations.

Table 2. Simulated warning times for major target sites from four scenario earthquakes

	Warning time for various sensor configurations (s)							
	TZNet		Onsite		Regional		Hybrid	
Scenario event magnitude	SBA	MBY	SBA	MBY	SBA	MBY	SBA	MBY
6.8	22	70	44	98	58	118	60	120
6.5	−35	20	−15	45	3	75	5	80
6.0	30	−15	50	10	68	26	70	28
5.9	25	−20	50	6	68	20	70	22

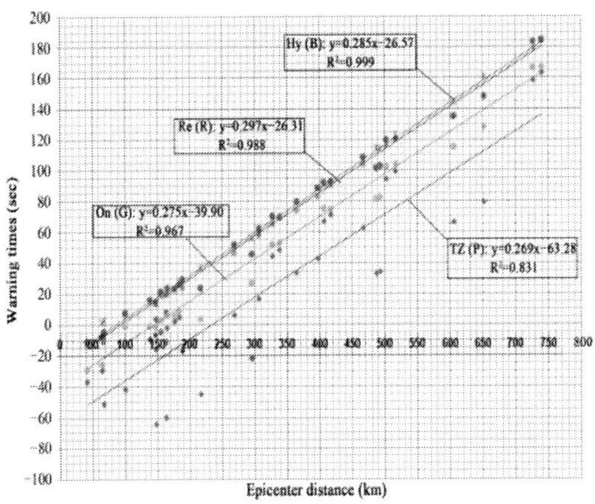

Figure 3. Available warning times at target sites for each sensor configuration.

5.2. PGA Variations across the Region

There is correlation between PGA values to the experienced damages, but not always absolute agreement since experiences and damage can be affected by many other factors, including the quality of earthquake engineering, type soil cover, as well as construction practices. Generally, PGA values of 0.001 g to 0.01 m/s^2 are perceptible by people, 0.02 g to 0.2 m/s^2 people lose their balance, and above 0.50 g are very high and only well-designed buildings can survive if the duration is short. For simplicity, in this study PGA values above 0.5 g are considered to cause severe ground shaking, PGA values between 0.5 g and 0.2 g are considered to cause strong ground shaking, PGA values below 0.2 g but above 0.01 are classified to cause weak ground shaking, and PGA values below 0.01 g are considered unfelt. The simulate PGA variations across the SWTZ region from the four scenario earthquakes considered in this study, are presented in Figure 4.

From Figure 4, all PGA curves simulate the decrease in PGA values as the distance from the event epicenter increase. Largest earthquake (magnitude 6.8) show large values of PGA at each grid point and attain PGA values than can be felt at larger distances. The decrease in PGA with distance for each case is large close to the epicenter and decrease as distance increases.

Considering PGA curve for magnitude 6.8 earthquakes, severe ground shaking is predicted to reach a distance of 160 km, strong shaking will be between 160 km and 360 km, weak ground shaking is between 360 km and 700 km, while the other areas will not feel the event.

Magnitude 6.5 earthquakes is simulated to cause severe ground shaking up to a distance of 110 km, strong ground shaking between 110 km and 230 km, weak ground shaking between 230 km and 480 km, while areas above 480 km from the epicenter will not feel the event.

Magnitude 6.0 is simulated with severe ground shaking that reaches 80 km, strong ground shaking between 80 km and 110 km, weak ground shaking between 110 km and 230 km and areas above 230 km will not feel the event.

Magnitude 5.9 is simulated to cause severe ground shaking that reaches about 43 km, strong ground shaking between 43 km and 90 km, weak ground shaking between 90 km and 200 km, while above 200 km the event will not be felt.

Thus, when the Sumbawanga Townships is predicted to be in severe ground shaking from magnitude 6.5 earthquakes, people dwelling in this town will be warned with only five seconds to move to safe positions. When the same city is predicted to fall under strong ground shaking of magnitude 6.8 earthquakes, there will be 22 seconds of warning time before the start of strong ground shaking at their location. Due to its location, a minimum of

164 km from the considered earthquakes, Mbeya city is simulated with strong shaking from magnitude 6.8 with about 60 seconds of warning times.

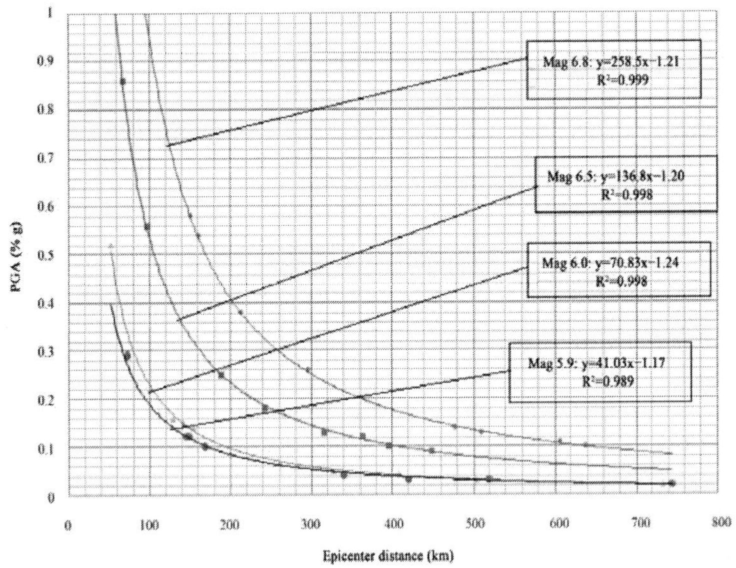

Figure 4. Predicted PGA values at each epicentral distance for various earthquake magnitudes.

6. Conclusion

EEW system can play an important role for earthquake hazard mitigation in Mbeya city and surrounding regions. The evaluation of warning capability in terms of warning time T_w is essential for system design to reflect the "ideal" performance of an CEEW system. The hybrid sensor network configuration appears to offer reasonable warning time to mitigate seismic hazards and should be considered for implementation of CEEWS to improve seismological observations in the region and for an early-warning purpose.

References

1. USGS Web Site 2000, 2005, 2009. www.earthquakes.usgs.gov/
2. B?se, M., Ionescu, C. and Wenzel, F. (2007) Earthquake Early Warning for Bucharest, Romania: Novel and Revised Scaling Relations. Geophysical Research Letters, 34.

3. Horiuchi, S., Negishi, H., Abe, K., Kamimura, A. and Fujinawa, Y. (2005) An Automatic Processing System for Broadcasting System Earthquake Alarms. Bulletin of the Seismological Society of America, 95, 347-353.
4. Hsiao, N.C., Wu, Y.M., Zhao, L., Chen, D.Y., Huang, W.T., Kuo, K.H., Shin, T.C. and Leu, P. L. (2011) A New Prototype System for Earthquake Early Warning in Taiwan. Soil Dynamics and Earthquake Engineering, 31, 201-208.
5. Mavonga, T. (2007) An Estimate of the Attenuation Relationship for the Strong Ground Motion in the Kivu Province, Western Rift Valley of Africa. Physics of the Earth and Planetary Interiors, 162, 13-21.
6. Mavonga, T.G. and Durrheim, R.J. (2009) Probabilistic Seismic Hazard Assessment for Democratic Republic of Congo and Surrounding Areas, Western Rift Valley of Africa. Special Africa Array Issue of South African Journal of Geology, 112, 329-342.
7. Espinoza-Aranda, H.M. and Rodriguez, F.H. (2003) The Seismic Alert System of Mexico City. In: Lee, W.H.K., Kanamori, H., Jennings, P.C. and Kisslinger, C., Eds., International Handbook of Earthquake and Engineering Seismology, Part B, Academic Press, Amsterdam, 1253-1259.
8. Macheyeki, A.S., Delvaux, D., De Batist, M. and Mruma, A. (2008) Fault Kinematics and Tectonic Stress in the Seismically Active Manyara—Dodoma Rift Segment in Central Tanzania—Implications for the East African Rift. Journal of African Earth Sciences, 51, 163-188.
9. UN Office for the Coordination of Humanitarian Affairs (2014) Tanzania-Earthquake OCHA Situation Report No. 1. Ref: OCHA/GVA-2000/0185. OCHA Situation Report No. 1 Tanzania-Earthquake 2 October 2000.
10. UN Office for the Coordination of Humanitarian Affairs (2014) Ref: OCHA/GVA-2000/0193. OCHA Situation Report No. 2 Tanzania-Earthquake 5 October 2000. Tanzania-Earthquake OCHA Situation Report No. 2.
11. UN Office for the Coordination of Humanitarian Affairs (OCHA) (2014) East Africa: Earthquake OCHA Situation Report No. 1. Ref: OCHA/GVA-2005/0217. 2005/12/05. OCHA SITUATION Report NO. 1. DRC, Tanzania, Burundi, Kenya, Rwanda, Uganda, Earthquake-Lake Tanganyika 5 December 1995.

CHAPTER 12

Comparison of Energy Dissipation, Stiffness, and Damage of Structural Oriented Strand Board (OSB), Conventional Gypsum, and Viscoelastic Gypsum Shearwalls Subjected to Cyclic Loads

David W. Dinehart [1,*] and Andrew S. Blasetti [2]

[1] Department of Civil and Environmental Engineering, Villanova University, 800 Lancaster Avenue, Villanova, PA 19085, USA

[2] Thornton Tomasetti, 744 Broad Street, Newark, NJ 07102, USA

Abstract

A key element in the seismic load resisting system of a wood framed structure is the shear wall which is typically sheathed on one side with plywood or oriented strand board (OSB) and gypsum on the other. The shear capacity of gypsum sheathed shear walls is typically neglected in high seismic areas due to the susceptibility of conventional drywall screw connections to damage caused by earthquakes. The earthquake resistance of an innovative viscoelastic (VE) gypsum shearwall is evaluated and compared to conventional structural and non-structural walls. Ten 8 ft × 8 ft wood framed wall specimens of three configurations [nailed-OSB, screw-gypsum, and VE polymer-gypsum] were subjected to a cyclic test protocol. The energy dissipation, stiffness, and damage characteristics of all shearwalls are reported herein. Testing results indicate the VE-gypsum walls can dissipate more energy than the OSB structural panels and 500% more energy that the conventional gypsum sheathed walls and contains a constant source of energy dissipation not seen in the structural and non-structural walls. The wall stiffness of the

OSB wall degrades at a far greater rate that the VE gypsum wall and at continued cycling degrades below the VE wall stiffness. Unlike both of the conventional wall types, the VE wall showed no visible or audible signs of damage when subjected to shear displacements up to 1.

Keywords: wood shearwall; seismic; gypsum; viscoelastic polymer; energy dissipation

1. Introduction

Light-frame wood structures are quite common throughout North America due to their high strength-to-weight ratio. Their ductility via inelastic deformation protects human life during moderate to large earthquakes, but leads to economic hardship afterwards. Nonstructural, structural, and secondary damage add considerable financial burden to home owners and insurance companies. In the United States, the devastating 6.7 magnitude Northridge earthquake damaged tens of thousands of residential homes. Over $20 billion in property damage and the loss of nearly 48,000 housing units was due to the inadequacy of wood-frame construction [1]. With today's need for performance-based design, the demand for new and innovative ideas for energy dissipation in wood-frame structures has grown considerably.

The major lateral load-resisting system for a light-frame wood structure consists of floor diaphragms, roof diaphragms, and vertical shear walls. Wood-frame shear walls have a critical role in preventing the destruction of residential buildings in earthquake regions—they transfer horizontal inertia forces from one floor to the next and ultimately to the foundation. Earthquakes inflict significant damage to wood shear walls. Sheathing-to-frame connections have been identified as the primary source of degrading performance and are now reinforced via reduced nail spacing, stronger nails, and thicker panels [2]. This guideline, however, is not as effective as one might expect due to the degradation of nailed sheathing connections subjected to dynamic loading [3]. As sheathing connections are lost, the wall damage continues and a reduction in stiffness and energy dissipation continues with each passing cycle of displacement.

Generally, nonstructural elements such as gypsum wall board and exterior finishes are not counted on in design of the lateral force resisting system. Attempts to quantify the value of nonstructural elements aiding in reducing seismic response have been made [4]. Shake table testing was conducted of a two-story, single family woodframe house with and without interior gypsum wall board and exterior wall finish materials. The structure with the wall finishes showed an increase in lateral stiffness and a substantial reduction in its seismic displacement response. The tests however, were unable to isolate the individual contribution of the gypsum wallboard and the exterior finish. Damage in the structure included diagonal cracking in the gypsum wallboard and stucco.

The damage threshold of nonstructural gypsum wallboard partition walls subjected to seismic loads was investigated with a series of 11 full-scale woodframe partitions sheathed on both sides with 13 mm gypsum [5]. Typical wall damage evident in all wall tests included fasteners pulling through the gypsum, cutting the gypsum, or tearing through the edge of the panel. Since the gypsum is not counted on for its structural use, this damage is considered nonstructural. Never the less, this damage caused during an earthquake can be extensive and expensive.

While gypsum wallboard is typically disregarded as a structural material, many experiments have shown that it does add to the structural capacity of the wall system. The shear capacity of shear walls sheathed with gypsum wallboard was investigated in thirty woodframe walls tests [6]. It was shown that wallboard orientation greatly affected strength and stiffness of the shear walls. Walls tested with panels oriented horizontally were over 40% stronger and stiffer than those with vertical panels. The results of this study allow structural engineers to consider the effects of gypsum wallboard when designing a structure to resist wind and seismic loads. In another study, the performance of shear walls sheathed with gypsum wallboard was compared to shear walls sheathed with plywood [7]. It was concluded that the contribution of gypsum sheathing to racking strength was significant compared to plywood, with gypsum wallboard on one side provided 38%–64% the racking strength produced by shear walls with plywood on one side, and 30%–39% the racking strength of shear walls sheathed with plywood opposite gypsum wallboard.

Shear walls with finish materials including gypsum were investigated in an effort to characterize their damage patterns [8]. Tested shear walls contained window and or door openings. At 0.2% drift, the gypsum wallboard sustained damage confined to small hairline cracks near the wall openings. Cracks were typically less than 0.002" wide. A small number of sheathing connections exhibited a bulging effect in the sheathing material near the edges of the panels. At 0.4% drift, the gypsum wallboard panels began to deteriorate and crack at the corners of the openings. Cracks ranged from 0.005" to 0.025" wide. Tearing along the taped joints also became evident. At 0.7% drift, the cracks ranged from less than 0.002" to 0.05" wide. Tearing along the taped joints and bulging near the sheathing connections were more pronounced.

As part of the NEESWood project there was a focus on a performance-based seismic design (PBSD) philosophy for woodframed structures. Experimental studies as part of this project included large scale shake table testing of a full-scale, two story, light frame wood townhouse building, designed according to modern U.S. engineered seismic design requirements [9] and a full-scale mid-rise, six story, light frame wood apartment building [10,11]. Additionally a seismic performance assessment of woodframed structures with energy dissipation systems was conducted [12,13].

The two story building structure was tested on two tri-axial shake tables operating simultaneously. The scope of this portion of project was to determine the dynamic characteristics and the seismic response of the structure under various base input intensities that were representative of ordinary and near-field ground motions typical in California. The structure was tested with and without gypsum wallboard finished on the interior and stucco finishes on the exterior. The experimental results clearly showed that the addition of gypsum wall board to the interior structural wood sheathed walls substantially improved the seismic response of the test structure. Additionally, the addition of stucco to the exterior added to the improved seismic response. The improvement was most noticeable in the longitudinal direction, where the shear response of low aspect ratio wall piers dominated the response. This work was a major step forward in noting the significance of gypsum and stucco in improving the seismic performance of the lateral-load resisting systems in light-frame wood construction [9].

The goals of the experimental testing program of the six story apartment building were to demonstrate that the performance-based design procedure developed as part of NEESWood worked on a full-scale building and to gain a better understanding of how mid-rise light-frame wood buildings respond to a major earthquake (2500 year event for Los Angeles, California). Additionally, the testing was intended to provide a baseline data set to the earthquake engineering community. An enormous amount of data was collected from numerous tests, including base shears, wall drifts, inter-story drifts, accelerations, hold-down forces and roof drifts that highlighted many significant findings; however, the information most relevant to the research reported herein focuses on the damage assessment of the test structure [10].

Another goal of the six-story shake table testing in the NEESWood project was focused on how the various construction elements influenced the seismic behavior of woodframed structures so that they could be better considered in the PBSD philosophy. To examine this issue several 50 year seismic tests were conducted at various stages of construction. The drywall panels were "normal" strength Japanese drywall panels. The panels were installed with tape and putty on all joints except the wall-to-ceiling joints and corners. The gypsum wall board was repaired between tests; however, the sill plates were not. Visual damage inspections were conducted after all tests. No edge tearout of the gypsum was evidenced due to the fact that large edge distances had been prescribed in this engineered building. Overall damage was characterized as non-structural damage. Most cracking was located at door and window openings and propagated diagonally outward, similar to damage that is typical for a moderate earthquake. Overall, the building performed very well during all earthquakes reviewed. There was no damage to any wood components in the building. The only damage found was attributed to the gypsum wallboard and noted to be non-structural [11].

While this recent information in regards to full-scale performance will aid design engineers, there is currently a lack of information regarding the dynamic effects of interior and exterior wall finishes on a structure. The extensive damage caused by earthquakes has led to the disposition of gypsum wallboard as a non-structural material. The initial response is to reduce the amount of allowable stress for gypsum-sheathed shear walls. In fact, these walls are typically neglected when designing a structure's lateral load-resisting system—especially in high seismic regions. However, simply neglecting the effects of a structural element can result in an overly conservative and possibly dangerous design. Generally, load acting upon a system is attracted to members with higher stiffness, so by neglecting an element's stiffness one might develop a misunderstanding of the overall structural system behavior. The attraction of force on certain walls during an earthquake may be considerably larger than anticipated, resulting in additional damage and potential collapse.

The objective of this research is to improve the seismic performance of wood-frame shear walls (*i.e.*, increase energy dissipation while reducing damage to the wall) by utilizing gypsum sheathing attached by a viscoelastic (VE) polymer. A wall that undergoes neither structural nor non-structural damage during earthquakes has the potential to save lives and billions of dollars worth in repairs for homeowners and insurance companies. The isolated performance of the VE polymer is relevant, but beyond the scope of this paper. Details related to the hysteretic behavior, stiffness and energy dissipation characteristics of the VE polymer have been well documented [14,15,16,17,18]. Described herein are the results of an experimental study investigating the dynamic behavior (hysteresis form, energy dissipation, stiffness, and damage pattern) of three shearwall types: 1. conventional structural walls (nailed—oriented strand board, OSB), conventional non-structural walls (screws/gypsum), and VE walls (VE polymer/gypsum).

2. Shearwall Specimens

Three types of wood-frame shear walls are presented in this study: walls sheathed with oriented strand board using conventional nailed sheathing connections (OSBwall), walls sheathed with gypsum wallboard using conventional screwed connections (GYPwall), and walls sheathed with gypsum wallboard using viscoelastic connections (VEwall). Most wood-frame shear walls are designed with either OSB or plywood sheathing; therefore, the OSBwall specimens are considered the structural walls.

The GYPwalls are conventional gypsum sheathed walls that are generally considered non-structural, while the VEwall specimens represent a modified shear wall design. The framing of all walls consisted of Spruce-Pine-Fir (SPF) 2 in. × 4 in. lumber. Four OSBwall specimens, shown in Figure 1, were constructed. The walls were sheathed with 7/16" OSB which was connected to the framing via Bostich 8d clipped-head cooler nails driven by a pneumatic Bostich nail gun. Nails were spaced 4" around the perimeter of each panel with an edge spacing of 3/4" along the end studs and sill plates and 3/8" along the center stud. Nails along the field studs were spaced every 12".

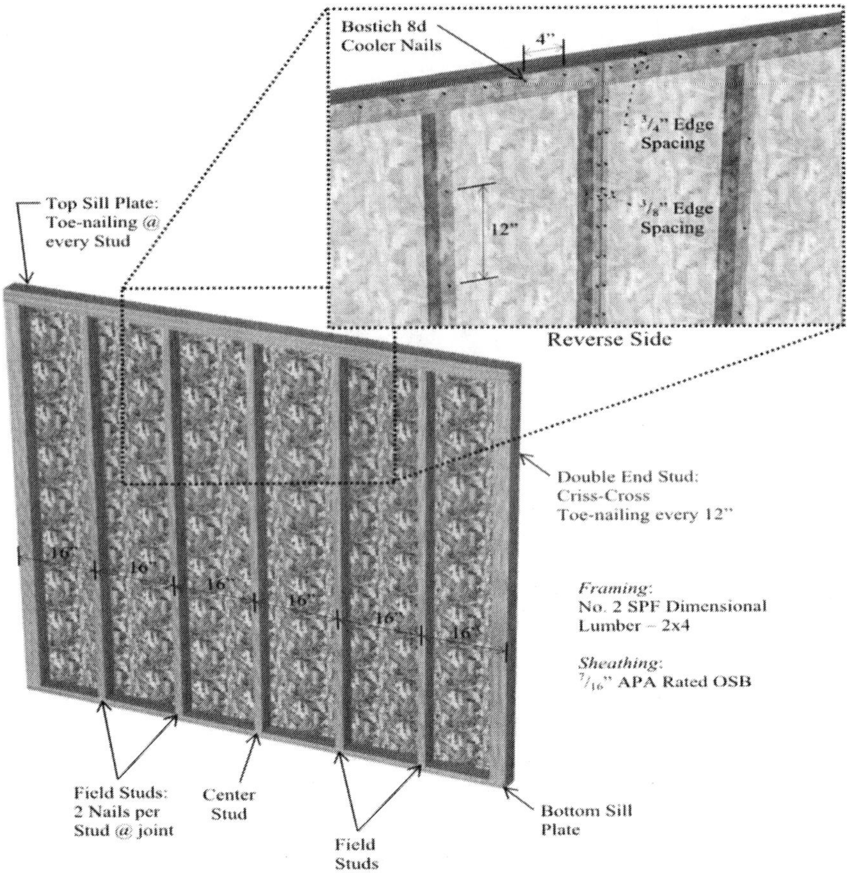

Figure 1. OSBwall (oriented strand board) Specimen.

Three 8 ft × 8 ft gypsum shear walls (GYPwall) and three VE gypsum shear walls (VEwall) were constructed and tested. The GYPwall and VEwall specimens, shown in Figure 2, consisted of two 4 ft × 8 ft gypsum panels 1/2 in. thick oriented vertically and separated by 0.125 in. The end studs and top sill plate are comprised of two studs nailed together, while all other studs as well as the bottom sill plate were made from one stud. Framing members were connected together with two 16d 3 + 1/2 in. nails at every joint and 1 + 1/4 in. drywall screws spaced at 12 in. along the perimeter and the field connected the gypsum to the frame.

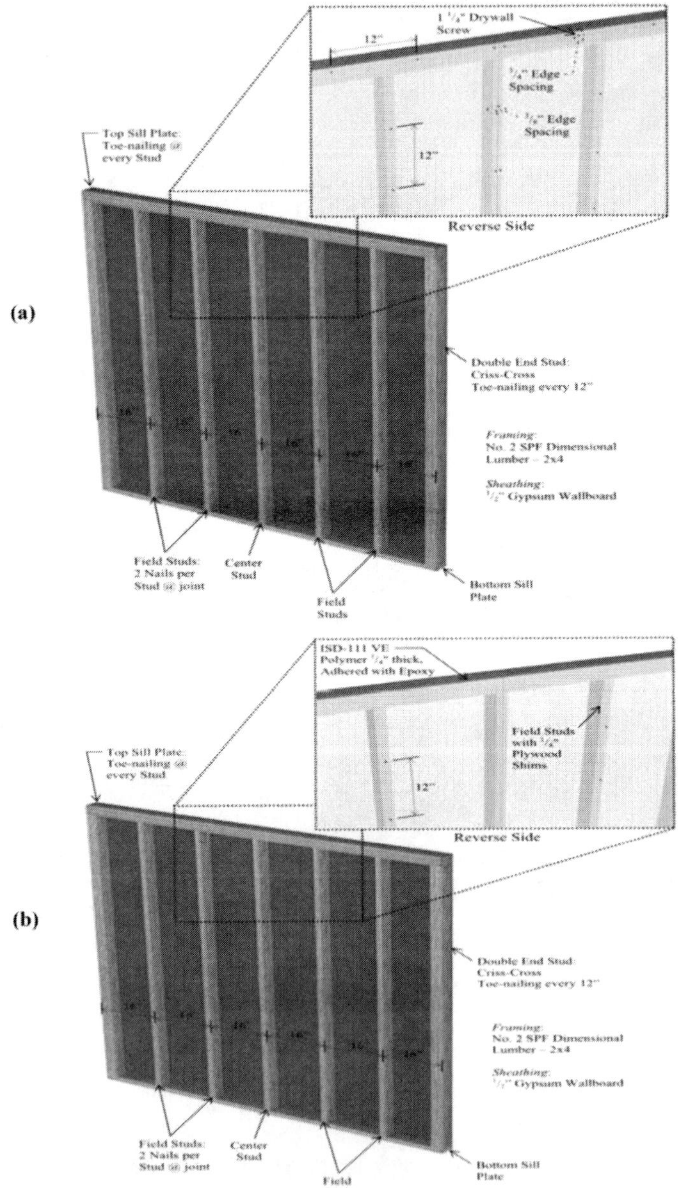

Figure 2. (**a**) GYPwall (gypsum) specimen; and (**b**) VEwall (viscoelastic) specimen.

The VEwalls used ISD-111 VE polymer manufactured by 3M that is isotropic and has properties that simultaneously exhibit those of a viscous fluid as well as a solid [14,15,16,17,18]. Therefore, its dynamic response is dominated by a stiffness force (function of displacement) and a damping force (function of velocity). The perimeter of the sheathing panels was fastened to the framing by adhering the VE material between the two. Viscoelastic material was located along the top and bottom sill plates as well as the end and center studs. Strips of VE polymer were initially adhered to the stud framing using high strength epoxy. The plastic backing was removed from the strips and epoxy was applied to the sheathing side of the polymer and a sheathing panel was laid on top of it. Screws were immediately installed along the perimeter with a spacing of 12" to produce a clamping effect as recommended by the epoxy manufacturer. After 24 hours, the perimeter screws were removed. Screwed sheathing connections along the field studs typically undergo little if any damage because the relative displacement between the sheathing and the field studs is typically small. Hence, a 1/4 in. shim was used in lieu of VE material. Joint tape and compound were applied between the two gypsum panels 24 hours prior to testing for both types of shear wall specimens.

3. Experimental Setup and Procedure

The wall specimen is installed between two load-resisting self-reacting frames, and is anchored to a lower support channel. Figure 3 shows the shear wall test frame with a GYPwall specimen installed. A horizontal displacement load was distributed along the top sill plate, while the bottom sill plate was fully braced with anchor bolts and cover plates. Lateral load is applied via a 20,000 lb hydraulic actuator with a guide beam that is bolted to the top of the specimen. Bearing rollers, oriented perpendicular to the direction of loading, were installed along the top sill plate as lateral bracing. Uplift was restrained at the end studs through the use of TDX5 hold downs manufactured by USP Connectors.

The data acquisition and control system simultaneously provided the hydraulic actuator with the displacement test sequence while recording measurements during its execution. Five channels of data were recorded including load from the actuator, lateral displacement of the actuator located at the top of the wall, lateral displacement at the other top corner of the wall via an independent cable potentiometer, and uplift displacements at each corner.

Figure 3. Shearwall test frame with specimen installed.

All shear wall specimens were subjected to the recommended CUREE displacement protocol for shear walls at a frequency of 1.0 Hz [19]. The maximum displacement was 2.86". The shear wall loading protocol is plotted in Figure 4. This protocol, designed to mimic the effects of an earthquake, is split into three components: initiation cycles, primary cycles, and secondary cycles. The initiation cycles simulate the cyclic loading that structures typically undergo. After the initiation cycles are sets of one primary cycle followed by several secondary cycles, which simulate earthquake loading. The primary cycles have larger amplitudes than their preceding secondary cycles, and simulate major events. Shear walls were subjected to a total of 37 cycles.

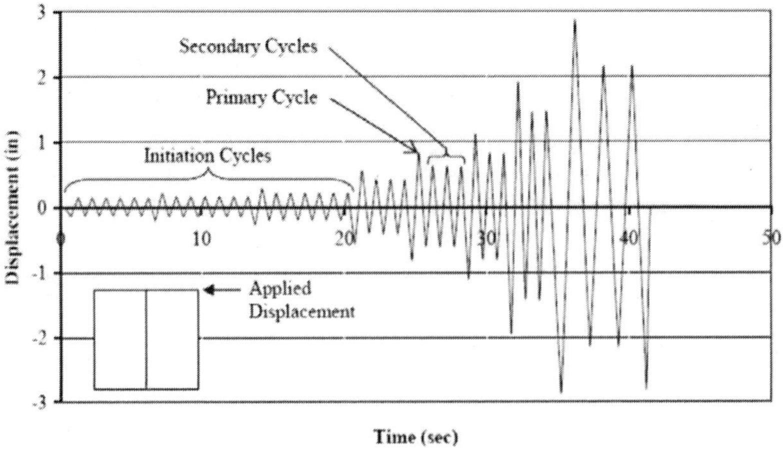

Figure 4. Shearwall loading protocol.

4. Hysteresis Results

A typical hysteresis (load-displacement) plot for the four OSBwalls tested is presented in Figure 5. The area inside a single cycle hysteresis is defined as the energy dissipated in that cycle. The slope of the loop is defined as the stiffness [20,21,22]. The OSBwall hysteresis is characterized by a pinched region at small displacements and load degradation at constant amplitude cycling.

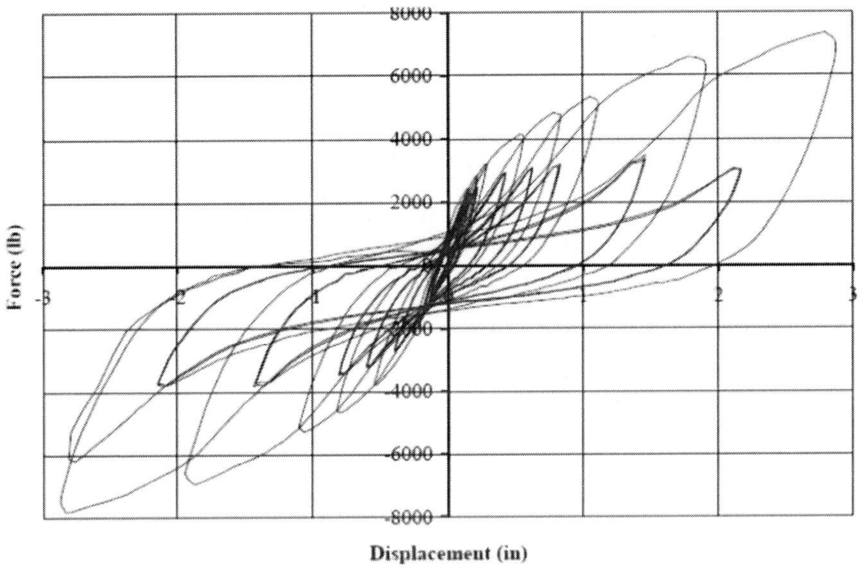

Figure 5. Hysteresis plot for OSBwall 1.

A typical hysteresis plot for one of the three GYPwall specimens tested is shown in Figure 6. Qualitatively, the hysteresis plots for the gypsum shear walls take on some but not all characteristics of the OSB walls. The most prominent characteristic is the extreme degradation between primary and secondary cycles, marked by reduced area of the hysteresis loops. The pinched region is also well-pronounced, and has virtually no thickness near the origin. However, certain characteristics of the hysteresis plots are unique to these shear walls. The primary cycles exhibit loops that are smooth and rounded, without sharp peaks or corners. This differs from both OSB shear walls and baseline gypsum connections and is an unanticipated response to the ramped loading defined by the recommended CUREE displacement protocol.

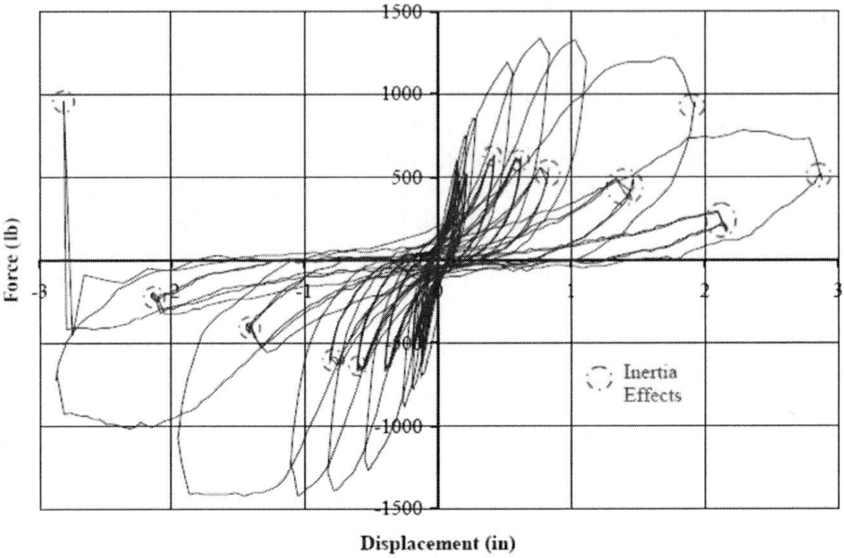

Figure 6. Hysteresis plot for GYPwall 3.

The sudden change in motion caused by ramped loading often produces abrupt changes in direction of the hysteresis plots, which are not present in the primary cycles of the gypsum shear walls. They are, however, present in the secondary cycles, which exhibit abnormalities at the extreme displacements. These abnormalities are most likely caused by inertia effects imposed by the mass of the load beam that rests atop the shear wall specimen during testing. These effects are highlighted by the red dashed circles in Figure 6. With ramped loading, the actuator changes direction very quickly, requiring maximum acceleration from the hydraulic system. This acceleration is applied to the steel load beam, which contains a significant mass and is bolted to the top of the shear wall. The inertia forces caused by the apparatus are typically neglected because they represent a slight percentage of the dynamic response of most shear walls. However, gypsum shear walls provide very weak responses, so the inertia effects represent a greater percentage of the total response.

Three VEwalls were tested with very similar results. A typical dynamic response is presented in Figure 7. Investigation of the hysteresis shows that the pinched region has been eliminated and there is very little degradation between cycles. Hysteresis loops for the VE gypsum shear walls appear more as parallelograms and not ovals, similar to a hysteresis loop of a VE damper. This can be attributed to the loading protocol rather than the type of specimen. The ramped loading used for the shear wall protocol consists of regions of constant velocity. When a damper is subjected to a constant velocity it produces a constant damping force, producing a hysteresis loop that resembles a rectangle. If a spring is assembled in parallel to the damper, then the rectangular hysteresis loop would be projected onto the linear response of the spring. It is significant to note that all three VEwall specimens failed after 31 cycles.

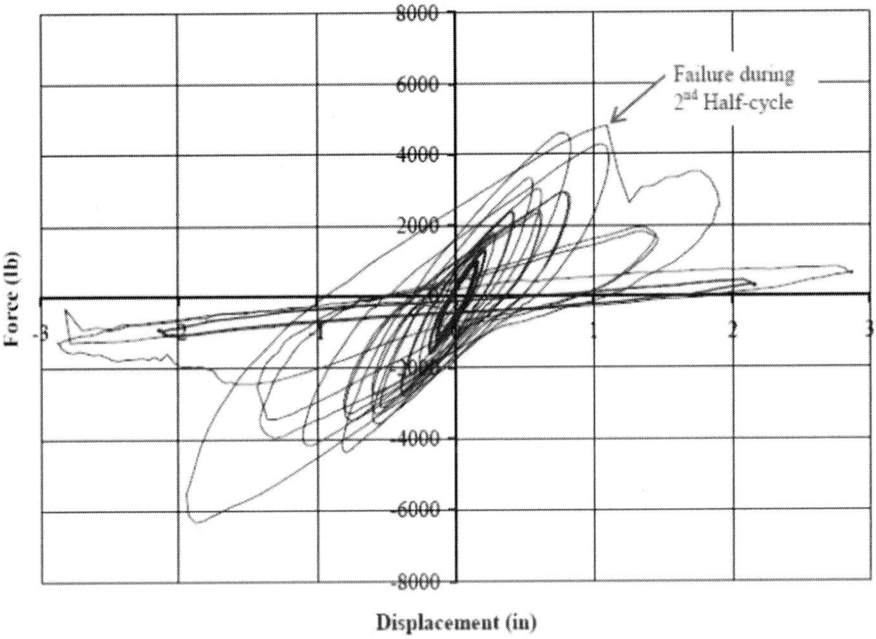

Figure 7. Hysteresis plot for VEwall 3.

Figure 5–Figure 7 provide a detailed view of the cyclic response for each wall. The displacement scales for each plot are identical; however, to view the detail of the plot the load scales are significantly different. Figure 8 presents the three hysteresis responses shown in Figure 5–Figure 7 plotted to the same load scale. It is clear that the response of each of these wall systems to the simulated earthquake cyclic loading is dramatically different. In comparing the GYPwall with the OSBwall, both have similar forms; however, the GYPwall clearly provides substantially less load resistance than the OSBwall. The VEwall has a drastically different response as indicated by the shape of the hysteresis loops and also has a much greater cyclic load resistance when compared to the GYPwall. A general comparison of the cyclic behavior of these connections and walls has been reported [20]. Based on these significant differences in behavior between the three wall types an in- depth analysis of the energy dissipation and stiffness characteristics on a per cycle and cumulative basis is warranted. This detailed analysis is provided in the following sections.

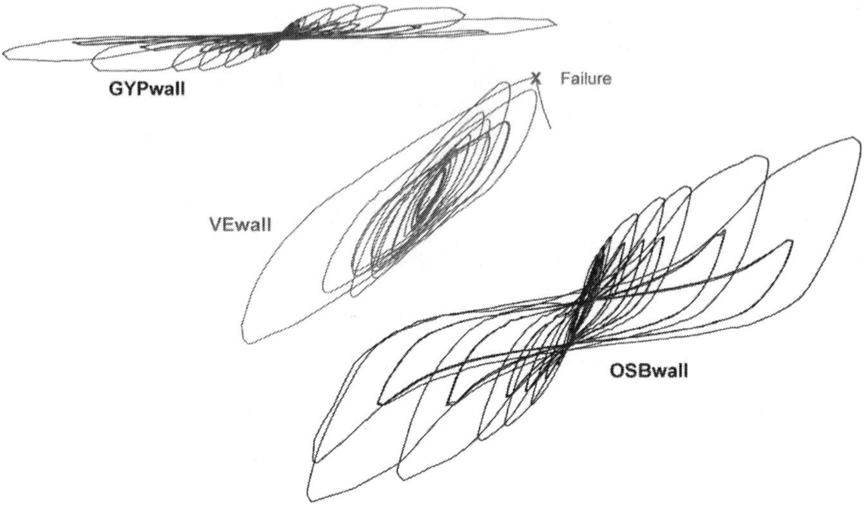

Figure 8. Representative hystereses plotted to the same scale.

5. Energy Dissipation Results

The energy dissipation, E_D, of the wall is calculated per cycle as the area enclosed in one loop. Table 1–Table 3 present the energy dissipation data for the OSBwalls, the GYPwalls and VEwalls respectively. The cycle number, lateral wall displacement, individual test results, averages, and standard deviations are included. Cycles are grouped in sets according to cycle displacement. The first cycle within each set is the primary cycle, and the following cycles of that set are the secondary cycles.

The relationship between energy dissipation and cycle number for all wall specimens is presented in a graphical format inFigure 9. Trend lines, representing the average energy dissipation, are plotted for each type of wall. Error bars are plotted along the trend lines to illustrate the standard deviation within the data at each cycle. Figure 10 presents the total energy dissipated through a given cycle for each wall type. Similar to Figure 9, trend lines, representing the average, are plotted for each wall specimen and error bars are plotted along the trend lines to illustrate the standard deviation within the data at each cycle.

Table 1. Energy Dissipation (lb-in) of OSBwalls.

Cycle	Displacement (in)	OSBwall1	OSBwall2	OSBwall3	OSBwall4	Average	Standard deviations
1	0.14	124.4	174.8	163.8	109.1	**143.1**	*31.3*
2	0.14	127.9	157.5	126.4	100.2	**128**	*23.4*
3	0.14	125.5	153.5	127.7	98.9	**126.4**	*22.3*
4	0.14	122.6	145.5	108	96.6	**118.2**	*21.1*
5	0.14	121.4	151.4	113.1	97.7	**120.9**	*22.6*
6	0.14	119.9	146.1	111.5	92.1	**117.4**	*22.4*
7	0.21	291	313.6	266	244.1	**278.7**	*30.1*
8	0.15	163.9	192.9	141.1	136.1	**158.5**	*25.9*
9	0.15	158.9	185	137.8	130.3	**153**	*24.5*
10	0.15	155.5	169.6	134.1	128.8	**147**	*19*
11	0.15	157.2	174.9	133.5	129.2	**148.7**	*21.4*
12	0.15	155.2	178.4	131.9	131.6	**149.3**	*22.4*
13	0.15	152.6	179.3	136.5	127.2	**148.9**	*22.8*
14	0.28	478.4	486.3	428.3	435.4	**457.1**	*29.5*
15	0.2	275.1	297	216.1	246.3	**258.6**	*35.1*
16	0.2	266.2	273.9	215.9	233.8	**247.5**	*27.3*
17	0.2	262	272.2	223.8	231.6	**247.4**	*23.4*
18	0.2	261.7	273.7	199.4	228.6	**240.8**	*33.6*
19	0.2	262.6	263.5	207.9	225.9	**240**	*27.6*
20	0.2	258.4	267.7	224.4	229.8	**245.1**	*21.2*
21	0.55	1813.5	1714.2	1457.2	1898.7	**1720.9**	*191.3*
22	0.41	807.6	749.1	679.9	838	**768.7**	*69.7*
23	0.41	786.4	730.6	631.1	802.7	**7373.7**	*77.5*
24	0.41	751.4	742.7	604.7	792.7	**722.9**	*81.7*
25	0.83	3392.8	3074.6	2631.6	3536.3	**3158.8**	*401*
26	0.61	1458.4	1288.4	1113.5	1580.3	**1360.1**	*203.4*
27	0.61	1392.2	1265.9	1041.1	1473.1	**1293.1**	*188.4*
28	0.61	1362.2	1266.2	965.8	1440.3	**1258.6**	*207.8*
29	1.11	5021.2	4576	3962.3	5135	**4673.6**	*532*
30	0.82	2206.9	1991.7	1654	2307.2	**2040**	*289*
31	0.82	2090.5	1981.8	1488.1	2167.5	**1932**	*305.5*
32	1.93	13439.5	13016	11687.7	13513.9	**12914.3**	*846.6*
33	1.44	4990.1	4683	3954.7	4763.9	**4597.9**	*448.1*
34	1.44	4582.3	4497	3785.2	4394.2	**4314.7**	*361.2*
35	2.87	21142.2	21448.7	19598.9	20691.9	**20720.4**	*809.7*
36	2.15	7625.3	7598.2	6475.9	6583.5	**7070.7**	*626.4*
37	2.15	7050.5	6979.5	6065.7	5811.9	**6476.9**	*630.6*
Total		83953.5	82060.2	71344.6	81184.4	**79635.7**	*5646.9*

Table 2. Energy Dissipation (lb-in) of GYPwalls.

Cycle	Displacement (in)	GYPwall1	GYPwall2	GYPwall3	Average	Standard deviation
1	0.14	69.7	101.0	95.0	**88.6**	*16.6*
2	0.14	49.0	60.1	63.4	**57.5**	*7.5*
3	0.14	43.7	52.4	61.7	**52.6**	*9.0*
4	0.14	40.9	49.8	55.3	**48.6**	*7.3*
5	0.14	40.6	48.3	55.4	**48.1**	*7.4*
6	0.14	37.6	45.0	50.3	**44.3**	*6.4*
7	0.21	105.6	112.4	122.7	**113.6**	*8.6*
8	0.15	46.4	50.3	65.5	**54.1**	*10.1*
9	0.15	40.7	54.7	55.1	**50.2**	*8.2*
10	0.15	43.1	49.9	56.9	**50.0**	*6.9*
11	0.15	38.4	44.6	54.7	**45.9**	*8.2*
12	0.15	39.4	49.0	57.5	**48.6**	*9.0*
13	0.15	40.4	48.1	52.4	**47.0**	*6.1*
14	0.28	162.2	158.9	180.0	**167.0**	*11.3*
15	0.2	65.1	74.5	84.3	**74.6**	*9.6*
16	0.2	60.7	67.3	79.2	**69.1**	*9.3*
17	0.2	55.7	64.8	74.6	**65.1**	*9.4*
18	0.2	55.4	62.8	73.6	**63.9**	*9.2*
19	0.2	53.8	61.6	71.6	**62.3**	*8.9*
20	0.2	54.0	63.3	71.1	**62.8**	*8.5*
21	0.55	628.8	616.7	621.9	**622.4**	*6.1*
22	0.41	160.8	169.9	191.0	**173.9**	*15.5*
23	0.41	142.2	150.3	166.9	**153.1**	*12.6*
24	0.41	137.0	149.6	155.3	**147.3**	*9.4*
25	0.83	900.4	967.0	959.3	**942.2**	*36.4*
26	0.61	227.4	286.1	275.2	**262.9**	*31.2*
27	0.61	178.1	224.6	230.8	**211.1**	*28.8*
28	0.61	167.2	210.7	208.1	**195.3**	*24.4*
29	1.11	1056.6	1216.1	1200.7	**1157.8**	*88.0*
30	0.82	260.9	329.2	310.5	**300.2**	*35.3*
31	0.82	217.6	264.3	267.7	**249.8**	*28.0*
32	1.93	2492.0	2608.4	2851.5	**2650.6**	*183.4*
33	1.44	377.3	500.5	544.3	**474.0**	*86.6*
34	1.44	271.8	358.6	417.6	**349.3**	*73.3*
35	2.87	2307.8	2188.0	2713.5	**2403.1**	*275.4*
36	2.15	378.1	474.8	510.9	**454.6**	*68.6*
37	2.15	322.8	370.5	385.7	**359.7**	*32.8*
Total		11369.3	12403.7	13490.9	12421.3	*1060.9*

Table 3. Energy Dissipation (lb-in) of VEwalls.

Cycle	Displacement (in)	VEwall1	VEwall2	VEwall3	Average	Standard. deviation
1	0.14	147.7	150.0	135.4	**144.4**	*7.9*
2	0.14	150.6	150.9	136.0	**145.8**	*8.5*
3	0.14	146.8	151.1	138.5	**145.5**	*6.4*
4	0.14	148.4	147.8	139.7	**145.3**	*4.8*
5	0.14	145.5	147.9	140.0	**144.5**	*4.0*
6	0.14	149.3	152.2	136.6	**146.0**	*8.3*
7	0.21	325.8	332.7	311.4	**323.3**	*10.9*
8	0.15	192.6	195.2	182.9	**190.2**	*6.5*
9	0.15	185.0	187.0	178.0	**183.3**	*4.7*
10	0.15	188.1	187.6	175.6	**183.8**	*7.1*
11	0.15	183.6	183.8	178.2	**181.9**	*3.2*
12	0.15	186.0	188.7	175.4	**183.4**	*7.0*
13	0.15	188.5	185.4	176.7	**183.5**	*6.2*
14	0.28	563.4	575.8	546.3	**561.8**	*14.8*
15	0.2	335.4	337.1	313.7	**328.7**	*13.0*
16	0.2	326.1	325.7	315.5	**322.4**	*6.0*
17	0.2	324.9	324.1	311.7	**320.2**	*7.4*
18	0.2	324.7	328.0	312.9	**321.9**	*8.0*
19	0.2	323.4	322.0	312.3	**319.2**	*6.1*
20	0.2	331.5	332.4	319.3	**327.7**	*7.3*
21	0.55	2114.5	2307.8	2021.7	**2148.0**	*146.0*
22	0.41	1187.6	1151.1	1158.1	**1165.6**	*19.4*
23	0.41	1171.7	1136.6	1145.4	**1151.2**	*18.3*
24	0.41	1200.4	1157.4	1174.9	**1177.6**	*21.6*
25	0.83	4561.3	4328.4	4249.7	**4379.8**	*162.1*
26	0.61	2198.9	2382.1	2298.5	**2293.2**	*91.7*
27	0.61	2070.4	2344.7	2209.4	**2208.2**	*137.1*
28	0.61	2084.5	2350.8	2043.0	**2159.4**	*167.0*
29	1.11	6062.3	7072.2	5931.2	**6355.3**	*624.4*
30	0.82	3535.6	3709.7	3477.2	**3574.2**	*121.0*
31	0.82	3518.3	3595.1	3435.1	**3516.2**	*80.0*
32	1.93	**FAILED**	**FAILED**	**FAILED**	–	–
33	1.44	–	–	–	–	–
34	1.44	–	–	–	–	–
35	2.87	–	–		–	–
36	2.15	–	–		–	–
37	2.15	–	–		–	–
Total		34573.3	36441.4	33780.1	**34931.6**	*1366.3*

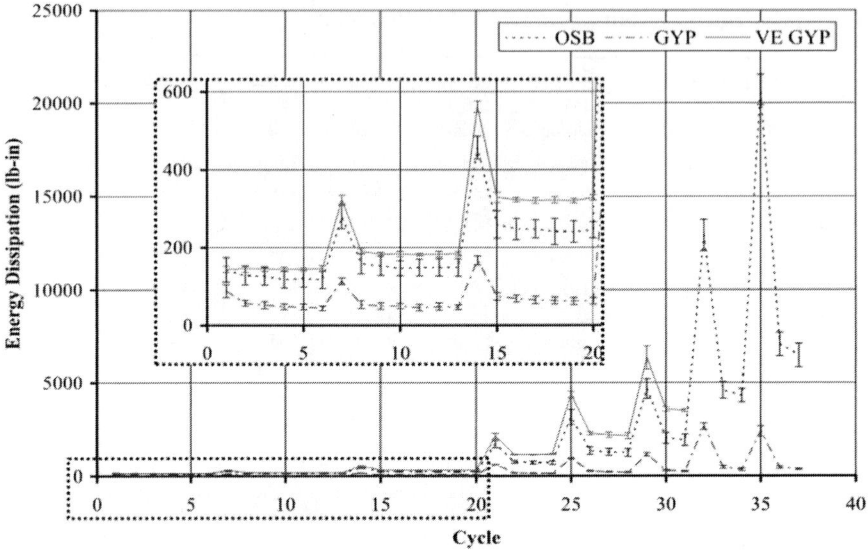

Figure 9. Average energy dissipation *vs.* cycle number for all walls.

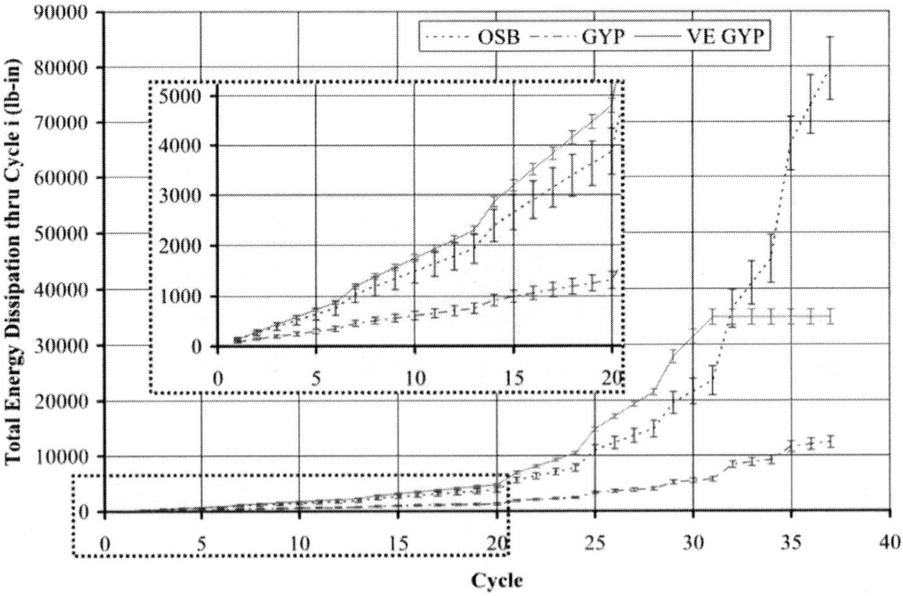

Figure 10. Total energy dissipated for all wall types.

The most prominent characteristic of the energy dissipation trends for the OSBWalls is the dramatic decrease in energy dissipation between primary cycles and secondary cycles. This decrease is somewhat deceiving because it is not entirely due to degradation. From Table 1 it is clear that for the CUREE loading protocol used, the primary cycles have larger displacement amplitudes than their subsequent secondary cycles. This has a significant effect on the area of the hysteresis loops and energy dissipation of the secondary cycles. Had the secondary cycles been given the same amplitude as their preceding primary cycles, their hysteresis loops would be longer, their areas would be greater, and their energy dissipation would be higher. Though, it is evident from Figure 5 that there is still significant degradation after primary cycles because the pinched regions are extended for the secondary cycles. Energy dissipated during secondary cycles of constant amplitude remained fairly constant throughout the first 20 cycles. Cycle 1 (0.14") produced 124.4 lb-in of energy dissipation, and gradually declined to 119.9 lb-in for cycle 6. Energy dissipated during cycle 8 (0.15") was 163.9 lb-in, compared to 152.6 lb-in for cycle 13. Cycle 15 (0.20") generated 275.1 lb-in of energy dissipation, and decreased to 258.4 lb-in. The total energy dissipated by OSBwall1 was 83,953.5 lb-in. The greatest standard deviation was 20.2% of the average and occurred at cycle 28 (0.61"). No failures occurred among the four OSB shear walls tested.

For the GYPwalls energy dissipated during secondary cycles of constant amplitude decreased slightly throughout the first 20 cycles. Cycle 1 (0.14") produced an average energy dissipation of 88.6 lb-in, and gradually declined to 44.3 lb-in for cycle 6. Energy dissipated during cycle 8 (0.15") was 54.1 lb-in on average, compared to 47.0 lb-in for cycle 13. Cycle 15 (0.20") generated an average energy dissipation of 74.6 lb-in, and decreased to 62.8 lb-in.

Among the three gypsum shear wall specimens, GYPwall3 dissipated the most total energy (13,490.9 lb-in), and GYPwall1 dissipated the least total energy (11,369.3 lb-in). The average total energy dissipation was 12,421.3 lb-in with a standard deviation of 1,060.9 lb-in (8.5% of the average). Generally, GYPwall3 produced the highest values for energy dissipation at each cycle and GYPwall1 produced the lowest values. While consistent, these values are considerably low for a shear wall (approximately 14.8% of OSBwall1), indicating that gypsum shear walls are not suitable for a major lateral force resisting system. The secondary cycles produced very low energy dissipation such that a single moderate sized seismic event would effectively render a gypsum shear wall incapacitated during future seismic events.

As shown in Table 3 and Figure 9, the values for energy dissipation were very similar for all three VEwall specimens. VEwall2 dissipated the most total energy (36,441.4 lb-in), and VEwall3 produced the lowest total energy dissipation (33,780.1 lb-in). The three specimens averaged a total energy dissipation of 34,931.6 lb-in with a standard deviation of 1,366.3 lb-in (3.9% of the average). Energy dissipated during secondary cycles of constant amplitude remained virtually constant throughout the first 20 cycles. Cycle 1 (0.14") produced an average energy dissipation of 144.4 lb-in, and increased to 146.0 lb-in for cycle 6. Energy dissipated during cycle 8 (0.15") was 190.2 lb-in on average, compared to 183.5 lb-in for cycle 13. Cycle 15 (0.20") generated an average energy dissipation of 328.7 lb-in, and decreased to 327.7 lb-in.

The standard deviation was as low as 1.6% at cycle 23 (0.41") and as high as 14.0% at cycle 28 (0.61"). Energy dissipation throughout the first 31 cycles was greater for the VE gypsum walls than for the baseline gypsum walls and OSB walls combined. By comparison, the OSB walls averaged a total energy dissipation of 23,540.7 lb-in (67.4% of VEwalls) and the baseline gypsum walls averaged a total energy dissipation of 5,729.9 lb-in (16.4% of VEwalls) throughout the first 31 cycles of displacement. The greatest increases occurred at secondary cycles because there was no inelastic deformation during primary cycles. Since all of the energy was dissipated by the viscoelastic polymer, the VE gypsum shear walls were able to provide a constant, non-degrading supply of energy dissipation.

6. Stiffness Results

The wall stiffness is defined as the slope of the hysteresis plot and is calculated per cycle. Table 4–Table 6 present the stiffness data for the OSBwalls, the GYPwalls and VEwalls respectively. Similar to the energy dissipation data presented previously, the cycle number, lateral wall displacement, individual test results, averages, and standard deviations are included. Cycles are grouped in sets according to cycle displacement with the first cycle within each set defined as the primary cycle, followed by the secondary cycles. Figure 11 presents the graphical representation of stiffness per cycle for all wall types and Figure 12 highlights the percentage of total stiffness loss per cycle. For Figure 11, Figure 12 trend lines, representing the average, are plotted for each wall type and error bars are plotted along the trend lines to illustrate the standard deviation within the data at each cycle.

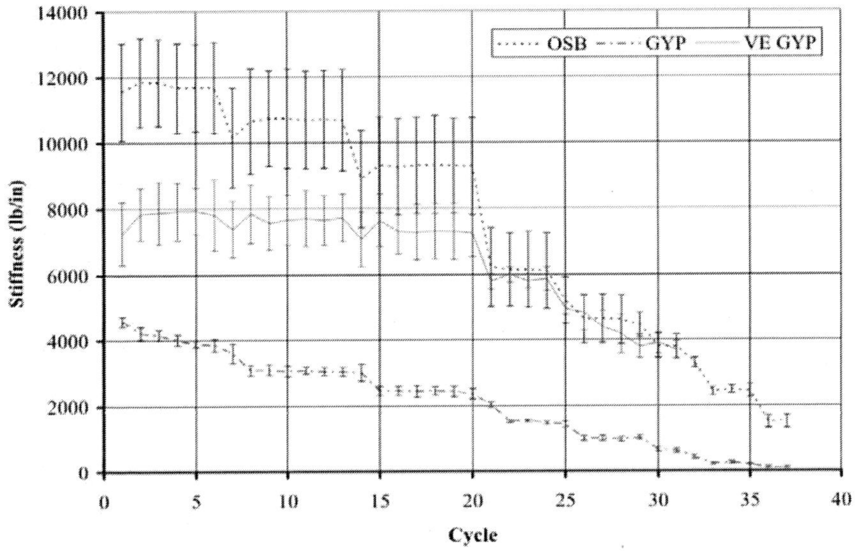

Figure 11. Average stiffness *vs.* cycle number for all walls.

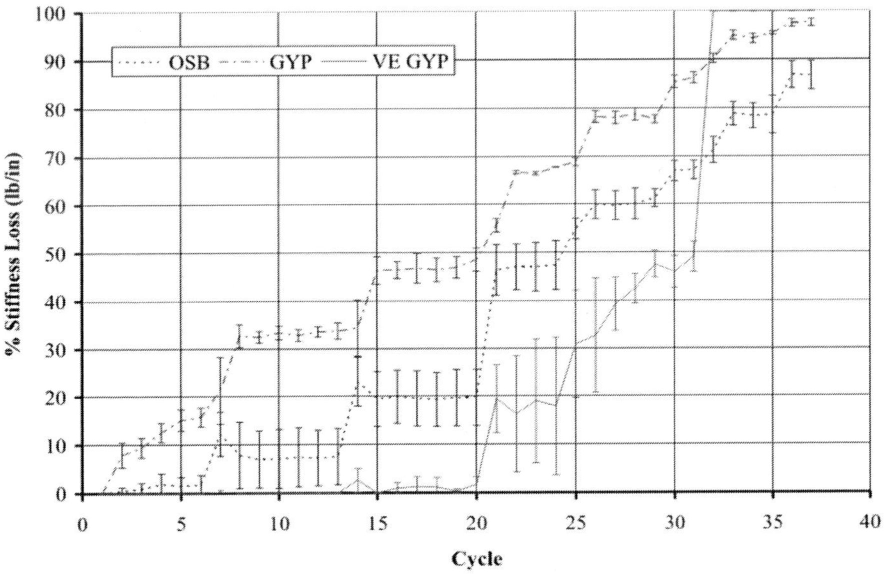

Figure 12. Average stiffness *vs.* cycle number for GYPwalls.

Table 4. Stiffness (lb/in) of OSBwalls.

Cycle	Displacement (in)	OSBwall1	OSBwall2	OSBwall3	OSBwall4	Average	Standard. deviation
1	0.14	11532.2	10906.9	10175.0	13619.0	**11558.3**	*1481.5*
2	0.14	11914.3	11916.8	10125.3	13416.6	**11843.3**	*1346.3*
3	0.14	12020.0	11985.7	10070.6	13273.7	**11837.5**	*1321.6*
4	0.14	11878.3	11914.3	9803.2	13087.2	**11670.8**	*1365.8*
5	0.14	11735.3	11914.3	9926.4	13130.8	**11676.7**	*1321.3*
6	0.14	11771.0	11983.7	9818.1	13166.7	**11684.9**	*1387.7*
7	0.21	10693.3	9735.6	8322.5	11877.9	**10157.3**	*1504.7*
8	0.15	10947.0	10870.3	8491.5	12345.2	**10663.5**	*1598.8*
9	0.15	11100.0	10845.3	8783.0	12276.5	**10751.2**	*1452.7*
10	0.15	11131.6	10813.4	8714.3	12313.0	**10743.1**	*1498.5*
11	0.15	11100.0	10781.8	8682.7	12240.4	**10701.2**	*1484.2*
12	0.15	11004.2	10813.7	8746.3	12340.1	**10726.1**	*1484.3*
13	0.15	11158.6	10631.7	8650.8	12341.4	**10695.6**	*1539.3*
14	0.28	9567.9	8332.1	7182.0	10574.0	**8914.0**	*1474.4*
15	0.2	9934.1	9094.6	7414.8	10851.5	**9323.8**	*1460.9*
16	0.2	9897.4	8988.3	7397.5	10803.9	**9271.8**	*1452.8*
17	0.2	9982.3	9012.1	7417.2	10830.2	**9310.5**	*1464.5*
18	0.2	9982.3	9010.0	7441.1	10925.5	**9339.7**	*1487.8*
19	0.2	10006.1	9012.1	7417.2	10740.9	**9294.1**	*1437.9*
20	0.2	9920.9	9053.9	7366.7	10822.1	**9290.9**	*1472.0*
21	0.55	6984.3	5419.0	5015.6	7479.7	**6224.7**	*1192.0*
22	0.41	6837.5	5588.8	4873.1	7288.5	**6147.0**	*1112.7*
23	0.41	6876.0	5532.8	4846.5	7340.7	**6149.0**	*1158.3*
24	0.41	6837.4	5501.5	4818.1	7286.9	**6111.0**	*1148.0*
25	0.83	5555.2	4842.8	4428.6	5996.5	**5205.8**	*703.1*
26	0.61	5064.8	4379.4	3730.8	5359.0	**4633.5**	*728.5*
27	0.61	5123.9	4315.1	3794.7	5382.7	**4654.1**	*731.4*
28	0.61	5096.7	4309.0	3721.8	5340.6	**4617.0**	*741.6*
29	1.11	4579.9	4354.7	3998.6	4897.8	**4457.7**	*378.6*
30	0.82	4055.1	3739.1	3316.6	4132.4	**3810.8**	*370.8*
31	0.82	4019.4	3684.9	3314.6	4132.9	**3787.9**	*368.4*
32	1.93	3326.7	3375.8	3087.8	3399.0	**3297.3**	*142.9*
33	1.44	2461.6	2566.2	2275.3	2422.9	**2431.5**	*120.5*
34	1.44	2512.1	2621.1	2345.2	2454.9	**2483.3**	*115.0*
35	2.87	2491.4	2625.6	2483.5	2168.4	**2442.2**	*193.9*
36	2.15	1565.7	1674.1	1497.4	1250.5	**1496.9**	*179.7*
37	2.15	1570.3	1709.3	1507.0	1237.0	**1505.9**	*198.2*

Table 5. Stiffness (lb/in) of GYPwalls.

Cycle	Displacement (in)	GYPwall1	GYPwall2	GYPwall3	Average	Standard deviation
1	0.14	4460.4	4533.8	4752.6	**4582.3**	*152.0*
2	0.14	4186.1	4043.2	4436.4	**4221.9**	*199.0*
3	0.14	4114.7	4006.3	4331.6	**4150.9**	*165.7*
4	0.14	3971.4	3862.5	4186.1	**4006.7**	*164.7*
5	0.14	3900.0	3792.2	3971.6	**3887.9**	*90.3*
6	0.14	3790.0	3720.8	4079.1	**3863.3**	*190.1*
7	0.21	3841.4	3272.0	3696.6	**3603.3**	*296.0*
8	0.15	3078.2	2925.7	3244.0	**3082.6**	*159.2*
9	0.15	2989.6	3021.5	3275.9	**3095.7**	*156.9*
10	0.15	2926.1	2989.6	3244.0	**3053.2**	*168.2*
11	0.15	3053.1	2989.6	3186.1	**3076.3**	*100.3*
12	0.15	2989.6	2957.7	3186.4	**3044.6**	*123.9*
13	0.15	3014.1	2916.9	3174.7	**3035.3**	*130.2*
14	0.28	3199.1	2718.5	3102.8	**3006.8**	*254.3*
15	0.2	2519.8	2303.5	2551.8	**2458.4**	*135.1*
16	0.2	2432.6	2337.9	2599.7	**2456.7**	*132.5*
17	0.2	2466.0	2252.5	2599.7	**2439.4**	*175.1*
18	0.2	2495.7	2314.0	2551.8	**2453.8**	*124.3*
19	0.2	2408.6	2290.0	2599.7	**2432.8**	*156.2*
20	0.2	2351.7	2207.7	2519.5	**2359.6**	*156.1*
21	0.55	2034.4	1942.0	2109.2	**2028.5**	*83.7*
22	0.41	1498.2	1510.0	1564.8	**1524.3**	*35.6*
23	0.41	1505.7	1529.4	1576.7	**1537.3**	*36.2*
24	0.41	1443.2	1462.7	1533.9	**1479.9**	*47.8*
25	0.83	1363.2	1400.6	1527.7	**1430.5**	*86.3*
26	0.61	920.4	1045.4	1045.3	**1003.7**	*72.1*
27	0.61	926.5	1059.2	1049.2	**1011.6**	*73.9*
28	0.61	896.8	1031.6	1007.8	**978.7**	*72.0*
29	1.11	957.9	1029.3	1092.9	**1026.7**	*67.5*
30	0.82	589.5	715.6	708.6	**671.2**	*70.9*
31	0.82	561.4	675.8	661.9	**633.1**	*62.5*
32	1.93	394.4	443.0	510.4	**449.3**	*58.2*
33	1.44	176.5	258.1	257.5	**230.7**	*46.9*
34	1.44	204.7	288.9	288.6	**260.7**	*48.5*
35	2.87	191.6	225.6	208.6	**208.6**	*17.0*
36	2.15	97.3	153.9	90.5	**113.9**	*34.8*
37	2.15	99.5	149.3	86.0	**111.6**	*33.4*

Table 6. Stiffness (lb/in) of VEwalls.

Cycle	Displacement (in)	VEwall1	VEwall2	VEwall3	Avergae	Standard. deviation
1	0.14	7137.6	8262.5	6376.9	**7259.0**	*948.7*
2	0.14	7587.1	8729.9	7227.2	**7848.1**	*784.6*
3	0.14	7692.2	8908.8	7048.4	**7883.1**	*944.8*
4	0.14	7656.7	8908.8	7227.2	**7930.9**	*873.7*
5	0.14	7728.1	8729.9	7366.5	**7941.5**	*706.3*
6	0.14	7227.6	9071.6	7155.7	**7818.3**	*1086.0*
7	0.21	7089.4	8346.0	6718.3	**7384.6**	*853.1*
8	0.15	7378.5	8882.1	7308.6	**7856.4**	*889.0*
9	0.15	7251.5	8491.5	6965.1	**7569.4**	*811.3*
10	0.15	7442.0	8491.5	7060.6	**7664.7**	*741.0*
11	0.15	7410.4	8650.8	7060.6	**7707.3**	*835.6*
12	0.15	7378.5	8491.5	7092.5	**7654.2**	*739.1*
13	0.15	7501.0	8523.8	7150.7	**7725.2**	*713.4*
14	0.28	6791.9	8015.3	6443.3	**7083.5**	*825.5*
15	0.2	7436.4	8518.6	6982.8	**7645.9**	*789.0*
16	0.2	7083.3	8084.8	6773.1	**7313.7**	*685.5*
17	0.2	6875.9	8251.8	6749.1	**7292.3**	*833.4*
18	0.2	6892.3	8278.7	6773.1	**7314.7**	*837.0*
19	0.2	7083.3	8251.8	6606.3	**7313.8**	*846.7*
20	0.2	6910.8	8110.6	6790.6	**7270.7**	*729.9*
21	0.55	5765.2	6055.3	5580.6	**5800.4**	*239.3*
22	0.41	6249.2	5796.1	5956.8	**6000.7**	*229.7*
23	0.41	5933.0	5548.2	5910.3	**5797.2**	*215.9*
24	0.41	6153.8	5462.2	5986.9	**5867.6**	*360.9*
25	0.83	4872.5	4819.6	5149.2	**4947.1**	*177.0*
26	0.61	4205.5	5126.0	5168.2	**4833.2**	*544.0*
27	0.61	3975.8	4947.9	4252.1	**4391.9**	*500.9*
28	0.61	3856.7	4848.7	3817.1	**4174.1**	*584.5*
29	1.11	3619.8	4209.9	3549.8	**3793.2**	*362.6*
30	0.82	3762.7	4270.1	3691.1	**3908.0**	*315.7*
31	0.82	3564.1	4001.2	3474.9	**3680.1**	*281.7*
32	1.93	FAILED	FAILED	FAILED	-	-
33	1.44	-	-	-	-	-
34	1.44	-	-	-	-	-
35	2.87	-	-	-	-	-
36	2.15	-	-	-	-	-
37	2.15	-	-	-	-	-

The initial stiffness of OSBwall1 was 11,532.7 lb/in and decreased significantly during primary cycles. This indicates that the transition to inelastic deformation occurs due to crushing of the wood around the sheathing fasteners, which only happens during primary cycles. Stiffness remained fairly constant within cycles of constant amplitude, implying that the majority of sheathing connections were not subjected to fatigue. The stiffness produced during cycles 1–6 (0.14"), cycles 8–13 (0.15") and cycles 15–20 (0.20") was approximately 11,900 lb/in, 11,100 lb/in and 9,950 lb/in, respectively. The final stiffness of OSBwall1 at cycle 37 (2.15") was 1,570.3 lb/in (a loss of 86.4%). The greatest standard deviation was 21.6% of the average and occurred at cycle 23 (0.41").

For the GYPwalls the stiffness initially ranged from 4,460.4 lb/in to 4,752.6 lb/in and decreased almost immediately, indicating a small range of elasticity (less than 0.14") followed by inelastic deformation. The three shear wall specimens averaged an initial stiffness of 4,582.3 lb/in with a standard deviation of 152.0 lb/in (3.3% of the average), and decreased to an average final stiffness of 111.6 lb/in with a standard deviation of 33.4 lb/in (29.9% of the average). Cycle 1 (0.14") produced an average stiffness of 4,582.3 lb/in, and declined almost linearly to 3,863.3 lb/in for cycle 6. Stiffness during cycle 8 (0.15") was 3,082.6 lb/in on average, compared to 3,035.3 lb/in for cycle 13. Cycle 15 (0.20") generated an average stiffness of 2,458.4 lb/in, and decreased to 2,359.6 lb/in. The standard deviation for the shear wall specimens was as low as 2.3% at cycle 22 (0.41"). GYPwall2 produced the lowest values of stiffness throughout most of the test, while GYPwall3 produced the highest values. The average loss in stiffness was 4,470.7 lb/in, or 97.6%. As with the energy dissipation results, these values for stiffness are consistent but very low for an 8' × 8' shear wall. By comparison, OSBwall1 produced an initial stiffness of 11,532.2 lb/in and a stiffness of 1,570.3 lb/in during cycle 52 (2.15").

The three VE gypsum shear wall specimens averaged an initial stiffness of 7,259.0 lb/in, which remained fairly constant throughout the first 20 cycles (≤0.28"). Stiffness decreased approximately 500 lb/in during primary cycles (1, 7 and 14). The standard deviation for the first cycle stiffness was 948.7 lb/in (13.1% of the average). At cycle 21 (0.55"), the average stiffness dropped to 5,800.4 lb/in. Then, at cycle 24 (0.41"), stiffness began to drop almost linearly with respect to cycle number until cycle 32 (1.93"), during which all specimens failed. Therefore, by the end of the test, all VE gypsum shear walls sustained a total loss in stiffness. Had the gypsum wallboard panels remained intact and the stiffness continued to decline linearly, the final stiffness would have been approximately 2,000 lb/in.

The decrease in stiffness for the VE gypsum shear walls resembles the behavior of the VE gypsum connections, which exhibited a decrease in stiffness after 32 cycles of shear displacement. They too showed a linear decrease in stiffness after a period of constant stiffness. As with the VE gypsum connections, the VE gypsum shear walls did not exhibit signs of yielding or permanent damage prior to the sheathing failure. Therefore, it is likely that the decrease in stiffness was due to a change in the temperature, due to heat buildup, within the viscoelastic polymer. The loss of stiffness due to temperature change is well documented [15,16,18].

7. Wall Performance and Damage Results

Damage to the OSBWalls was restricted to localized deformations near the sheathing connections and included nail withdrawal, nail pull-through and flaking of the OSB, as shown in Figure 13–Figure 15. In some areas, the nail withdrawal allowed the sheathing to pull away from the frame. Damage to the sheathing connections occurred along the center stud, top sill plate and double end stud (actuator side). Crushing of the wood sheathing and bending of the nails were observed similar to sheathing connections deformed in a manner similar to the nailed OSB connection specimens reported in a previous connection study [3].

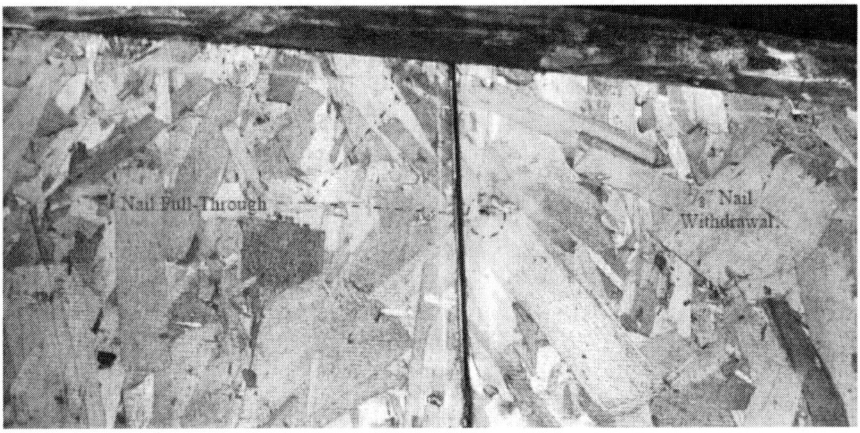

Figure 13. Damage along the center stud and top sill plate.

Figure 14. Nail pull-through along the center stud.

Figure 15. Nail pull-through along the center stud.

The wood frame remained intact throughout the test, as did the OSB sheathing panels. Sheathing connections along the field studs were also in good condition. However, the degradation of stiffness suggests significant inelastic behavior and permanent deformation. Had this wall been located in an actual structure during an earthquake, it would be in need of significant repair—specifically, re-nailing along the perimeter of each sheathing panel would be required.

Global damage sustained by the baseline GYPwalls included separation of the wallboard panels from the wood frame, as shown in Figure 16. The separation was more severe at the top of the walls (over 3/4"), whereas the bottom of the walls did not separate more than 1/4". The separation was most likely due to localized damage sustained at the sheathing connections, which was worse at the top half of the wall (see Figure 16). The wood frames remained intact throughout all three tests.

The most severe localized damage was observed near the corners of the gypsum wallboard panels. Portions of the panels were crushed and torn by the dowel-type fasteners, as shown in Figure 17. This occurred near the corners and along the bottom sill plate, where the sheathing separation was minimal. Bulging of the wallboard was also observed near the fasteners. The heads of the drywall screws, which were slightly over-driven, slid just beneath the outer veneer of the wallboard when displaced, creating a bulging effect on the sheathing surface. This occurred along the field studs and top sill plate, where fastener pull-through was more evident as shown in Figure 18. Overall, the damage observed was substantial, which can be attributed to using rigid dowel-type fasteners with easily crushable gypsum sheathing panels. Had these walls been located in an actual structure during an earthquake, they would certainly be in need of repair—specifically, total replacement of the drywall panels.

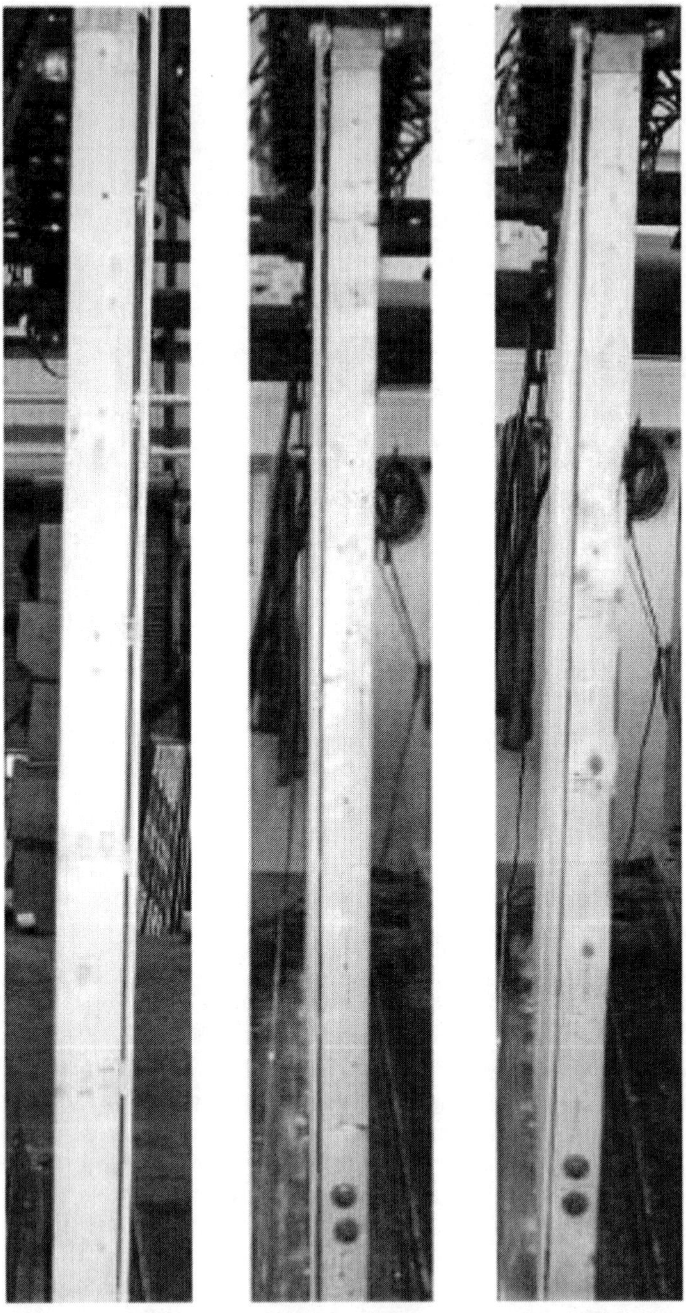

Figure 16. Separation of Gypsum Wallboard from Wood Frame.

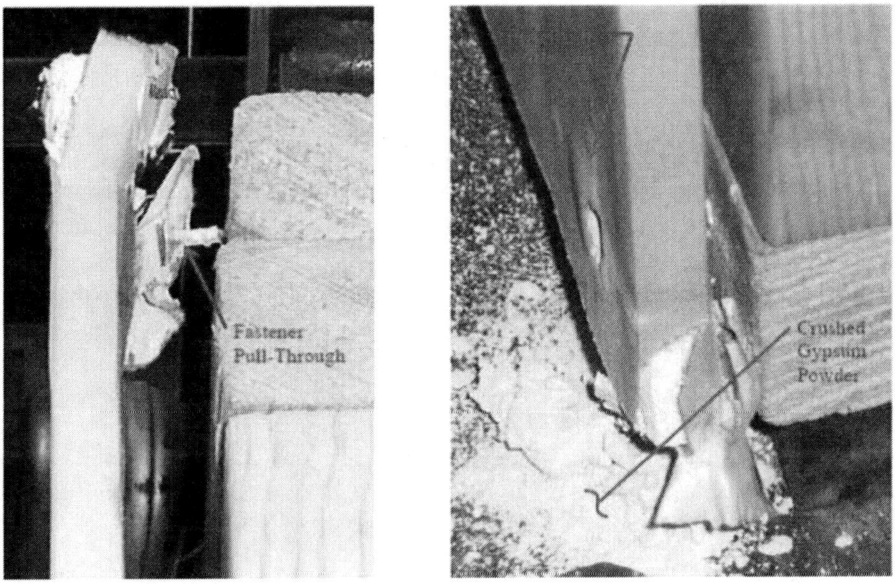

Figure 17. Typical Damage to Gypsum Wallboard Corners.

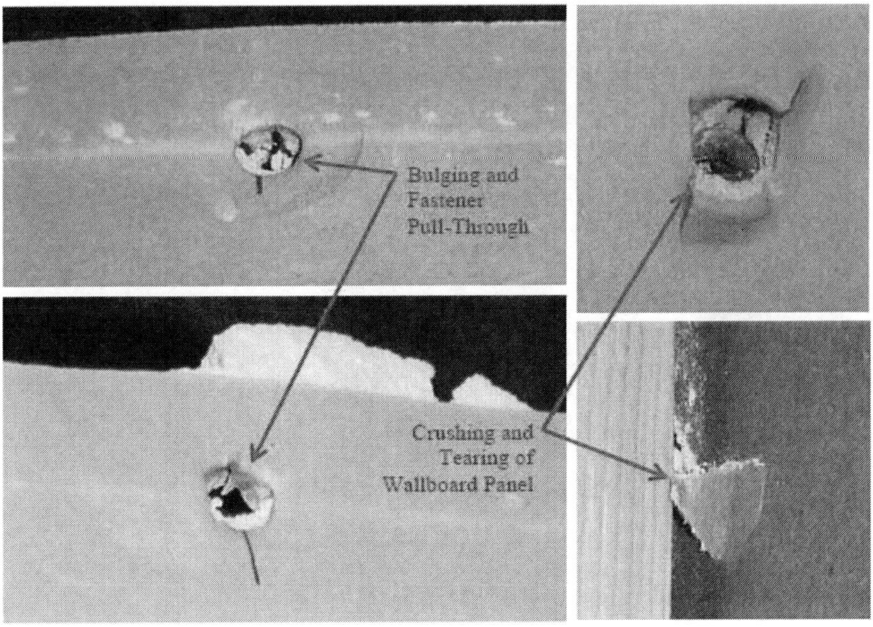

Figure 18. Typical Damage to Gypsum Sheathing Connections.

For all VEwalls during the first 31 cycles of displacement, there were no visible or audible signs of damage to the wood, gypsum, or VE polymer of all three specimens. All components of the VE gypsum connections remained completely adhered even when forces exceeded 6,000 lbs. The only indication of possible damage is the decrease in stiffness beginning at cycle 21 (0.55"). Though the decrease in stiffness is not expected to be a result of permanent damage, the possibility can not be ruled out.

Immediately following cycle 31, all three walls failed due to excessive tensile stresses in the gypsum wallboard, leading to permanent damage of the sheathing panels, as shown in Figure 19. The damage consisted of long, diagonal cracks through the wallboard panels oriented perpendicular to the direction of the principal tensile stresses. Cracks were generally located on the lower and upper corners of the sheathing panels, and propagated along the center stud and sill plates. The failures were brittle, as there were no indications that a failure was imminent. Essentially, the integrity of the VEwalls relied on the capacity of the paper veneers on either side of the wallboard. Figure 19–Figure 23 illustrate the damage sustained by the three specimens. While the failures were unpredicted, it is essential to note that prior to a target displacement of 1.9 in, there was no visual or audible sign of damage.

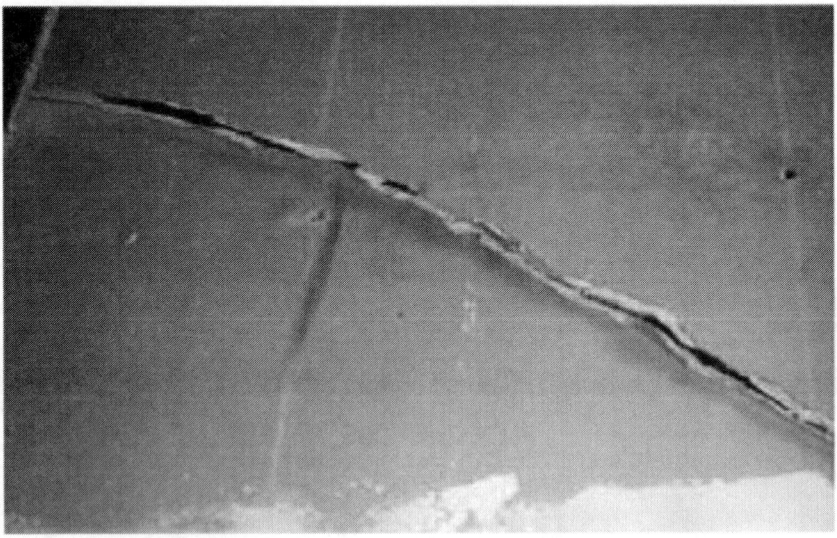

Figure 19. Tension failure of gypsum wallboard.

Figure 20. Damage to upper half of VEwall 1.

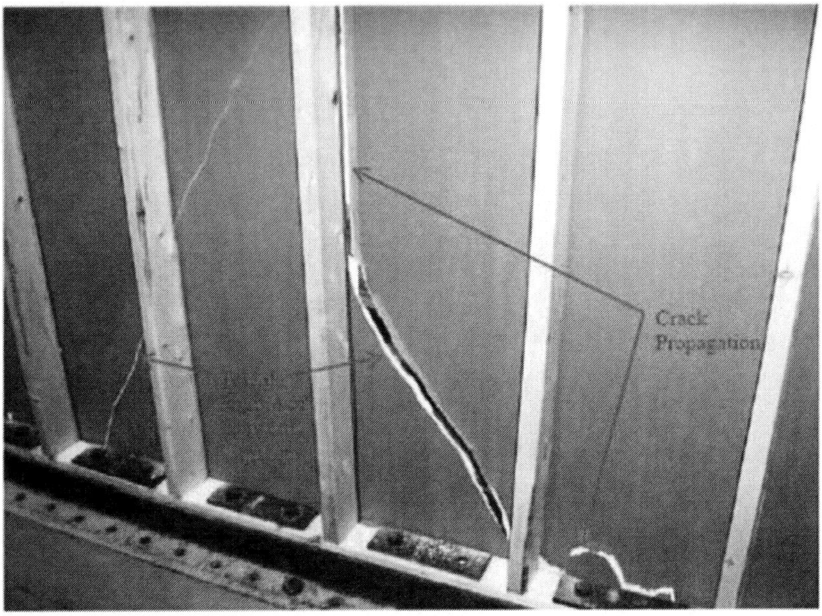

Figure 21. Damage to lower half of VEwall 1.

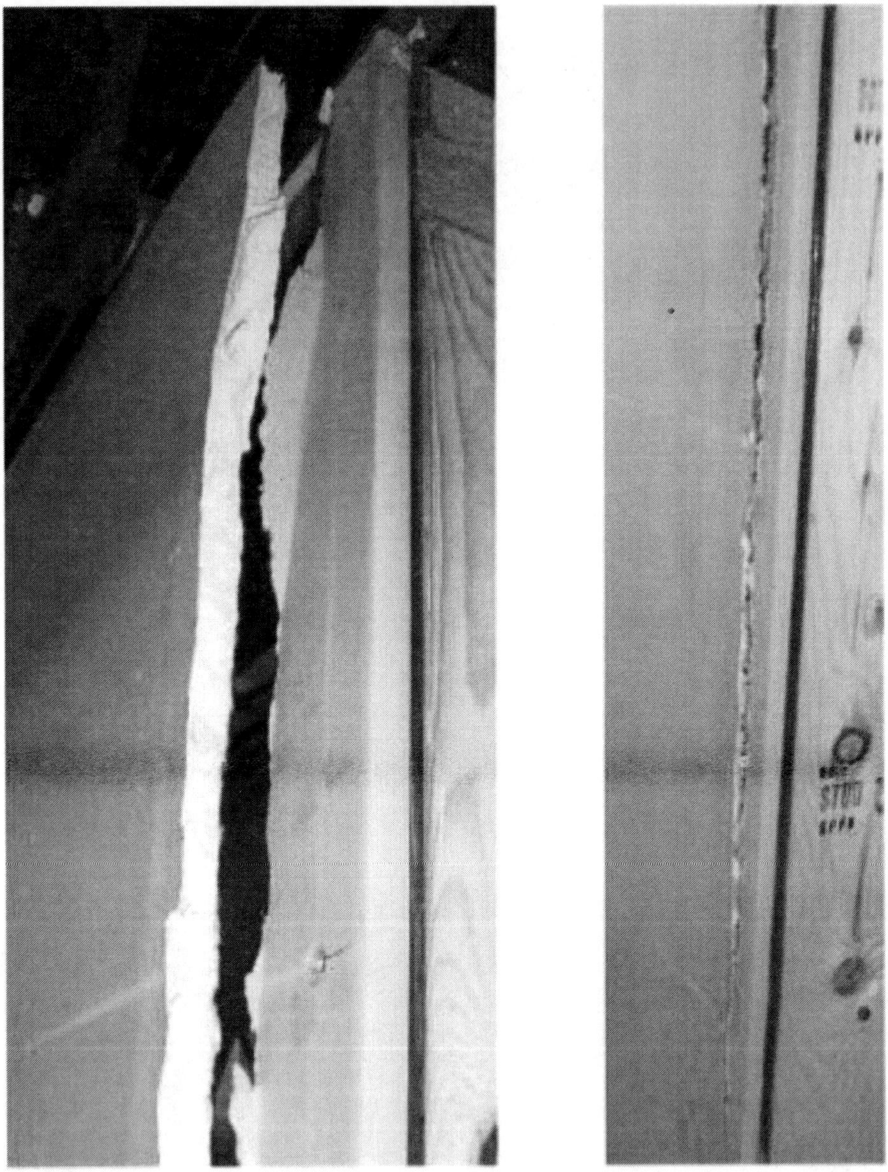

Figure 22. Crack propagation along the edge of VEwall 2.

Figure 23. Damage to lower half of VEwall 3.

8. Summary

Three conventional gypsum shear walls and three viscoelastic shear walls were tested. The baseline walls used conventional screwed sheathing connections, whereas the VE shear walls utilized VE material along the perimeter of the sheathing panels. The walls were loaded at frequency of 1.0 Hz using a displacement-based load protocol recommended by CUREE. Table 7 provides a summary of the performance.

Table 7. Summary of wall performance.

Type of shear wall	ED at cycle 29 (lb-in)	Total ED to cycle 31 (lb-in)	Initial K (lb/in)	K at cycle 31	Damage summary
OSBwall	4,673.6	30,177.0	11,558.3	4,019.4	Nail withdrawal Nail pull-through OSB flaking
GYPwall	1,157.8	5,730.0	4,582.3	633.1	Sheathing separation from frame Fastener pull-through Bulging of gypsum near fasteners
VEwall	6,355.3	34,931.6	7,259.0	3,680.1	No damage < 1.1 inches of displacement Long. diagonal cracks due to sheathing Tension failure Sheathing separation from frame

9. Conclusions

Based on the results of the experimental program described, the following conclusions with respect to energy dissipation, stiffness, and wall damage can be made:

9.1. Energy Dissipation

1. The OSB sheathed walls and gypsum sheathed walls dissipate energy via inelastic deformation of the sheathing connections, whereas the VE walls dissipate energy via viscous damping of the VE polymer;

2. At a wall displacement of 1.11 inches (cycle 29), the energy dissipation of the VE gypsum walls is 36% greater than OSB walls and over 400% greater than the non-structural conventional walls with the same sheathing;

3. The VE gypsum walls dissipated more energy than the OSB walls through 31 cycles of displacement and over 5 times as much energy than the conventional gypsum walls;

4. The average total energy dissipated by the VE gypsum shear walls through 31 cycles of displacement was greater than that dissipated by the OSB walls;

5. The average total energy dissipated by the VE walls was 5 times greater than that dissipated by the non-structural gypsum walls;

6. Viscoelastic gypsum shear walls dissipate energy with greater consistency. The standard deviations (percentage) for total energy dissipated by each type of shear wall are as follow: 3.9% for VE gypsum, 7.1% for OSB, 8.5% for gypsum.

9.2. Stiffness

1. Viscoelastic gypsum shear walls produce an initial stiffness of 7,259 lb/in, 37% less than the OSB shear walls and 58% greater than the gypsum shear walls;
2. Gypsum and OSB shear walls lose stiffness immediately when displaced, whereas VE gypsum shear walls do not lose stiffness until cycle 21 (0.55"). The gypsum and OSB shear walls average a 50% loss of stiffness by cycles 21 (0.55") and 25 (0.83"), respectively. On average, the VE gypsum shear walls do not lose 50% stiffness prior to failure;
3. Cycle 31 (0.82") is the last cycle for which all VE gypsum shear wall specimens remained intact. During this cycle, the VE gypsum shear walls average 3,680 lb/in, an increase of 481% over the conventional gypsum shear walls. By comparison, the OSB shear walls averaged a stiffness of 3,788 lb/in during cycle 31. All three VE gypsum shear walls failed during cycle 32 (1.93"), resulting in a total loss of stiffness.

9.3. Wall Damage

1. Damage to the OSB and gypsum shear walls consists of localized deformation near the sheathing connections and the corners of the sheathing panels. Separation between the sheathing and the frame is also present (3/4" for the gypsum shear walls and 1/8" for the OSB shear wall). These findings are similar to that seem in other studies;
2. The VE gypsum shear walls do not display any visual signs of damage to the sheathing or the sheathing connections prior to cycle 32 (1.93"). During cycle 32, all three specimens sustain a brittle failure due to excessive tensile stresses in the gypsum wallboard panels, resulting in long, diagonal cracks propagating through the panels and a loss of capacity;
3. Viscoelastic material offers gypsum shear walls the potential to be used as substantial structural members, while reducing the risk of costly damage during seismic events.

Acknowledgments

The National Science Foundation (NSF) and Partnership for Advancing Technology in Housing (PATH) is acknowledged for partial funding of this work (Grant No. CMS-0229724). The academic and technical mentoring of Harry "Tripp" Shenton is greatly appreciated by the authors.

References

1. Seible, F.; Filiatrault, A.; Uang, C.M. Proceedings of the Invitational Workshop on Seismic Testing, Analysis and Design of Woodframe Construction; California Universities for Research in Earthquake Engineering: Berkeley, CA, USA, 1999.
2. American Wood Council, *Manual for Engineered Wood Construction*; American Forrest and Paper Association: Washington, DC, USA, 2001.
3. Foley, D. An Experimental Investigation of Viscoelastic Material in Wood Connections and Shearwalls. Master's Thesis, Villanova University, Villanova, PA, USA, 2005..
4. Filiatrault, A.; Fischer, D.; Folz, B.; Uang, C. Seismic testing of two-story woodframe house: Influence of wall finish materials. *ASCE J. Struct. Eng.* **2002**, *128*, 1337–1345.
5. McMullin, K.M.; Merrick, D.S. Seismic damage thresholds for gypsum wallboard partition walls. *J. Archit. Eng.* **2007**,*13*, 22–29.
6. Wolfe, R.W. *Contribution of Gypsum Wallboard to Racking Resistance of Light-Frame Walls*; United States Department of Agriculture: Washington, DC, USA, 1983.
7. Patton-Mallory, M.; Gutkowski, R.M.; Soltis, L.A. *Racking Performance of Light-Frame Walls Sheathed on Two Sides*; Forest Products Laboratory: Madison, WI, USA, 1984.
8. Arnold, A.E.; Uang, C.; Filiatrault, A. *Cyclic Behavior and Repair of Stucco and Gypsum Woodframe Walls: Phase I*; Consortium of Universities for Research in Earthquake Engineering: Richmond, CA, USA, 2003.
9. Filiatrault, A.; Christovasili, I.P.; Wanitkorkul, A.; van de Lindt, J.W. Experimental seismic response of a full-scale light-frame wood building. *ASCE J. Struct. Eng.* **2010**, *136*, 1262–1272.
10. Van de Lindt, J.W.; Pei, S.; Pryor, S.E. Construction and experimental seismic performance of a full-scale six-story light-frame wood building. In Proceedings of the 12th East Asia-Pacific Conference on Structural Engineering and Construction, Hong Kong, China, 26–28 January 2011.
11. Van de Lindt, J.W.; Gupta, R.; Pei, S.; Tachibana, K.; Araki, Y.; Rammer, D.; Isoda, H. Damage assessment of a full-scale six-story wood-frame building following triaxial shake table tests. *ASCE J. Perform. Construct. Facil.* **2012**, *25*, 17–25.
12. Shinde, J.K.; Symans, M.D.; Hongyan, L.; van de Lindt, J.W. Seismic performance assessment of woodframed structures with energy dissipation systems. In Proceedings of the 18th Analysis and

computation Specialty Conference—Structures Congress, Vancouver, BC, Canada, 24–26 April 2008.

13. Van de Lindt, J.W.; Hongyan, L.; Symans, M.D.; Shinde, J.K. Seismic performance and modeling of a half-scale base-isolated wood frame building. *J. Earthq. Eng.* **2011**, *15*, 469–490.
14. Blasetti, A.S.; Hoffman, R.M.; Dinehart, D.W. Simplified hysteretic finite element model for wood and viscoelastic polymer connections for the dynamic analysis of shear walls. *ASCE J. Struct. Eng.* **2008**, *134*, 77–86.
15. Joye, D.D.; Dinehart, D.W. Improved viscoelastic damping for earthquake resistant wood structures. *Int. J. Polym. Mater.* **2007**, *56*, 55–64.
16. Dinehart, D.W.; Leese, K.A.; Patterson, S. Cyclic performance of a viscoelastic connection for use in wood-frame and steel-frame gypsum shearwalls. In Proceedings of the 11th World Conference on Timber Engineering, Riva del Garda, Italy, 20–24 June 2010.
17. Dinehart, D.W.; Joye, D.D.; de Mendoza, G.E. Viscoelastic polymers improve the cyclic performance of wood connections. In Proceedings of the 8th World Conference on Timber Engineering, Lahti, Finland, 14–17 June 2004.
18. Dinehart, D.W.; Lewicki, D.E. Viscoelastic material as a seismic protection system for wood-framed buildings. In Proceedings of the 2001 Structures Congress & Exposition, Washington, DC, USA, 21–23 May 2001.
19. Krawinkler, H.; Parisi, F.; Ibarra, L.; Ayoub, A.; Medina, R. *Development of a Testing Protocol for Woodframe Structures*; Consortium of Universities for Research in Earthquake Engineering: Richmond, CA, USA, 2001.
20. Dinehart, D.W.; Shenton, H.W., III. Comparison of the static and dynamic response of timber shear walls. *ASCE J. Struct. Eng.* **1999**, *125*, 686–695.
21. Shenton, H.W., III; Dinehart, D.W.; Elliott, T.E. Stiffness and energy degradation of wood frame shear walls. *Can. J. Civ. Eng.* **1998**, *25*, 412–423.
22. Dinehart, D.W.; Blasetti, A.S.; Shenton, H.W., III. The experimental cyclic performance of viscoelastic gypsum connections and shear walls. *J. Struct. Eng.* **2008**, *134*, 87–95.

Index

P

R

S

V

W

Y